RESOURCE RECOVERY FROM WASTEWATER

Toward Sustainability

RESOURCE RECOVERY FROM WASTEWATER

Toward Sustainability

Edited by

Veera Gnaneswar Gude, PhD

AAP | APPLE ACADEMIC PRESS

First edition published 2022

Apple Academic Press Inc.
1265 Goldenrod Circle, NE,
Palm Bay, FL 32905 USA

4164 Lakeshore Road, Burlington,
ON, L7L 1A4 Canada

CRC Press
6000 Broken Sound Parkway NW,
Suite 300, Boca Raton, FL 33487-2742 USA

2 Park Square, Milton Park,
Abingdon, Oxon, OX14 4RN UK

Library and Archives Canada Cataloguing in Publication

Title: Resource recovery from wastewater : toward sustainability / edited by Veera Gnaneswar Gude, PhD.

Names: Gude, Veera Gnaneswar, 1978- editor.

Description: Includes bibliographical references and index.

Identifiers: Canadiana (print) 20200314572 | Canadiana (ebook) 20200315013 | ISBN 9781771889285 (hardcover) | ISBN 9781003055501 (ebook)

Subjects: LCSH: Sewage—Purification. | LCSH: Sewage—Purification—Technological innovations. | LCSH: Sewage—Purification—Environmental aspects. | LCSH: Water reuse.

Classification: LCC TD745 .R49 2021 | DDC 628.3—dc23

Library of Congress Cataloging-in-Publication Data

Names: Gude, Veera Gnaneswar, 1978- editor.

Title: Resource recovery from wastewater : toward sustainability / edited by Veera Gnaneswar Gude, PhD.

Description: First edition. | Palm Bay, FL : Apple Academic Press, Inc. ; Boca Raton, FL : CRC Press, 2021. | Includes bibliographical references and index. | Summary: "This informative volume provides comprehensive knowledge on various aspects of wastewater resource management from the point of process sustainability and resource recovery. This authoritative compendium is crucial for developing resource-efficient and sustainable wastewater treatment technologies and management strategies for both small (decentralized) and large (centralized) communities. Traditional wastewater systems have become increasingly energy-consuming and cost-intensive while also not meeting the increasing standards for nutrient removal and sustainable development. This book incorporates the latest developments in pollutant removal and resource recovery schemes in wastewater treatment. It highlights advances that have been made in microbiological processes; design of treatment methods; process configurations ; energy conservation and efficiency improvement schemes ; nutrient removal; recovery, reclamation, and recycling ; beneficial uses of wastewater ; and bioenergy and biochemical production from wastewater and sludge streams. Waste-to-energy technologies, especially wastewater treatment as a potential biofuel energy alternative through bioelectrochemical and other processes, are also discussed in this book. Basic and advanced principles of wastewater treatment are covered with theoretical development, illustrations, samples and examples, and case studies. Eminent scholars around the world have shared their knowledge here, making this volume a valuable resource for novice researchers as well as practicing and experienced professionals in biofuel research and industrial sectors. The scientific experimental methods and analytical procedures and interpretation of the results capture the state-of-the-art knowledge in the literature and future trends in this area"-- Provided by publisher.

Identifiers: LCCN 2020037577 (print) | LCCN 2020037578 (ebook) | ISBN 9781771889285 (hardcover) | ISBN 9781003055501 (ebook)

Subjects: LCSH: Sewage--Purification--By-products. | Sewage--Recycling.

Classification: LCC TD765 .R474 2021 (print) | LCC TD765 (ebook) | DDC 628.3/8--dc23

LC record available at https://lccn.loc.gov/2020037577

LC ebook record available at https://lccn.loc.gov/2020037578

ISBN: 978-1-77188-928-5 (hbk)
ISBN: 978-1-77463-791-3 (pbk)
ISBN: 978-1-00305-550-1 (ebk)

About the Editor

Veera Gnaneswar Gude, PhD
Kelly Gene Cook, Sr. Endowed Chair and Associate Professor of Civil and Environmental Engineering, Richard A Rula School of Civil and Environmental Engineering, Mississippi State University, Mississippi State, Mississippi, USA

Veera Gnaneswar Gude, PhD, is Kelly Gene Cook, Sr. Endowed Chair and Associate Professor of Environmental Engineering in the Department of Civil and Environmental Engineering at Mississippi State University (MSU). He is also an affiliate research faculty at the Energy Institute of Louisiana and the Department of Chemical Engineering at the University of Louisiana at Lafayette. His academic research and industry design experiences focus on the cutting-edge research areas of biofuels, desalination, and wastewater treatment scientific and technological development.

Dr. Gude has published over 150 scientific research articles on biofuels, desalination, and water and wastewater treatment research and his research work is well utilized in the literature, with over 5,500 citations. He has published two books in biofuels research and three books in desalination research and one book in sustainable water management, as well as more than 18 invited book chapters, 50 conference proceedings papers, 20 technical reports, several popular press articles, and media releases, and two patents in microalgae biofuels and low-temperature desalination technologies respectively.

Dr. Gude has delivered over 50 invited lectures, including panelist talks and over 150 scientific research and educational presentations. He has organized many workshops on water-energy-environment nexus topics on national and international platforms. He was the chair and board representative for clean energy and water division of ASES between 2011 and 2016. He serves on numerous scientific advisory boards and task committees across the world, including AAEES, ASCE-EWRI, ASEE, ASES, and AWWA.

He is a member of several editorial boards and editor for many scientific journals, including *ASCE Journal of Environmental Engineering, ASCE Journal of Hazardous, Toxic and Radioactive Waste, Desalination, and Water Treatment; Nature npj Clean Water, Renewable Energy;* and *Water*

Environment Research (Water Environment Federation) Journal. His research is supported by NSF, USEPA, USGS, USDA, and many industrial and international agencies. He has received much recognition for research, teaching, and service activities from regional, national, and international organizations, conferences, and professional societies (ASCE, ASEE, ASES, and Chi Epsilon National Civil Engineering Honor Society). He also received the 2017 Albert Nelson Marquis Lifetime Achievement Award and the 2017 ASCE MS Engineer of the Year Award.

Dr. Gude received a BS degree in Chemical Engineering Technology from Osmania University and worked for Du Pont Singapore. He received MS and PhD degrees in Environmental Engineering from the National University of Singapore and New Mexico State University, respectively. His research expertise includes biofuel synthesis using sustainable chemistry principles and process intensification, and advanced wastewater treatment for resource recovery. He is a licensed professional engineer, a board-certified environmental engineer (also known as diplomat environmental engineer, DEE), a board-certified diplomat water resources engineer (D.WRE), and an elected Fellow of ASCE (American Society of Civil Engineers).

Contents

Contributors

Syed Muhammad Hassaan Ali
University of Pannonia, Veszprém, Hungary

Andrea Arias
Department of Chemical Engineering, Universidade de Santiago de Compostela – 15782 Santiago de Compostela, Spain

Francisco Corona
CARTIF Technology Center, Parque Tecnológico de Boecillo – 205, 47151, Boecillo, Valladolid, Spain

Nancy Diaz-Elsayed
Research Assistant Professor, University of South Florida, 4202 E. Fowler Avenue, Tampa, Florida – 33620, USA, Phone: 760-220-8167, Fax: (813) 974-2957, E-mail: nancyd1@usf.edu

Endre Domokos
University of Pannonia, Veszprém, Hungary, E-mail: drdomokosendre@gmail.com

Gumersindo Feijoo
Department of Chemical Engineering, Universidade de Santiago de Compostela – 15782 Santiago de Compostela, Spain, E-mail: gumersindo.feijoo@usc.es

Hannah Feldman
Process and Systems Engineering Center (PROSYS), Department of Chemical and Biochemical Engineering, Technical University of Denmark, Building – 229, 2800 Kongens Lyngby, Denmark

Xavier Flores-Alsina
Process and Systems Engineering Center (PROSYS), Department of Chemical and Biochemical Engineering, Technical University of Denmark, Building – 229, 2800 Kongens Lyngby, Denmark

Oscar Franchi
Facultad de Ingeniería y Ciencias, Universidad Adolfo Ibáñez, Avenida Padre Hurtado 750, Viña del Mar – 2562340, Chile, Phone: +56964695194, E-mail: oscar.franchi@edu.uai.cl

Krist V. Gernaey
Process and Systems Engineering Center (PROSYS), Department of Chemical and Biochemical Engineering, Technical University of Denmark, Building – 229, 2800 Kongens Lyngby, Denmark, E-mail: kvg@kt.dtu.dk

Veera Gnaneswar Gude
Richard A Rula School of Civil and Environmental Engineering, Mississippi State University, Mississippi State, MS – 39762, USA, E-mail: gude@cee.msstate.edu

Dolores Hidalgo
CARTIF Technology Center, Parque Tecnológico de Boecillo – 205, 47151, Boecillo, Valladolid, Spain, Phone: +34 983 546504, E-mail: dolhid@cartif.es

Christian Hubert
Bundeswehr University Munich, Chair of Sanitary Engineering and Waste Management, Werner-Heisenberg-Weg – 39, 85577 Neubiberg, Germany, E-mail: christian.schaum@unibw.de

Steffen Krause
Bundeswehr University Munich, Chair of Sanitary Engineering and Waste Management, Werner-Heisenberg-Weg – 39, 85577 Neubiberg, Germany

Wenhai Luo
Beijing Key Laboratory of Farmland Soil Pollution Prevention and Remediation, College of Resources and Environmental Sciences, China Agricultural University, Beijing – 100193, China, E-mail: luowenhai@cau.edu.cn

Seyed Soheil Mansouri
Process and Systems Engineering Center (PROSYS), Department of Chemical and Biochemical Engineering, Technical University of Denmark, Building – 229, 2800 Kongens Lyngby, Denmark

Jesús M. Martín-Marroquín
CARTIF Technology Center, Parque Tecnológico de Boecillo – 205, 47151, Boecillo, Valladolid, Spain

Macarena Mellado
Facultad de Química y Biología, Universidad de Santiago de Chile, Avenida Libertador Bernardo O'Higgins, 3363, Santiago – 9170022, Chile, Phone: +56962490603, E-mail: macarena.mellado.m@usach.cl

Weiwei Mo
Assistant Professor, University of New Hampshire, 35 Colovos Road, Durham, New Hampshire – 03824, US, Phone: 603-862-2808, Fax: (603) 862-3957, E-mail: Weiwei.Mo@unh.edu

Maria Teresa Moreira
Department of Chemical Engineering, Universidade de Santiago de Compostela – 15782 Santiago de Compostela, Spain, E-mail: maite.moreira@usc.es

Elham Ramin
Process and Systems Engineering Center (PROSYS), Department of Chemical and Biochemical Engineering, Technical University of Denmark, Building – 229, 2800 Kongens Lyngby, Denmark

Pedram Ramin
Process and Systems Engineering Center (PROSYS), Department of Chemical and Biochemical Engineering, Technical University of Denmark, Building – 229, 2800 Kongens Lyngby, Denmark, E-mail: pear@kt.dtu.dk

Gideon Sarpong
Richard A Rula School of Civil and Environmental Engineering, Mississippi State University, Mississippi State, MS – 39762, USA, E-mail: maximatlove@gmail.com

Christian Schaum
Bundeswehr University Munich, Chair of Sanitary Engineering and Waste Management, Werner-Heisenberg-Weg – 39, 85577 Neubiberg, Germany, E-mail: christian.schaum@unibw.de

Viktor Sebestyén
University of Pannonia, Veszprém, Hungary

Viola Somogyi
University of Pannonia, Veszprém, Hungary, E-mail: somogyiv@uni-pannon.hu

Bettina Steiniger
Bundeswehr University Munich, Chair of Sanitary Engineering and Waste Management, Werner-Heisenberg-Weg – 39, 85577 Neubiberg, Germany, E-mail: bettina.steiniger@unibw.de

Bülent Topkaya
Akdeniz University, Department of Environmental Engineering, Antalya, Turkey,
E-mail: btopkaya@akdeniz.edu.tr

Isuru A. Udugama
Process and Systems Engineering Center (PROSYS), Department of Chemical and Biochemical Engineering, Technical University of Denmark, Building – 229, 2800 Kongens Lyngby, Denmark

Xinhua Wang
Jiangsu Key Laboratory of Anaerobic Biotechnology, School of Environmental and Civil Engineering, Jiangnan University, Wuxi – 214122, China

Zhicheng Xu
Beijing Key Laboratory of Farmland Soil Pollution Prevention and Remediation, College of Resources and Environmental Sciences, China Agricultural University, Beijing – 100193, China

Mustafa Yildirim
Antalya Water and Wastewater Administration General Directorate (ASAT), Antalya, Turkey,
E-mail: yildirim.mustafa@gmail.com

Qiong Zhang
Associate Professor, University of South Florida, 4202 E. Fowler Avenue, Tampa, Florida – 33620, USA, Phone: (813) 974-6448, Fax: (813) 974-2957,
E-mail: qiongzhang@usf.edu

Abbreviations

A	acidification
ABR	anaerobic baffled reactor
ABW	automatic backwash
AD	abiotic depletion
ADM1	anaerobic digestion model no. 1
ADs	anaerobic digesters
AFB	anaerobic fluidized bed reactor
AFMBR	anaerobic fluidized membrane bioreactor
AIWPS	advanced integrated wastewater pond systems
AnMBR	anaerobic membrane bioreactor
ASBR	anaerobic sequencing batch reactor
ASM1	activated sludge model no. 1
ASM2d	activated sludge model no. 2d
AST	activated sludge treatment
ATP	adenosine triphosphate
ATS	algal turf scrubber
BDO	increase biological oxygen
BES	bioelectrochemical systems
BNR	biological nutrient removal
BNRM1	biological nutrient removal model no.1
BOD	biochemical oxygen demand
BSM2	benchmark simulation model no. 2
BW	blackwater
CAS	conventional activated sludge
CC	climate change
CH_4	methane
CHP	combined heat and power
CO_2	carbon dioxide
COD	chemical oxygen demand
COP	coefficient of performance
CSTR	continuous stirred tank reactor
CTA	cellulose triacetate
CWA	clean water act
CWs	constructed wetlands

DCP	di-calcium phosphate
DHW	domestic hot water
DNA	deoxyribonucleic acid
DPR	direct potable reuse
E	eutrophication
EA	extended aeration
EBMUD	East Bay Municipal Utility District
EBPR	enhanced biological phosphorus removal
ED	electrodialysis
EDTA-2Na	ethylene-diamine-tetra acetic acid disodium salt
EGSB	expanded granular sludge bed
EIO-LCA	economic input-output LCA
EP	eutrophication potential
EPA	Environmental Protection Agency
ERA	environmental risk assessment
EU	European Union
FAO	Food and Agriculture Organization
FAST	facilities accelerating science and technology
FD	fossil depletion
FE	freshwater ecotoxicity
FL	Florida
FO	forward osmosis
FOG	fat-oil-grease
FTHFS	formyltetrahydrofolate synthetase
FU	functional unit
FWS	free water surface
GAC	granular activated carbon
GHG	greenhouse gas
GJJWWTP	Gloversville-Johnstown joint wastewater treatment plant
GW	greywater
GWP	global warming potential
HCWTP	Howard F. Curren advanced wastewater treatment plant
HDPE	high-density polyethylene
HE	heat exchangers
HF	horizontal flows
HFO	hydrous ferric oxide
HNO_3	nitric acid
HRAP	high rate algal pond
HRSD	Hampton roads sanitation district

HRT	hydraulic residence time
HSSF	horizontal subsurface flow
HT	human toxicity
IPR	indirect potable reuse
K	potassium
LCA	life cycle assessment
LCC	life cycle costing
LCCA	life cycle cost analysis
LCI	life cycle inventory
LEM	low energy mainline
Lh	hydraulic loading rate
LT	list of transformations
M&M	major and minor
MBBR	moving bed bioreactor
MBRs	membrane bioreactors
MCP	mono-calcium phosphate
MD	membrane distillation
ME	marine ecotoxicity
MEC	microbial electrolysis cell
MF	microfiltration
MFA	material flow analysis
MFCs	microbial fuel cells
MFZ	multi-family zones
Mg	magnesium
$MgCl_2$	magnesium chloride
MGD	million gallons per day
MRC	microbial recovery cell
MWRD	metropolitan water reclamation district
N	nitrogen
N_2O	nitrous oxide
NaCl	sodium chloride
NaOAc	sodium acetate
NEB	net environmental benefit
NF	nanofiltration
NGS	next-generation sequencing
NH_3	ammonia gas
NH_3SO_4	ammonium sulfate
NH_4^+	ammonium
NO_2^-	nitrite

NPR	non-potable reuse
NPV	net present value
NRC	National Research Council
NSC	nitrate short-cut
NSF	National Sanitation Foundation
NUE	nutrient use efficiency
NYSERDA	New York State Energy Research and Development Authority
O&M	operation and maintenance
O_2	oxygen
OD	ozone depletion
OLAND	oxygen limited autotrophic nitrification-denitrification
OMBR	osmotic membrane bioreactor
P	phosphorus
PABF	passively aerated biological filter
PAOs	phosphorus accumulating organisms
PE (p.e.)	population equivalent
PEI	potential environmental impacts
PHA	poly-β-hydroxyalkanoates
PMF	particulate matter formation
PO_4^{3-}	phosphate
POF	photochemical oxidation formation
PRR	partition-release-recover
PS	primary sludge
PVC	polyvinyl chloride
RBC	rotating biological contactor
RO	reverse osmosis
RPG	reactor performance indicator
RW	reclaimed water
SAS	soil absorption system
SBR	sequencing batch reactors
SCENA	short cut enhanced nutrient abatement
SCN	nitrate short-cut
SCP	struvite crystallization process
SDGS	small diameter gravity sewers
SFZ	single-family zones
SI	saturation index
SMS	shotgun metagenomic sequencing
SP	stabilization pond
SRB	sulfate-reducing bacteria

SRT	sludge retention time
SRWWTP	Sheboygan regional wastewater treatment plant
SS	surplus sludge
SSA	sewage sludge ashes
SSF	subsurface flow
ST	septic tank
T	toxicity
TA	terrestrial acidification
TCP	tri-calcium phosphate
t-DS	ton of dry solids
TE	terrestrial ecotoxicity
TFC	thin film composite
TN	total nitrogen
TRL	technology readiness levels
TS	total solids
TSS	total suspended solids
TUD	Technical University of Delft
UAC	uniform annual cash flow
UASB	up-flow anaerobic sludge blanket
UCT	University of Cape Town
UF	ultrafiltration
US	United States
UV	ultraviolet
UWWTD	Urban wastewater treatment directive
VF	vertical flows
VFAs	volatile fatty acids
VFD	variable frequency drives
VLT	vegetated land treatment
VSS	volatile suspended solids
WAGBR	water aeration grooves biofilm reactor
WD	water depletion
WRRF	water resource recovery facilities
WSP	waste stabilization ponds
WWSHP	wastewater source heat pump
WWTPs	wastewater treatment plants

Preface

Wastewater has to be properly treated prior to discharge into the environment to protect the receiving water bodies from adverse impacts. As the quantities increase with the population growth, the requirements for organic carbon and macronutrient removal also increased. There are many physical, chemical, and biological processes that help achieve this goal. The most commonly used technology to remove organic carbon is activated sludge process which requires aeration while nitrification/denitrification processes remove nitrogen. Phosphorous can be removed through sludge collection, chemical precipitation, and biological processes.

Wastewater treatment requires significant energy and chemicals and produces large quantities of sludge requiring further management. However, in recent years, wastewater has become a key player in managing the nutrient and water needs of many communities due to increasing water scarcity and water quality issues. Once considered as a nuisance and environmental burden, wastewater is now viewed as a resource in many communities around the world due to its potential to contribute to nutrient and water resource demands. However, many advances are essential to exploit and extract the resources available through wastewater. There is a strong urge for developing both scientific and practical knowledge required for transitioning the current resource-demanding wastewater treatment plants (WWTPs) into resource and revenue-generating water resource recovery facilities (WRRF). Towards this end, this book provides a comprehensive knowledge on various aspects of wastewater resource management from the point of process sustainability and resource recovery. This authoritative information is crucial for developing resource-efficient and sustainable wastewater treatment technologies and management strategies for both small (decentralized) and large (centralized) communities.

Chapter 1 provides an excellent overview of various resource recovery technologies and a framework for assessing the sustainability of various treatment schemes. Important methods used for environmental, economic, and social assessments are presented to help researchers and industry practitioners in their design and evaluation of sewer and wastewater systems. Chapter 2 discussed general platforms and modeling tools used in assessing the water resource recovery facilities. Different simulation approaches to evaluate and develop innovative configurations for future water resource recovery facilities are presented. These tools also can be used to predict the efficiency of these configurations. In Chapter 3, osmotic membrane bioreactors (OMBR) and hybrid systems are presented as a potential solution for efficient nutrient harvesting from wastewater. Major issues related to the process performance and practice applications are discussed in detail.

Energy recovery is an important aspect of developing sustainable water resource recovery facilities. Chapter 4 provides a comprehensive review of various energy recovery options in wastewater treatment plants. Wastewater treatment systems were categorized into three configurations based on their potential for achieving energy-neutral or energy-positive treatment. Mass and energy flow balances were presented for these configurations along with case studies. Some practical considerations to address the limitations associated with the proposed approaches were also discussed. Thermal energy embedded in wastewater sources presents many opportunities for its recovery and reuse in the plant operations. Chapter 5 delineates the sinks and sources of thermal energy in wastewater treatment plants. Detailed energy balances with and without anaerobic digesters (ADs) are developed to provide solutions for thermal energy recovery from wastewater. Best practices and case studies are used to discuss the design of thermal heat recovery systems.

Nutrient (nitrogen and phosphorous) recovery is critical for sustaining the quality of our environment. Chapter 6 explores various alternatives (processes and technologies) for recovering nitrogen from wastewater. Success stories at industry scale demonstrations are discussed to highlight opportunities available for closing nutrient cycles. Chapter 7 discusses opportunities for phosphorous recovery in water resource recovery facilities. Factors that affect the feasibility of phosphorous removal in full-scale operations are discussed. Energy recovery from anaerobic digesters is the first step to achieve resource recovery from the wastewater. Chapter 8 discusses the microbial factories that are responsible for biogas production in anaerobic digesters. Currently, available methods to characterize the microbial community in anaerobic digesters have been discussed in detail. In addition, methods to monitor the microbial communities to understand and to maintain the performance of anaerobic digesters are presented.

Small scale and decentralized wastewater treatment systems face a different set of challenges. Chapter 9 provides an analysis of different scenarios for the collection and treatment of wastewater from individual households and small communities, moving from centralized to decentralized approaches. These scenarios are compared from a life-cycle assessment perspective and implications for the reuse of treated effluents and the recovery of resources are discussed. Chapter 10 examines different wastewater treatment options for small scale communities. Technologies such as vegetated land treatment (VLT), constructed wetlands (CWs), rotating biological contactors (RBC), conventional activated sludge treatment, MBR, extended aeration (EA), and stabilization ponds (SP) by which effluents are discharged to sensitive and less sensitive areas are evaluated using life cycle assessment (LCA) approach. Influencing factors for efficient wastewater treatment are discussed.

CHAPTER 1

The Sustainability Dimensions of Resource Recovery from "Wastewater"

NANCY DIAZ-ELSAYED,[1] WEIWEI MO,[2] and QIONG ZHANG[1]

[1]*University of South Florida, 4202 E. Fowler Avenue, Tampa, Florida – 33620, USA*

[2]*University of New Hampshire, 35 Colovos Road, Durham, New Hampshire – 03824, USA*

ABSTRACT

Resource recovery provides a paradigm shift from viewing "wastewater" as an undesired waste that requires energy-intensive treatment to considering it as a product with valuable resources. However, identifying viable and sustainable resource recovery technologies can be challenging given the myriad of solutions available, the heterogeneous wastewater characteristics, and local conditions, and the lack of a consistent sustainability assessment framework. This chapter provides an overview of technologies for recovering water, energy, and nutrients from municipal wastewater at various implementation scales. Three distinctive system scales are discussed in this chapter: small (design flows of 17 m^3/day or less), medium (8 to 20,000 m^3/day), and large (3,800 m^3/day or more), each presenting a unique population coverage, treatment technologies and infrastructures, and regulations. Mature and emerging technologies are introduced along with their major benefits and challenges. Furthermore, the methods used for environmental, economic, and social assessments are presented to provide guidance for the sustainability evaluations of resource recovery. The emerging trends in sustainability assessments are also discussed to provide insights about the synergies and trade-offs among resource recovery technologies, which can aid researchers and industry practitioners in their design and evaluation of municipal sewer and wastewater systems.

1.1 INTRODUCTION

Wastewater is considered as "a combination of the liquid or water-carried wastes removed from residences, institutions, and commercial and industrial establishments, together with such groundwater, surface water, and stormwater as may be present" (Metcalf and Eddy, 2004). Due to the adverse impacts of organics, pathogens, nutrients, and other pollutants contained in untreated wastewater, contaminant removal is the primary objective of wastewater systems in order to protect human health and ecosystems. Driven by this objective, wastewater is considered as waste with an end-of-pipe treatment and subsequent discharge management approach. Approximately 65% of the US population is served by conventional centralized wastewater systems, which are categorized by an extensive sewer network for collection and a central location for treatment (Tchobanoglous and Leverenz, 2013). Such systems take advantage of economies of scale for densely-populated areas and better quality control due to centralized monitoring and management (Maurer et al., 2005); however, they face many challenges, such as potential diseconomies of scale for collection systems; potential water quality deterioration of receiving surface water bodies due to nutrients, presence of pharmaceuticals and other emerging contaminants; expensive treatment due to diluted wastewater; high vulnerability to natural disasters and terrorism; and high water and energy consumption (Wilderer and Schreff, 2000; Bakir, 2001; Means, 2004; Ho, 2005; Angelakis and Snyder, 2015). In addition, these systems have been facing the challenge of water quality standards becoming more stringent over time. Since the implementation of the clean water act (CWA) in the United States (US)inone various federal regulation have been established (e.g., Water Quality Act (WQA), total maximum daily load (TMDL)) to tighten the water quality standards. This was partly due to the increased understanding and awareness of long-term adverse effects caused by wastewater discharges.

To comply with these standards, wastewater treatment facilities have increased the degree of treatment to meet the additional treatment objectives. For example, the treatment objectives have been expanded from the reduction of biological oxygen demand (BOD), total suspended solids (TSS), and pathogens to the reduction of nitrogen and phosphorus. As a result, additional economic burden and resource requirements have been put into existing resource-limited wastewater treatment systems. Facing external stressors such as population growth, urbanization, climate change (CC), and resource shortages (especially water, energy, and phosphorus) (Zimmerman et al., 2008; Gleeson et al., 2012), internal stressors such as aging infrastructure (Copeland and Tiemann, 2010) and increasing energy costs (Means, 2004), as well as criticisms on the sustainability of the water and wastewater industry (Anastas, 2012), a paradigm shift has occurred that has transformed the perception of wastewater from a waste to a product with valuable resources over the last few decades (Guest et al., 2009).

Various studies have been conducted to review the wastewater treatment processes and associated costs for water reuse (Asano et al., 2007; Guo et al., 2014), source separation technologies (Larsen et al., 2009), resource recovery solutions at the large-scale (Mo and Zhang, 2013) and across various scales (Diaz-Elsayed et al., 2019). This chapter presents the technical, environmental, and social perspectives of resource recovery from "wastewater." Specifically, the recovery of water, energy, and nutrients are discussed in depth. Since there are substantial differences in the types of technologies used for resource recovery, their respective efficacy in recovering resources, the materials and energy required for implementation, and the sustainability of these systems are discussed considering their scale (e.g., size of implementation). The classification of the scale of a resource recovery system has been defined with respect to the design flow for wastewater (Diaz-Elsayed et al., 2019): small-scale systems with flows of 17 m^3 per day or less, medium-scale systems with flows of 8 to 20,000 m^3 per day, and large-scale systems with flows of 3,800 m^3 per day or more.

1.2 OVERVIEW OF RESOURCE RECOVERY TECHNOLOGIES

This section provides an overview of the mature and emerging technologies that can be used for the recovery of water, energy, and nutrients from wastewater.

1.2.1 WATER RECOVERY

1.2.1.1 NON-POTABLE REUSE (NPR)

Most water reuse occurs as non-potable reuse (NPR) applications. Small-scale NPR projects consist primarily of lawn or garden irrigation and toilet flushing, while larger-scale implementations include irrigation at higher volumes (e.g., golf courses or farms), industrial cooling, artificial wetlands, and recreational impoundments. The US Environmental Protection Agency (EPA) recommends secondary treatment and disinfection for most NPR applications, and the addition of a filtration process when direct human contact with the water can occur (e.g., unrestricted reuse applications or for the irrigation of food crops that can be consumed raw) (US EPA, 2012).

For small-scale systems, primary treatment can be accomplished by a septic tank, secondary treatment by a number of alternative processes (e.g., a membrane bioreactor (MBR), aerobic treatment unit, or sand/multi-media filtration), and disinfection by chlorination and/or ultraviolet (UV) light. The onsite technologies that have been approved by the National Sanitation Foundation (NSF) 350 standard rely primarily on these technologies (NSF, 2017). Onsite NPR systems primarily utilize greywater (GW), which includes wastewater collected from sinks, showers, baths, and washing machines, as opposed to blackwater (BW), which is usually generated from the toilet flushing. By separating BW from GW via source separation, the constituent loading is more predictable and a different treatment process can be used to target each stream of wastewater. For example, the Phipps Center for Sustainable Landscapes sought Living Building certification and treats GW by means of wetlands, sand filtration, and UV disinfection (ILFI, 2017). MBRs are commonly used at the small or medium-scale due to their relatively small footprint and high removal efficiency of solids and pathogens (Melin et al., 2006). However, they do have disadvantages: they are expensive to install and operate, must be operated carefully to avoid membrane fouling, and can be sensitive to chemicals (Melin et al., 2006).

Centralized wastewater treatment plants (WWTPs) are naturally equipped to accommodate NPR projects since secondary treatment is the minimum treatment required in the US (US EPA, 2009). The treatment trains implemented are highly variable due to the large spectrum of NPR applications, and the need to accommodate factors such as treatment costs, land space availability, and climate zone. When all or only a portion of the secondary effluent will be reused, advanced wastewater treatment processes can be implemented within the WWTP. For example, the William E. Dunn Water Reclamation Facility in Pinellas County, Florida produces reclaimed water (RW) primarily for irrigation, and implements the following processes: screening and grit removal, a 5-stage Bardenpho process (e.g., fermentation, oxidation, and aeration with recirculation), clarification, automatic backwash (ABW) sand filtration, and chlorine disinfection (Pinellas County, n.d.). Alternatively, a separate plant can be constructed for advanced treatment as is the case for the Green Acres Project in Orange County, California, which incorporates coagulation, flocculation, dual-media bed filtration (anthracite and sand), and chlorine disinfection (OCWD, 2019). The Integrated

Treatment Train Toolbox (IT^3), developed by Trussell et al. (2015), can be used as a design tool to identify alternative wastewater treatment options for all types of reuse.

1.2.1.2 INDIRECT POTABLE REUSE (IPR)

Indirect potable reuse (IPR) projects incorporate an environmental buffer (i.e., groundwater recharge or surface water augmentation) prior to treatment at a water treatment plant. The environmental buffer can help to improve public perception (e.g., by adding a natural component), improve the water quality (e.g., through attenuation processes or dilution), and increase the time period between wastewater treatment and use (NRC, 2012). IPR projects are typically at the medium or large scale, and some well-known projects include Singapore's NEWater facilities (PUB, 2019) and Orange County's Groundwater Replenishment System (OCWD, n.d.). IPR treatment trains are also variable, but generally incorporate filtration (e.g., multi-media filtration, membrane filtration, or ultra/microfiltration), ozonation, reverse osmosis (RO), and/or activated carbon prior to disinfection (Diaz-Elsayed et al., 2019). This is, in part, based on requirements for California IPR projects, which require ultra/microfiltration, RO, and UV disinfection prior to the environmental buffer (Gerrity et al., 2013).

1.2.1.3 DIRECT POTABLE REUSE (DPR)

Although no specific regulations have been drafted in the US concerning direct potable reuse (DPR), treatment trains that can achieve drinking water quality have been recommended (US EPA, 2012). A multi-barrier approach introduces numerous stages in the treatment process to facilitate water quality testing and promote the identification of hazards in case a fault should arise during the treatment process. Although small-scale DPR was implemented in the 1970s and early 1980s in Colorado in the US, the treatment system was not sustainable as the technology was too expensive to maintain (Tchobanoglous et al., 2011). Larger scale DPR implementations include Cloudcroft, New Mexico (380 m^3/day), Big Spring, Texas (7,570 m^3/day), and Windhoek, Namibia (21,000 m^3/day). The wastewater treatment trains often rely on a combination of secondary treatment, ultra/microfiltration, RO, and/or biological/granular activated carbon (GAC) with multiple forms of disinfection (Diaz-Elsayed et al., 2019). Thereafter, the effluent is blended with raw water prior to entering the water treatment plant for further treatment. While DPR avoids the construction and operation of a parallel water distribution pipeline (resulting in savings in costs and materials), there still remain several barriers to implementation including overcoming the "yuck factor" customers face with the notion of drinking treated wastewater, as well as the lack of well-defined treatment guidelines. Public outreach and education help overcome such barriers and a demonstration plant can be implemented to provide the community with an opportunity to learn about the proposed treatment train firsthand. Moreover, the demonstration plant helps plant operators monitor the water quality and fine-tune the processes and operating parameters prior to commissioning the new facility.

1.2.2 ENERGY RECOVERY

Although wastewater treatment is an energy-intensive process, wastewater itself contains nearly five times the amount of energy needed for treatment, primarily in the form of thermal energy (Tarallo et al., 2015). Other available forms of energy recovery include hydropower and options for biosolids management such as biogas production from anaerobic digestion or biosolids incineration.

1.2.2.1 THERMAL ENERGY

Thermal energy can be extracted with a heat exchanger and heat pump(s) when there is a temperature difference between the wastewater and another medium (e.g., the surrounding environment or incoming water); the heat can thereafter be used onsite or in local facilities. Since wastewater is typically warmer than the incoming potable water, the temperature difference would be greater in regions with cold climates where the incoming water has the opportunity to further decrease in temperature. Thermal energy recovery spans small-scale implementations at the household-level to large-scale implementations at centralized WWTPs. At the small scale, current technologies can achieve as much as 70% in energy reductions for water heating from GW energy recovery (Nexus, 2015). Heat recovery from shower drains is an emerging technology at the small scale, but the amount of heat that can be recovered varies significantly (30–75%) as it depends on several factors including the type of system implemented (Słyś and Kordana, 2014; Gabor et al., 2017). Another emerging application of wastewater heat is the preheating of outdoor air in mechanical ventilation with heat recovery systems to reduce the heat expended on defrosting such systems in cold climates (Nourozi et al., 2019). Presently, the largest wastewater heat recovery plant is located in Hammarbyverket, Sweden (25–114 MGD), which produces 1,235 GWh per year of heat for residential buildings (Mikkonen et al., 2013).

1.2.2.2 HYDROPOWER

The kinetic energy from high wastewater flows and head loss in raw or treated wastewater provides an opportunity for hydropower generation via turbines. This form of energy recovery is implemented at the medium or large scale since higher flow rates lead to a greater energy recovery potential. Examples include the Profray plant (8,600 m^3/day capacity) in Switzerland, which has a gross head of 449 m and generates 851 MWh per year (San Bruno et al., 2010), and San Diego's Point Loma WWTP (180 MGD average flow-rate), which can generate 1.35 MW from an ocean outfall that drops 27 m in elevation over 7.2 km (San Diego, 2017).

1.2.2.3 BIOSOLIDS-ANAEROBIC DIGESTION

Anaerobic digestion is a common biosolids management option in the US, and introduces the possibility of energy recovery from biogas production. Mesophilic digestion can be implemented at operating temperatures of 25–33°C, while thermophilic digestion requires a temperature of at least 55°C. The plant can capture the biogas produced by the digestion process to produce heat for local use or to generate electricity via combined heat and power (CHP). A plant size of 20,000 PE (person-equivalent) or greater is recommended for mesophilic anaerobic digestion with energy recovery via CHP with regards to economic feasibility (Nowak et al., 2004). For large WWTPs (>100,000 PE) with CHP, the ratio of electricity generated versus electricity used by the plant is expected to be 0.68 to 1.5 (Aarhus, n.d.; Bachmann, 2015). In contrast, a ratio of 0.37 is expected for smaller plants (<10,000 PE) (Bachmann, 2015). At the medium scale, the Delhi Charter Township WWTP (2.5 MGD) in Michigan can treat up to 30,000 GPD of biosolids with parallel treatment trains of two-phased anaerobic digesters (ADs) (Viswanathan, 2010). Since the plant's two 30 kWh micro-turbines (used for CHP) have a low electricity conversion efficiency (~30%), the site focuses on heat generation. Accordingly, the system can eliminate the need for nearly all outsourced heat and is capable of reducing electricity consumption by more than 40%. At a large scale, the Marselisborg WWTP in Denmark treats approximately 25 MGD for 200,000 PE and was able to optimize its treatment processes and enhance its biogas production from onsite digesters to the point where it became a net energy producer. In 2015, the plant consumed 0.18 kWh/m^3 for wastewater treatment and generated 0.28 kWh/m^3 (or ~150% of the energy needed for treatment) (Aarhus Vand, n.d.). However, this is not typical of most large-scale implementations, since the significant improvements achieved in energy efficiency helped make such savings feasible.

1.2.2.4 BIOSOLIDS-INCINERATION

The incineration of biosolids is becoming more popular as regulations become more stringent on disposal pathways, the availability of land becomes more constrained, and technologies for controlling harmful emissions become more prevalent. The heat from incineration can be used in a steam cycle power plant to produce electricity or it can be reused locally to provide heat for the incineration process. Not only does incineration offer reliable energy production for medium- and large-scale systems (Stillwell, 2010), but the process effectively destroys pathogens and can reduce the volume of wastewater solids by up to 95% (US EPA, 2003).

The dominant technologies for incineration are multiple hearth furnaces and fluidized bed furnaces. Since fluidized bed furnaces are "more efficient, more stable, and easier to operate," their installation has been more prevalent in the last few decades (US EPA, 2003). When the sludge is dewatered to ~30% solids (prior to incineration) and if it has an adequate heating value, the incineration process can be energy self-sufficient (US EPA, 2003). Although digestion provides the benefit of volume reduction for transportation, undigested sludge has a higher heating value: anaerobically digested primary sludge (~12.8 GJ per ton of dry solids (t-DS)), combined primary and waste activated sludge (11.6–23.3 GJ/t-DS), and primary sludge (18.6–30.3 GJ/t-DS) (US EPA, 2003). Hong et al. (2009) report electricity consumption of 304.8 kWh/t-DS and electricity generation from waste heat of 928.5 kWh/t-DS for the mono-combustion of biosolids. Co-combustion of biosolids (e.g., with coal, wood chips, municipal solid waste, or in cement kiln) resulted in a net energy gain of 0.58 to 5.0 kWh/kg-DS in the study by Cartmell et al. (2006). Incineration limits the recovery of nutrients since biosolids are converted to ash, but an emerging technology for combined incineration and phosphorus recovery, the Bio-Con process, was proposed by Lundin et al. (2004).

1.2.2.5 BIOSOLIDS-PYROLYSIS AND GASIFICATION

Pyrolysis and gasification output synthetic gas (syngas) and bio-oil from biosolids by means of thermochemical processes. In pyrolysis, biosolids are heated under anaerobic conditions and thermally decomposed to combustible gases, charcoal, and bio-oil (USDA, 2017). In gasification, sludge is thermally decomposed to syngas and ash in aerobic conditions at high temperatures. Ultra-high-temperature gasification (1100–1500°C) was tested on biosolids at a WWTP in Munich, Germany, and Gikas et al. (2014) estimated that a net electrical energy output of ~15 MJ/kg dry solids could be achieved at a 20 MGD facility. More information on the outcomes of demonstration plants and large-scale implementations of gasification technology in Germany, Japan, Canada, and the US can be found in a 2012 technology assessment report to the US EPA (GGTC, 2012).

1.2.2.6 EMERGING TECHNOLOGIES

The emerging technologies for energy recovery from wastewater include microalgae production, microbial fuel cells (MFCs), and nitrous oxide production. Energy can be recovered from microalgae grown in nutrient-rich wastewater by means of digestion, gasification, or extraction and transesterification. The energy yield depends on a number of factors including the type of process implemented and the type of energy recovered. For example, Caporgno et al. (2015) reported 0.15 to 0.34 L CH_4/g VS from microalgae digestion, whereas Aresta et al. (2005) found that the net energy of 9,500 MJ/dry ton can be produced via microalgae gasification. A pilot forward osmosis (FO) system was tested by NASA where municipal wastewater and membrane enclosures were used to grow microalgae offshore (Buckwalter et al., 2013; Wiley et al., 2013). The salinity gradient extracts water from the membrane enclosures and condenses algae culture; while the heat capacity of the ocean water-cooled, the photobioreactors and the wave energy provide supplementary mixing. Current challenges of microalgae production include contamination, unstable biomass production, and the harvesting of small cells and diluted cultures (Cai et al., 2013).

While microalgae production can yield biogas and/or bio-fuel, MFCs facilitate the production of electricity. In a review of the operating power density, Ge et al. (2014) found that small systems using less than 50 mL of wastewater generally resulted in higher power densities (>500 W/m^3wastewater) relative to larger systems of <2,000 mL (100–500 W/m^3) and >2,000 mL (<30 W/m^3). Energy losses and costs remain high, so significant improvements are needed to make MFCs a competitive form of energy recovery (McCarty et al., 2011).

The production and use of nitrous oxide (N_2O) to enhance methane combustion have been investigated in experimental studies. Specifically, Scherson et al. (2014) produced N_2O from the oxidation of ammonia (NH_4^+) and reduction of nitrite (NO_2^-) during the side stream treatment of ADcentrate. By replacing oxygen (O_2) with N_2O in a biogas-fed internal combustion engine, the power output increased by 6–7%. As with other emerging technologies, pilot-scale or commercial installations would aid in the evaluation of the feasibility and performance of these technologies from a technical, economic, and environmental perspective.

1.2.3 NUTRIENT RECOVERY

This section focuses on the recovery of nitrogen (N) and phosphorus (P) from wastewater. Nutrient recovery mechanisms exist across all scales of implementation and result in varying degrees of nutrient concentration.

1.2.3.1 URINE SOURCE SEPARATION

Urine contributes ~80% of the nitrogen (N), ~70% of the potassium (K), and ~50% of the phosphorus (P) in municipal wastewater (Wilsenach et al., 2003). This is because undiluted urine contains high concentrations of nutrients amounting to 6500–9200 mg/L (total N), 300–980 mg/L (total P), and 1400–2200 mg/L (K) (Mitchell et al., 2013). Conventional toilets dilute urine with water, but by separating urine from the main wastewater stream with a urine-diverting toilet and a separate pipeline (urine source separation), the nutrient-rich fluid could be used for fertilization. Proper maintenance of these systems is essential as mineral build-up and scaling could occur if they are not properly maintained. Processes such as evaporation, drying, or struvite precipitation can be implemented to increase the concentration of nutrients and facilitate transportation to another site (Wilsenach et al.,

2003). Prior to being land applied, the urine needs to be diluted with water to avoid overloading the plants with nutrients. Urine diverting toilets have been installed in Germany, Switzerland, and Australia in pilot case studies to evaluate factors contributing to their efficacy such as social considerations for their installation and operation and opportunities for nutrient recovery. Urine has been shown to produce similar or higher crop yields than commercial fertilizer for some crops (e.g., barley, cabbage, outdoor cucumber, red beets, and lettuce), but has resulted in lower yields for other crops (e.g., outdoor pumpkins and tomatoes in a greenhouse) (Heinonen-Tanski et al., 2010; Mitchell et al., 2013).

1.2.3.2 COMPOSTING TOILETS

Composting toilets are another technology that can be used in homes or buildings (small- or medium-scale systems) to capture the nutrients from BW with the production of compost. The compost can be used directly at the site (e.g., the Morris and Gwendolyn Cafritz Foundation Environmental Center in Accokeek, Maryland) or can undergo further treatment prior to land application to meet local regulations (e.g., the Bullitt Center in Seattle, Washington) (ILFI, 2017).

1.2.3.3 FERTIGATION

By keeping beneficial nutrients in treated wastewater for plant irrigation (fertigation), the need for potable water and fertilizer can be simultaneously offset. Fertigation can be implemented across all scales. A study conducted in Monterey County evaluated the efficacy of irrigating with well water and compared it to RW with higher nutrient concentrations (Asano et al., 2007). While some crops exhibited higher crop yields when irrigated with RW (e.g., broccoli, and celery), the yield of other crops (such as artichoke) showed no significant improvement (Sheikh et al., 1990).

1.2.3.4 BIOSOLIDS MANAGEMENT FOR FERTILIZER PRODUCTION

In 2004, about 49% of US biosolids were land applied (74% of which was used for agriculture), 45% were disposed of (e.g., landfilled, incinerated) or used as a daily cover in a landfill, and the remaining 6% underwent other practices such as long-term storage (NEBRA, 2007). The land application of biosolids offers several advantages including: offsetting the production and cost of commercial fertilizer, a lower probability of leaching into water supplies since nutrients are released at a slower rate, and a relatively inexpensive option for biosolids management (US EPA, 2000). Prior to land application, the biosolids must be stabilized. Typical treatment processes include thickening, dewatering, alkaline stabilization, digestion, composting, and/or heat drying (US EPA, 2000). Biosolids treated by digestion or alkaline stabilization can be used as a soil amendment or daily landfill cover, composting produces highly organic and soil-like biosolids for horticultural, nursery, and landscaping uses, and heat-dried biosolids can be directly used as fertilizer (Mo and Zhang, 2013).

Two examples of facilities that land applies stabilized biosolids include the William E. Dunn Water Reclamation Facility in Pinellas County, Florida, and the Metro Biosolids Center in San Diego, California. Specifically, the biosolids from the William E. Dunn Water Reclamation Facility are transformed into fertilizer pellets by undergoing the following processes: aeration in the sludge holding tanks, thickening via polymer addition, dewatering via belt filter presses, and thermal drying at a larger facility (Pinellas County, n.d., 2018). The process implemented at the Metro Biosolids Center relies on degritting, centrifugal dewatering, screening, blending, anaerobic digestion (with electricity generation), storage, and mixing with biosolids from another plant, and centrifugal dewatering (San Diego, 2018).

Challenges of land applying biosolids include labor-intensive processes, public opposition, and seasonal constraints (US EPA, 2000). In seasons when biosolids cannot be land applied, an alternative disposal pathway or storage mechanism is needed. In the case of the Beaver Dam WWTP in Wisconsin, digested biosolids with a 2% solids concentration are land applied in local farms throughout the warm months (Beaver Dam, 2014). But when the land is frozen, biosolids are stored after dewatering and polymer addition to achieve a 12% solids concentration. Since smaller plants risk producing biosolids with higher concentrations of hazardous substances, effective programs to monitor the quality of biosolids is of critical importance (Nowak et al., 2004).

1.2.3.5 ENHANCED NUTRIENT RECOVERY

Enhanced nutrient recovery from wastewater can be achieved by crystallization/precipitation, wet chemical processes, or thermochemical processes. In a review of struvite precipitation, Rahman et al. (2014) found that 67–94%

of the phosphorus and 20–65% of the nitrogen in wastewater from varied sources (i.e., swine, municipal, landfill leachate, and synthetic) could be recovered. The molar ratio of magnesium (Mg), phosphate (PO_4), and ammonium (NH_4) in pure struvite are 1:1:1. Since nitrogen and phosphorus concentrations in wastewater are higher than magnesium, the addition of a magnesium source (e.g., $MgCl_2$, $MgSO_4$, MgO, or $Mg(OH)_2$) is needed for the precipitation of struvite (Rahman et al., 2014). Other metal salts can enable phosphorus precipitation (e.g., calcium, aluminum, iron) as well. However, struvite precipitate and, secondly, calcium phosphate have shown the greatest potential for use as a fertilizer substitute (De-Bashan and Bashan, 2004; Egle et al., 2016).

Based on a material flow analysis (MFA), Egle et al. (2016) estimated that phosphorus could be recovered at a rate of about 10–30% (relative to the phosphorus in the influent of the WWTP) via crystallization or precipitation using digester supernatant. The recovery rates using wet chemical, thermo-chemical, and other recovery processes (wet oxidation, thermo-electrical, etc.) were much higher, reaching rates of 40–50% when applied to sewage sludge and 60–90% for sewage sludge ash. Detailed information concerning the application of wet chemical and thermo-chemical processes for phosphorus removal can be found in the review by Cornel and Schaum (2009).

1.3 SUSTAINABILITY ASSESSMENTS

The three dimensions of sustainability consist of the environment, economy, and society. Accordingly, this section discusses the sustainability of resource recovery systems with respect to each dimension.

1.3.1 THE ENVIRONMENTAL DIMENSION

To assess the environmental impacts of recycled water schemes, studies have applied a variety of methodologies including MFA, environmental risk assessment (ERA), life cycle assessment (LCA), and integrated approaches (Chen et al., 2012). Integrated models combine approaches to leverage the benefits of different assessment methods while accommodating their weaknesses; nonetheless, "the investigations on the selection and implementation of appropriate integrated models for particular recycling schemes are still limited and not well documented" (Chen et al., 2012). Material Flow Analyses quantify the flow of materials (e.g., treatment chemicals) or substances (e.g., pharmaceuticals, heavy metals, organic matter, nutrients, etc.) using a systems-based approach. For example, an MFA of Amsterdam's wastewater chain facilitated the design of four strategies for resource recovery that targeted the production of alginic acid, the production of bioplastics, the recovery of cellulose, and the recovery of phosphorus (van der Hoek et al., 2016). In contrast, ERAs evaluate the possibility that a "harmful consequence will occur as a result of an action or condition...[while considering] the interaction of agents or hazards, humans, and ecological resources" (Muralikrishna and Manickam, 2017). The potential risks for water reclamation projects alone are vast and include the displacement of populations, adverse effects on public lands, aquatic life, endangered species or their habitats, as well as the degradation of "ambient air quality, noise levels, surface or ground-water quality or quantity" (Urkiaga et al., 2008). LCA can be used to evaluate the environmental impacts of a system, technology, or process. The methods for implementation are outlined in the ISO 14040 standard (ISO, 2006), which consists of: (1) Defining the goal and scope of the assessment; (2) Conducting an inventory analysis; (3) Conducting the impact assessment; and (4) Interpreting the results. Since LCA is a standard methodology for the evaluation of the environmental dimension of sustainability, it will be discussed in further detail in this sub-section.

The approaches used to conduct an LCA fall into three categories: a process LCA, economic input-output LCA (EIO-LCA), or a hybrid approach. A process LCA captures individual inputs and outputs for a system; Figure 1.1 provides a general overview of the inputs and outputs that could be considered for the LCA as well as life cycle phases. The data can be obtained from numerous sources including existing systems or literature references, and LCA software applications like SimaPro and Gabi incorporate process models that have been previously developed and make them available to users to facilitate the assessment. Accordingly, process LCAs can provide a very accurate representation of the environmental impacts, but it can also be a time consuming and, consequently, expensive process. The EIO-LCA method takes a high-level approach to estimating environmental impacts based on material and energy requirements for activities within individual economic sectors. A web-based EIO-LCA tool that was developed by researchers at Carnegie Mellon University is available online for free, non-commercial use (CMU, 2008). It incorporates models specific to the US, Canada, China, Germany, and Spain. EIO-LCA database have also been integrated into LCA software applications such as SimaPro and Gabi. Finally, a hybrid LCA uses a combined approach by leveraging detailed process-level data and filling gaps in the model with an economic input-output dataset.

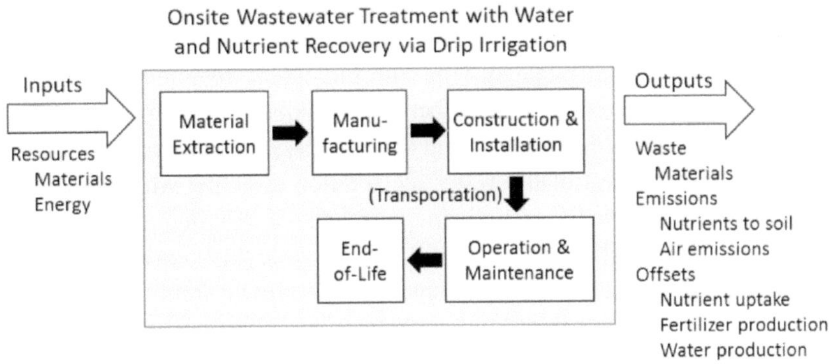

FIGURE 1.1 An example of the life cycle phases and inputs and outputs to the model that could be considered for an onsite resource recovery system.

Several characterization methods have been developed for conducting an LCA. For example, CML, ReCiPe, IMPACT, and TRACI are commonly used characterization methods, and provide a pathway to normalize the outcomes of the study with respect to the average impact of a person in the US, Europe, or the world. If LCA software is used to conduct the analysis, the characterization method can easily be selected once the system has been modeled, and a multitude of impact categories (sustainability indicators) can thereafter be calculated simultaneously.

Carbon footprint is most commonly evaluated in LCAs of resource recovery systems. The other environmental indicators considered vary significantly, but their selection can be based on prior studies or what is expected to have a significant contribution (or relative savings). For example, in case studies evaluating wastewater management systems, the eutrophication potential (EP) is usually also considered since the quality of the treated wastewater being discharged to the environment has a significant influence on this indicator. In the case of biosolids incineration, the direct emissions from the process can be significant. So in addition to carbon footprint, human toxicity (HT) and ecotoxicity should be evaluated as well (Suh and Rousseaux, 2002). Since the outcomes of an LCA depend on a variety of factors including the study scope and boundaries, system type, energy mix, and scale, several LCA examples considering resource recovery from wastewater will be discussed.

The case study by Cornejo et al. (2016), for example, evaluated small- to large-scale resource recovery systems across multiple sustainability indicators (embodied energy, carbon footprint, and EP). They found that the EP was largely dependent on the treatment schemes implemented and the type of reuse implemented as the scenarios with advanced treatment and/or targeted nutrient removal processes (nitrification and denitrification) had a lower EP (3.6–4.4 kg PO_4/m^3) than the small-scale system with biological treatment (primary treatment, aerobic treatment, and disinfection) intended for onsite irrigation (10.5 kg PO_4/m^3). Energy consumption and direct emissions were major contributors to the carbon footprint of the water reclamation systems, which is consistent with other water reuse LCAs with energy-intensive treatment processes (Tangsubkul et al., 2005; Muñoz et al., 2009). The breakdown of energy consumption resulting from different NPR, IPR, and DPR treatment trains are exhibited in Figure 1.2 for centralized systems ranging in scale from 1 to 100 MGD. It is evident that RO and MBRs are very energy-intensive, which would lead to higher environmental impacts for energy-centric indicators such as fossil fuel depletion and carbon footprint. LCAs that incorporated RO were naturally dominated by electricity consumption (Tangsubkul et al., 2005; Muñoz et al., 2009). Chemicals were another major contributor to water reuse LCAs, especially in the evaluation of tertiary treatment processes (Theregowda et al., 2015).

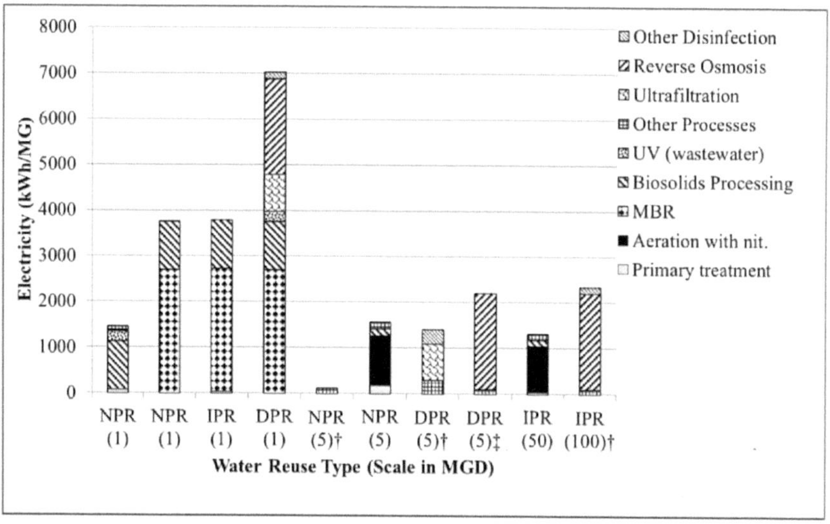

FIGURE 1.2 Specific electricity consumption for unit treatment processes commonly implemented in non-potable reuse (NPR), indirect potable reuse (IPR), and direct potable reuse (DPR) treatment schemes. Biosolids processing includes thickening (gravity belt thickener), aerobic or anaerobic digestion, and dewatering (screw-press). Unit electricity consumption data was sourced from (EPRI and WRF, 2013) and data for reverse osmosis was taken as the average for that listed for advanced wastewater treatment (Raucher and Tchobanoglous, 2014). †Treatment of secondary effluent; ‡Treatment of tertiary effluent; nit: nitrification; MBR: membrane bioreactor; UV: ultraviolet.

Fortunately, energy recovery technologies and process optimization can lower electricity and other energy requirements. For example, the Marselisborg WWTP in Denmark invested €3 million (with an expected 5-year payback) to optimize its treatment processes and improve biogas production via on-site digesters (Karáth, 2016). The plant treats 25 MGD, consumes 0.32 kWh per m^3 of wastewater treated, and generated more than 150% of its energy requirements in 2015 (Vand, n.d.; Danfoss, 2014). The Marselisborg WWTP is not alone; however, as Gu et al. (2017) highlighted 14 WWTPs that achieved 94% or more energy self-sufficiency (9 out of 14 were 100% self-sufficient). These facilities relied primarily on anaerobic digestion as well, and were large facilities ranging in capacity from 4 to 300 MGD.

Compared to alternative water sources, water reuse can result in significant savings. Meneses et al. (2010) showed that agricultural reuse with nutrient recovery and urban reuse had the lowest environmental impacts relative to conventional potable water production and desalination. Specifically, the carbon footprint amounted to 0.142 (agricultural reuse) and 0.219 (urban reuse) kg CO_2-e/m^3 for the water reuse scenarios and 0.277 and 3.42 kg CO_2-e/m^3 for potable water production and desalination, respectively. However, factors influencing the energy consumption of the RW distribution need to be considered on a case-by-case basis as hilly terrain, long distances, and low population densities lead to higher energy requirements and greater environmental loads. Kavvada et al. (2016) varied treatment scale and processes, and found that decentralized NPR could achieve 28% and 29% in energy and GHG emissions savings, respectively, relative to centralized treatment in locations that were very elevated and far away from the treatment plant. In contrast, when an end-user was close to a treatment plant and at a lower elevation, the decentralized scenario could consume up to 85% more energy and emit 49% more GHG emissions than centralized treatment at the plant.

With respect to energy recovery from wastewater, most LCAs focused on biosolids management processes and commonly evaluated the carbon footprint and energy consumption of the systems assessed. Accordingly, nutrient recovery was often also incorporated into the LCAs by substituting treated biosolids for fertilizer. The major contributors to the carbon footprint consisted of direct emissions (treatment processes and end-users), emissions from energy consumption (fuels and electricity), and opportunities for offsets (carbon sequestration and fertilizer production) (Brown et al.,2010). Although LCAs of biosolids management options often focused on the operational phase (Soda et al., 2010), energy recovery offsets increased the relative impact of components that would otherwise be small or negligible, such as construction (Stokes and Horvath, 2010) or transportation (Brown et al., 2010).

In the study by Brown et al. (2010), the municipality that implemented dewatering and incineration with heat recovery resulted in the highest emissions (1.48 Mg CO_2-e/ton of dry matter (t-DM)) relative to municipalities that relied primarily on thickening, digestion, dewatering, and/or composting for landfilling or land applying biosolids (-0.26 to 0.46 Mg CO_2-e/t-DM). However, when the incineration temperature was increased from 760°C to 880°C, N_2O emissions could be nearly eliminated and result in a reduction of 1.63 Mg CO_2-e/t-DM in the carbon footprint.

In a comparative LCA evaluating urine source separation at the building-scale to centralized wastewater treatment, Ishii, and Boyer (2015) found that urine source separation with high phosphorus recovery resulted in significantly lower environmental impacts. For the conventional scenario, the electricity for wastewater treatment, electricity for potable water production (extra water is needed in the conventional case for toilets), and the nutrient discharge dominated the life cycle impacts. The urine source separation scenario was also dominated by these inputs, in addition to the construction of the infrastructure and the manufacture of chemicals needed for struvite precipitation. Interestingly enough, Tillman et al. (1998) found similar results that favored urine source separation to centralized treatment for a village in Hamburgsund in Sweden, however, the ability to recover a significant amount of energy in a large-scale WWTP for the Bergsjön suburb resulted in prominent reductions in fossil fuel consumption and methane emissions relative to urine source separation. With respect to phosphorus recovery technologies applied to WWTPs, the study by Amann et al. (2018) revealed a large variation in emissions and energy demand. Nonetheless, the recovery of phosphorus from the liquid phase and sewage sludge ash generally had lower impacts than processes applied to sewage sludge, but the need for the efficient recovery of phosphorus warranted a preference for processes applied to sewage sludge ash because of the relatively high recovery rate (~60–85%).

1.3.2 THE ECONOMIC DIMENSION

Life cycle cost analysis (LCCA) can be used to evaluate the cost of a technology or process over its lifetime. While the phases considered can consist of transportation, manufacturing, operation, maintenance, and end-of-life, as with LCAs, the most prominent phases are often selected to facilitate the data acquisition process. The economic indicators evaluated are most commonly the net present value (NPV), uniform annual cash flow (UAC), and payback period. Note that costs can also be presented with respect to a functional unit (FU) such as the volume of wastewater processed (m^3) or the mass of biosolids (t-DM).

The NPV accumulates the variable costs and revenues over the functional life of a system, and represents the net cost (or profit) as a monetary value at the time of acquisition (the present). It accounts for the opportunity to accrue interest over time, which is why future payments become "less expensive" when converted to the present time period. Eqn. (1.1) represents the NPV calculation, where C_t is the cost during a particular time period, t; R_t is the revenue; L is the functional life of the system; and $i$represents the interest rate. The UAC represents the equivalent net annual cost (or profit) of the system, and is useful for comparing systems that have different life spans. The UAC can be obtained from the NPV using Eqn. (1.2). Finally, the payback period provides visibility into how long it takes to recover the costs of a system; it is calculated by dividing the investment cost by the net annual revenue (the revenue subtracted by the annual costs). In this section, costs will be provided for various water, energy, and nutrient recovery projects/case studies, but this section is not meant to be a comprehensive review of life cycle cost analyses to date.

$$NPV = \Sigma_{t=0}^{L} \frac{C_t - R_t}{(1+i)^t} \tag{1.1}$$

$$UAC = \frac{NPV * i * (1+i)^L}{(1+i)^L - 1} \tag{1.2}$$

Guo et al. (2014) developed cost models for wastewater treatment processes commonly used in DPR, including activated sludge, GAC, MBR, RO, and ultrafiltration (UF). The capital and operation and maintenance (O&M) costs were modeled separately and took on a logarithmic form as expressed by Eqn. (1.3) where C is the capital or O&M cost and Q is the flow rate of water processed. Most processes realized economies of scale (reduced costs per unit of water processed) as the flow rate increased. These models did not include the distribution costs of RW, which is significantly influenced by the population density and topography of the site. Guo and Englehardt (2015) analyzed distributed DPR systems for different terrains with varied population densities, and confirmed that DPR for rural (less populated) areas would be more expensive that alternative water/wastewater treatment systems. However, having a DPR facility in urban and suburban areas only needed 10,000 people served per facility to be economically competitive to conventional water/wastewater systems.

$$\log(C) = a(\log(Q))^b + c \tag{1.3}$$

As with electricity consumption, the cost to treat water was found to increase as the water quality improved. For example, O&M costs for the Green Acres Project (3.18 MGD average demand in the fiscal year 2016–17, 7.5 MGD capacity) in Orange County, California, and the William E. Dunn Water Reclamation Facility (6.5/9 MGD) in Pinellas County, Florida amount to about $0.40/m^3 (fiscal year 2014–2015) and $0.53/m^3, respectively (Pinellas County, n.d., 2013; Water Issues Committee, 2016; OCWD, 2019); the range of these NPR project costs is exhibited in Figure 1.3. A water reclamation study that considered fertigation (NPR with nutrient recovery) in Monterey County, California estimated the costs for a 114,000 m^3/day plant to be ~$0.06/m^3 for filtered effluent (i.e., coagulation, flocculation, filtration, and chlorination) and $0.09–0.13/m^3 for tertiary effluent (i.e., coagulation, flocculation, sedimentation, filtration, and chlorination) (Sheikh et al., 1990). Assuming that these estimates were calculated in USD (1990), they would amount to $0.11 and $0.17–0.24 per m^3 in USD (2016), respectively. The Groundwater Replenishment System in Orange County, California produces 100 MGD at a cost of $0.69/m^3 (or $0.43/m^3 with subsidies) (OCWD, n.d.). Considering that this is the largest IPR project in the world, the project likely leverages economies of scale, so smaller systems may have a higher per unit cost. For comparison, the cost of delivering water to Los Angeles has been estimated to be $0.45–$1.33 for NPR and $1.25–$2.14 for IPR (Porse et al., 2018).

The US EPA's water reuse guidelines recommend that drinking water quality guidelines are followed for DPR (US EPA, 2012). Accordingly, DPR would have the highest expected water quality. Advanced treatments for DPR applications include advanced membrane treatment with oxidation, for which costs would amount to $0.57 to $0.97 per m^3 as estimated by the National Research Council (NRC) (2012). The DPR facility in Windhoek, Namibia (4.0/5.4 MGD) had treatment costs amounting to $0.55/m^3 (USD, 2013) in the first year of operation in 2002 and total costs (with capital) of $0.99/m^3 (USD, 2013) (Du Pisani, 2006). One of the main advantages of DPR is that the main water distribution network can be used, thus avoiding a parallel distribution network for RW. Parallel distribution of RW treated to a tertiary level can be expensive and amount to more than treatment costs. The NRC (2012) estimates parallel distribution to cost between $0.32 and $1.70 per m^3, which can easily exceed the cost of treating wastewater for reclamation as exhibited in Figure 1.3.

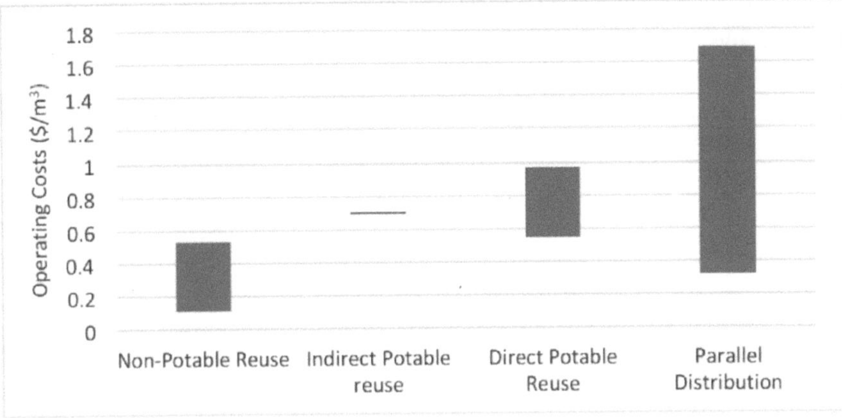

FIGURE 1.3 The range of operating cost estimates and existing case studies for non-potable reuse, indirect potable reuse, direct potable reuse, and the parallel distribution of reclaimed water.

In an evaluation of alternative energy recovery schemes, Chae et al. (2013) estimated the installation cost for a hydropower project to be $13,913/kW for a 10 kW system at the KiheungRespia WWTP. Typically, the installation costs for micro-hydropower generation are lower and range from $1,300–$8,000 per kW, while O&M costs amount to 1–4% of the installation costs (Power et al., 2014). Although the limited availability of local resources resulted in higher initial costs for the KiheungRespia WWTP, renewable energy certificates offered by the government ($0.138/kWh) resulted in a promising payback period for hydropower of 7.4 years. In contrast, the thermal energy recovery option could produce more energy (3.65% vs. 0.75% of the plant's energy) at a lower installation cost of $1,978/kW, and although this option did not qualify for renewable energy certificates, the payback period of 6.8 years still made this option economically viable.

At the household-level, Słyś and Kordana (2014) evaluated thermal energy recovery from a shower drain and estimated a payback period of 2.5 to 10 years. The ability to produce energy, however, was largely dependent on the amount and rate of water flow. Accordingly, reducing the flow rate and increasing the number of inhabitants and length of showers led to an increase in the energy recovery potential and a beneficial reduction in the NPV and payback period. The NPV was also sensitive to the electricity rate and capital expenditures, which influenced the possible savings from the heat recovery system and the initial costs, respectively. Nonetheless, the risk of investment failure decreased with greater water use.

With respect to large-scale biosolids management, conventional, energy, and/or nutrient recovery options are often compared to one another since they offer alternatives for processing this byproduct of wastewater treatment. For example, the study by Lundin et al. (2004) revealed that the agricultural reuse option (680 SEK/t-DM) was less expensive than co-incineration with household waste (2,510 SEK/t-DM) because an additional investment would be needed to expand the capacity of the existing incineration facility. The costs were dominated by transportation costs for agricultural reuse with nutrient recovery and capital investments for co-incineration with energy recovery. Novel options were also presented for incineration with phosphorus and energy recovery, and fractionation with phosphorus recovery, which amounted to 1,410 and 1,760 SEK/t-DM, respectively. These results and those from other life cycle cost analyses are summarized in Figure 1.4.

Li et al. (2017) compared biosolids management options in China and found that only biogas production from anaerobic digestion for high-organic-content sludge (volatile solids to total solids (TS) ratio of 0.7) resulted in a profit over the systems' operational life. In general, the NPV was lowest for the anaerobic digestion (about 90 to 190 million CNY), followed by co-incineration at a cement kiln (about ~130 to 190 million CNY) and coal-fired power plant (-340 to -420 million CNY), and incineration (-400 to -580 million CNY). Anaerobic digestion also resulted in the lowest costs relative to other energy recovery options (via biosolids management) in the study by Murray et al. (2008). Moreover, when digestion is paired with other treatment processes (e.g., thickening/dewatering, composting, drying, incineration, or melting) it reduces the extra resources that would otherwise be expended without it and can lower the life cycle costs (Hong et al., 2009). It should be noted, however, that local conditions can significantly influence the outcomes of the cost analysis, so assessments and recommendations should account for location-specific conditions, e.g., expensive landfill costs in resource-scarce locations, the availability of government subsidies, the dry matter concentration in the biosolids, the energy sources available, etc., (Murray et al., 2008; Hong et al., 2009; Liu et al., 2013; Li et al., 2017).

The largest nutrient recovery facility in the world is located in Chicago, Illinois, and processes the biosolids from the Stickney Water Reclamation Plant (1,440 MGD plant that treats 700 MGD) (MWRDGC, n.d.; Hawthorne, 2016). The nutrient recovery facility opened in 2016 and implements Ostara's enhanced biological nutrient removal (BNR) process paired with struvite precipitation to produce up to 10,000 tons of Crystal Green® per year, a slow release 5-28-0 fertilizer (Hawthorne, 2016; Ostara, 2017). Ostara's technology is capable of achieving up to 50% phosphorus removal and will reduce phosphorus discharges at this facility by 30%. Life cycle cost estimates were not readily available, but the project costs amounted to $31 million, revenue from fertilizer sales of up to $2 million could be generated annually, and a relatively short payback period of 3 to 5 years is expected (Haddaway, 2015; Hawthorne, 2016).

Small-scale implementations, unfortunately, lose the economies of scale that large-scale systems can leverage, but the savings from resource recovery can be enough to lead to improvements in life cycle costs and environmental impacts. For example, in a study comparing the centralized treatment of wastewater with urine source separation for nutrient recovery with struvite precipitation, infrastructure costs to equip a building with conventional toilets amounted to $20,700 whereas the cost of source separating toilets, piping, and storage tanks for urine amounted to $79,710 or nearly four times the costs (Ishii et al., 2015). Ultimately, however, the life cycle costs were similar and the environmental impacts for urine source separation with phosphorus recovery were actually lower than the impacts for centralized treatment. Such an outcome highlights the importance of conducting a comprehensive sustainability analysis, since the economies of scale of a centralized system may not outweigh the benefits of decentralized resource recovery solutions.

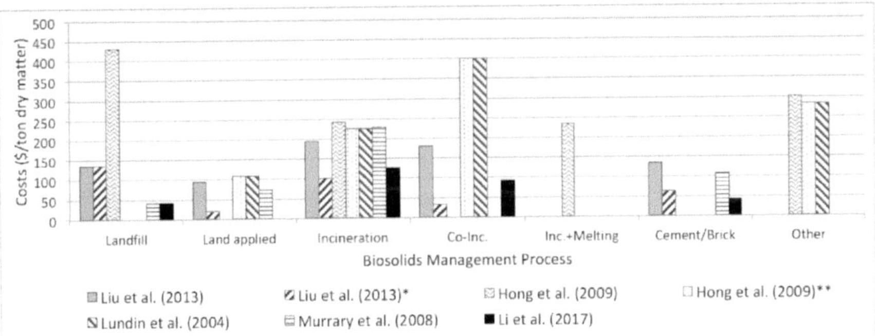

FIGURE 1.4 The net present value (per ton of dry matter) of biosolids management processes (treatment and end-use/disposal) that implement energy and/or nutrient recovery. Costs have not been converted to a specific year as most studies did not mention the year of their cost analysis. Conversion rates of 6.25 SEK/USD, 105 Y/USD, and 6.32 CNY/USD were assumed. *includes available government subsidies; **with digestion; Inc.: incineration.

1.3.3 THE SOCIAL DIMENSION

Resource recovery and human society are interconnected. On the one hand, the successful implementation of resource recovery technologies is often dependent on socio-cultural factors such as awareness and acceptance, institutional infrastructure, and available expertise (Balkema et al., 2002). On the other hand, increased awareness and adoption of resource recovery technologies may alter people's understanding and behavior towards resource consumption and sustainability. Previous studies on the social dimension of resource recovery are primarily focused on understanding the acceptance of certain recovered resources (e.g., RW, source separated urine) and/or its associated technologies (e.g., Hartley (2006), Ross et al. (2014), and Quezada et al. (2016)). Efforts have also been made to identify the key barriers and enablers of societal acceptance (e.g., Dolnicar et al. (2010); Quezada et al. (2016)). These studies usually utilize surveys, case studies, interviews, workshops, or news media data mining (Ching, 2010) to obtain qualitative and quantitative information related to stakeholders' perceptions/acceptance of the proposed resource recovery, willingness to pay for the new technologies or management schemes, along with their socioeconomic characteristics (e.g., job position, income, education, political affiliation) and environmental literacy. Survey instruments are also frequently combined with statistical data analyses to identify the key personal characteristics that enable acceptance or adoption (Ross et al., 2014). These studies have strived for both spatial and temporal heterogeneity (Garcia-Cuerva et al., 2016). Collectively, knowledge and information emanated from these studies could help guide and inform the design, planning, and implementation of resource recovery.

Some resource recovery techniques are more subject to social and institutional barriers than others, especially those that require high level public support for their successful implementation. Water reclamation and urine diversion are among these technologies. The "yuck factor," defined as emotional discomfort generated from close contact with certain unpleasant stimuli, is often cited as the primary contributor to the opposition of water reclamation programs. Efforts have been made to understand the source of the yuck factor as well as ways to reduce its effect and increase acceptance.

Participants' knowledge about water shortage, alternative technology, and its associated risks has been constantly reported to play an important role in the level of acceptance. A review of public acceptance studies conducted in the 1960s and 1970s on water reclamation found that the participants' knowledge about the nature of the water supply, the perception of its adequacy, income, and education greatly influence their acceptance of RW (Baumann, 1983). Similarly, a recent survey of 1000 respondents found that providing information about the water production process greatly increases respondents' willingness to use recycled water (Dolnicar et al., 2010). As a reflection of this study, the authors suggested to use factual information as opposed to persuasive campaigns to garner the public support of water augmentation projects. Fielding and Roiko (2014) further investigated how different types of information could affect participants' response to RW. A social experiment was conducted in which each participant would randomly receive one of the following treatments: (1) no information; (2) basic information regarding the recycling process and the safety of recycled water; (3) basic information plus pollutant concentrations after treatment; and (4) basic information plus an analogy of how much water you need to drink to get a 100 mg dose of a pharmaceutical. It was found that while providing basic information is generally effective at increasing acceptance, additional pollutant information could potentially decrease the level of acceptance.

The end-use and source of the RW also have a significant effect on acceptance (Garcia-Cuerva et al., 2016). Hartley (2006) found that public acceptance of water reclamation is higher when the degree of human contact was minimal, protection of public health was clear, and the perception of wastewater as the source of RW was minimal. Non-potable uses are generally more acceptable than potable uses regardless of the actual water quality (Marks, 2004; Hartley, 2006). Furthermore, people tend to have a higher acceptance of water generated from their own wastewater rather than communal sources (Jeffrey, 2002). One underlying reason for the public's preference of non-potable uses and self-generated wastewater sources is their perception of health risk (Marks et al., 2008). The health risk associated with non-potable uses is perceived to be smaller than potable uses. Similarly, the health risk associated with self-generated wastewater sources is perceived to be smaller than communal sources. Nevertheless, the perceived health risks do not necessarily reflect the actual health risks. Such perceptions are often strongly driven by cultural norms governing ideas of water cleanliness (Marks et al., 2008).

This can be changed by creating greater public awareness and encouraging public participation in decision-making of water reclamation initiatives (Radcliffe, 2006). The more community members perceived that the water authority used fair procedures (e.g., consulting with the community and providing accurate information); the greater sense of shared identity with the water authority is created (Ross et al., 2014). This is found to improve community's trust of the competence of the water authority and the image that the water authority has the community's interest at heart. Higher levels of trust lead to lower perceptions of risk and higher acceptance. In addition, an assessment of previously successfully implemented potable reuse projects in California found that legitimation of the water reclamation programs from pragmatic, moral, and cognitive perspectives greatly increase acceptance (Harris-Lovett et al., 2015). Such legitimation strategies include establishment of a portfolio of standards, procedures, and possibly new institutions and government support (Quezada et al., 2016).

From a socio-institutional perspective, some of the barriers have been identified as lack of financial incentives and accounting for externality benefits, fragmented industry structure, and the limited knowledge regarding the long-term O&M requirements of decentralized systems (Brown and Farrelly, 2009; Farrelly and Brown, 2011; Sharma et al., 2012; Ross et al., 2014). From an asset investment perspective, Marlow et al. (2013) identified the possible conflict of alternative water sources with utility revenues as an important socio-institutional barrier. Furthermore, there is currently no well-established economic framework for sharing costs and benefits of recovered resources, especially at the decentralized scale (Sharma et al., 2010). On the other hand, the adoption of alternative systems creates institutional benefits such as overcoming lack of available services, deferring upgrades to existing infrastructure, protecting sensitive waterways, showcasing sustainable development, improving landscape amenity, and demonstrating innovation (Sharma et al., 2013).

Unlike water reclamation, social acceptance of the concept of urine diversion is relatively high (Lienert and Larsen, 2009). A focus group study in Switzerland found that participants were not only willing to have a urine-diverting toilet in their home, but also willing to eat food crops that had been fertilized with urine (Pahl-Wostl et al., 2003). Similarly, a survey conducted in the University of Hawaii found a high acceptance of urine source separation (Lamichhane and Babcock, 2013). Another survey study conducted at the University of Florida found 84% acceptance of urine source separation in residence halls and there was little concern reported about the use of urine-based fertilizers (Ishii and Boyer, 2016). A survey of 2700 respondents across Europe later found that the pioneering concept of NoMix-technology for urine diversion was well accepted: 80% liked the idea, 75–85% were satisfied with design, hygiene, smell, and seating comfort of NoMix-toilets, 85% regarded urine-fertilizers as a good idea (50% farmers), and 70% would purchase such food (Lienert and Larsen, 2009). Nevertheless, 60% of the participants encountered problems during daily use of the NoMix-toilets (Lienert and Larsen, 2009). Furthermore, the public are generally unwilling to pay or to make additional efforts to maintain urine diversion infrastructures (Lamichhane and Babcock, 2013; Ishii and Boyer, 2016). Blume and Winker (2011) found that users are more likely to comply with sitting on toilet seats if a disinfection spray were made available in Germany.

1.4 SYNERGIES AND TRADE-OFFS

1.4.1 SYNERGIES IN RESOURCE RECOVERY

It is important to investigate the synergies and tradeoffs of resource recovery from wastewater. Synergy is a concept related to systems thinking that views a system as a whole with interacting parts instead of merely a sum of individual parts. The behavior of a system is the result of the interaction of the parts that comprise the system. When the parts work together in a way to produce greater effect than the sum of their individual effect, they are considered to have synergy among them. In the area of resource recovery, when a technology or strategy can recover more than one type of resource or can address more than one aspect of sustainability, the synergies are assumed to occur since the overall resource input or environmental burden will be lower compared to the recovery of resources separately. The following sections give some examples of strategies to recover more than one type of resource and to address more than one sustainability issue.

1.4.1.1 TECHNOLOGIES RECOVERING MULTIPLE RESOURCES

A variety of technologies are capable of recovering more than one resource. For instance, treated wastewater often contains nutrients that could be directly utilized by plants/crops when used for irrigation. Table 1.1 shows the water quality for large-scale NPR in several existing cases. For these facilities, no additional treatment is needed to remove nutrients in effluent water. This could simultaneously avoid the extra material and energy input required for advanced treatment to remove the residual nutrients. When the wastewater treatment plant is close to the location where irrigation water is needed, fertigation should be encouraged, especially to align the nutrient content in effluent water with the nutrient requirement for specific crops.

TABLE 1.1 Water Quality Achieved for Large-Scale Non-Potable Reuse Case Studies

Case Studies	Capacity (MGD)	COD (mg/L)	TSS (mg/L)	TN (mg/L)	TP (mg/L)	Turbidity (NTU)	Fecal (Coli/ 100 ML)
Denver	30	4–8^{TOC}	NS	5–30	0.04–0.4	0.15	<1
William E. Dunn	9.0	NS	NS	1.77^{NO3}	0.22	NS	NS
South Cross Bayou	33	NS	NS	0.96^{NO3}	0.27	NS	NS
South Cary	12.6	0–0.2	0.8–2.1	1.8–3.6	0.4–2.1	NS	1–9

NS: not specified.

Source: Adapted from Diaz-Elsayed et al. (2019).

Table 1.2 shows the representatives of the top five profitable crop groups in Florida as an example. Comparing Tables 1.1 and 1.2, it is evident that for some crops the nitrogen content from RW might be able to completely meet the nutrient requirement. As in the case of sugarcane, the nitrogen content could be met by the RW produced in Denver. However, the nitrogen requirement for peanuts is zero and RW with high TN concentration might cause nutrient emission. Typically, the phosphorus requirement for crops has to be supplemented by fertilizer.

TABLE 1.2 The Nutrient Requirement for the Five Most Profitable Crop Groups in Florida

	Irrigation Requirement (2015 FL)			Nutrient Requirement			
Group	**Water (MGD/ Year)**	**Irrigated Land (A)**	**MG/A/ Year**	**N (lb/A)**	**Concen–tration* (mg/L)**	**P_2O_5 (lb/A)**	**Concen–tration* (mg/L)**
Oranges	689	567,741	0.44	160	43	80	22
Strawberries	51	32,155	0.58	150	31	100	21
Tomatoes	337	220,880	0.56	200	43	100	22
Peanuts	92	132,909	0.25	0	0.00	40	19
Sugarcane	616	567,431	0.40	90	27	40	12

A: acreage; FL: Florida; MGD: a million gallons per day; *: concentration is calculated by dividing nutrient requirement by irrigation requirement.

Source: Data from The Balmoral Group, LLC (n.d.); University of Georgia (2019).

Another example is anaerobic digestion used for sludge treatment, which recovers both energy and nutrients. Energy is recovered in the form of biogas and nutrients in the form of stabilized biosolids. For example, the Howard F. Curren Advanced Wastewater Treatment Plant (HCWTP) in Tampa, FL captures the biogas produced from ADs and generates electricity onsite through engine generators. A portion of biosolids is heat dried to produce a fertilizer product and the rest of the biosolids are applied to agricultural land as a soil amendment (Mo and Zhang, 2012).

Urine diversion could synergistically reduce the amount of nutrients and water entering WWTPs. This is likely to ease the treatment of wastewater, and hence reduce the amount of energy needed by centralized systems. Replacement of petrochemical-based fertilizers by urine products could also synergistically reduce the transboundary inflow of nutrients into a region as well as the energy and water associated with producing conventional fertilizers. The study by Ishii and Boyer (2015), first introduced in Section 1.3.1, conducted a LCA to compare diverting urine for struvite precipitation with the conventional centralized wastewater treatment in a university setting. Urine diversion with MgO addition was found to reduce life cycle nitrogen emissions by one-half and life cycle energy consumption by two-thirds. However, the benefits of urine diversion are highly technology dependent. In this same study, urine diversion with MgO and Na_3PO_4 additions to achieve a higher P and N recovery was found to result in a higher life cycle nitrogen emission and energy consumption than conventional wastewater treatment (Ishii and Boyer, 2015). Scale and design are other factors that influence the performance of urine diversion. A spatial analysis in San Francisco, California showed that decentralized urine source separation has lower life cycle cost, unit energy, and GHG emissions than centralized nitrogen management (Kavvada et al., 2017).

GW recycling, on the other hand, allows synergistic energy (primarily through residual heat recovery) and water recovery. In the past, the potential of energy and water recovery from GW recycling were investigated separately. However, more analyses on the integrated recovery of both water and energy from GW have started to emerge. Meggers and Leibundgut (2011), for example, evaluated the potential of heat recovery from water immediately after its use, and the system investigated could be used for both energy and water recovery. Another study of a typical residential house in New York, New York found that recovering both heat and water from a GW system could reduce 33.9% of the household's energy consumption for space heating and cooling and hot water heating, and 27.2% of the potable water supply (Ni et al., 2012).

1.4.1.2 *TECHNOLOGIES ADDRESSING MULTIPLE SUSTAINABILITY ISSUES*

Some resource recovery technologies address multiple issues related to the different dimensions of sustainability (i.e., environmental, economic, and social). For example, anaerobic digestion not only has environmental benefits related to reducing the discharge of nutrients and organics, and the recovery of resources, it also provides social benefits in terms of protecting public health by reducing pathogens in the sludge. Sludge contains a variety of pathogens, such as *Salmonella* and *E. coli*, depending on the source of sludge (Sahlström, 2003). The pathogen reduction in anaerobic digestion is influenced by many factors, such as bacterial species, operation mode, temperature, solid retention time, pH, and volatile fatty acids (VFA) (Sahlström, 2003; Avery et al., 2014; Scaglia et al., 2014). Among these parameters, it is generally agreed upon that the most important factor is temperature. Hence, the pathogen reduction has to be carefully evaluated before the application of biosolids for agriculture land.

Similarly, urine diversion could potentially have both environmental and economic benefits, although it could be case dependent. Direct application of urine, for instance, has been practiced for a long time in many developing countries. Such practice creates am extremely low cost source of fertilizer without having to redevelop or reinvent the existing infrastructure. This is because often very simple collection and application methods are utilized. However, it is more difficult to address the economic benefits of urine diversion in many developed countries, as more sophisticated collection, treatment, and application infrastructure and methods are often expected or required. Some of the research effort has been directed to minimize the cost for integrating urine diversion into the existing wastewater infrastructures (Remy, 2010; Wood et al., 2015).

1.4.2 TRADEOFFS IN RESOURCE RECOVERY

Tradeoffs in resource recovery refer to instances where recovering one resource requires the consumption of other resources from a life cycle perspective; or addressing one dimension of sustainability requires sacrifice of other dimensions of sustainability. The following sections highlight some of these tradeoffs in the current resource recovery technologies.

1.4.2.1 *TECHNOLOGIES WITH RESOURCE TRADEOFFS*

While operation and construction of resource recovery technologies inherently require certain levels of energy and material inputs, the consumption of these additional resources could sometimes become an issue of concern if it is significant enough to create another environmental problem. Using biosolids as fertilizer, for example, often requires heat drying the solids to achieve the desired quality for marketing and application purposes. Nevertheless, the heat drying process itself could be energy and carbon intensive. Taking the aforementioned HCWTP for example, the system currently utilizes residual heat from a nearby power plant to heat dry the digested biosolids for subsequent uses. When taking the energy reduction from producing conventional fertilizers into account, the system is net energy negative, meaning more is saved than consumed (Mo and Zhang, 2012). However, the HCWTP had a proposal of installing a rotary kiln powered by natural gas to increase the fertilizer production rate. If this becomes reality, the system will have significant net positive energy consumption and carbon emissions, in exchange for a higher nutrient recovery. Such decisions need to be carefully made with all environmental impacts of concern evaluated.

Another case study was conducted to investigate different types of water reuse in Lakeland, Florida (Rezaei et al., 2019). The study found trade-offs between the degree of treatment for water reuse and the value of RW, infrastructure requirement, and potential benefits. The cost (as annual specific NPV) associated with wastewater treatment increases with the improved water quality as \$1,667/MG, \$2,220/MG, and \$3,561/MG for urban reuse, IPR, and DPR. The value of water also increases with the improved water quality at \$173/MG, \$1,500/MG, and \$3,500/MG for urban reuse, IPR, and DPR, respectively. In this case, the additional cost associated with improving water quality from urban reuse to IPR is minimal; however, the value of water increases by one order of magnitude (Rezaei et al., 2019).

The scale of implementation is an important factor to select the technology for treatment and resource recovery. For example, at the household scale, a septic tank with aerobic treatment is a common technology for wastewater treatment and fertigation. Advanced water treatment is typically applied at the community and city scale. Cornejo et al. (2016) evaluated the impact of scale of implementation on the environmental sustainability of resource recovery-oriented wastewater treatment. It was found that there are benefits associated with resource recovery from large-scale wastewater treatment with respect to the embodied energy and carbon footprint due to economies of scale. In addition to scale, other factors such as topographical conditions, system layout, and end-use application all play a significant role in technology selection and consequently the environmental impacts.

1.4.2.2 TECHNOLOGIES WITH TRADEOFFS IN SUSTAINABILITY DIMENSIONS

The tradeoffs between environmental and economic dimensions are prominent in many of the aforementioned resource recovery techniques. Cost minimization and economic incentives for resource recovery are ongoing research topics in the field. Some other technologies do provide both environmental and economic benefits. For instance, DPR has often been noted as having a lower cost and requiring less energy than other drinking water treatment alternatives (Leverenz et al., 2011; Rezaei et al., 2019). However, the lack of public acceptance is a major social barrier to implement such projects.

1.4.2.3 SOLUTIONS TO TRADEOFFS

For widespread implementation of wastewater resource recovery, the portfolio of regulation, procedures, new governance, and incentives have to be in place. In addition to educational campaigns, communication strategies, and outreach, other legitimation strategies should be adopted such as stakeholder engagement over the project life, data transparency and information sharing (Harris-Lovett et al., 2015). For a specific project, the selection of the optimal approach has to consider a variety of factors, including existing infrastructure, water demands, topographical, and climate conditions, available water and energy sources, local nutrient management regulations, scale of implementation, and utility governance structure. Often times, the decision has to be made on a case-by-case basis. There are no one-size-fits-all solutions. The potential synergies and tradeoffs have to be identified and evaluated under the given context. From a research perspective, development of decision support tools that take account of multiple factors to support the comprehensive assessment of the local condition and selection of technologies could be of great value.

1.5 RECOMMENDATIONS FOR FUTURE RESEARCH

The evaluation of resource recovery technologies through the lens of sustainability relies on the availability and abundance of high quality inventory data from real-world practices. Such data, however, are often not directly available to research communities for reasons such as safety concerns and the protection of business interests, among others. Published data are often scattered in the literature, and sometimes are not available with sufficient details to be reproducible. Hence, developing broad partnerships with industries and utilities, and creating data sharing protocols are imperative. An example of such partnerships is the FAST (facilities accelerating science and technology) Water Network initiated by multiple federal agencies in the US. This network connects researchers, new technology providers, water treatment facilities, and other innovators in the water resource recovery industry to allow testing of innovative technologies and the generation of standardized high-quality data (Snyder and Simon, 2017; Mihelcic et al., 2017).

Sustainability assessments can provide powerful insights about the tradeoffs that arise between design considerations and the technology's performance across the dimensions of sustainability. While many life cycle studies have been conducted for conventional processes such as water reclamation treatment technologies, end users (non-potable vs. potable), and biosolids management (anaerobic digestion, incineration, land application, etc.), life cycle evaluations for energy recovery technologies other than biosolids management processes are scarce. Furthermore, developing technologies, such as source separation systems, are much less investigated. From an impact assessment perspective, sustainability is a broad term that encompasses several dimensions. Previous studies in the field are primarily focused on the assessment of the environmental or economic dimension, while the social dimension or a combination thereof, as well as the tradeoffs among the three dimensions are much less discussed. One approach to understand and address such tradeoffs is to engage stakeholders from the very beginning of the technology design and assessment phases. Such engagement would allow prioritization of the different sustainability dimensions and create innovative case-specific solutions. Methods to engage stakeholders are widely available from social science-related disciplines, which include collaborative modeling tools, workshops, charrettes, and focus group approaches.

Additionally, the development of decision support tools that take into account multiple factors to support the comprehensive assessment of technologies could be of great value. Such tools need to allow the integration and generalization of site and technology-specific data as well as inform proactive decision-making under uncertain future conditions. This would require future sustainability assessments to be able to incorporate spatial and temporal variabilities. A shift from the currently dominant deterministic and static models to stochastic and dynamic models is hence necessary. As the desired analyses become increasingly complex, it is also important to develop multi-disciplinary collaborations to tackle these problems using advanced and systematic approaches.

Last but not least, the improvement of sustainability indicators also relies on effective policy-making, stakeholder communication, and (eventually) a wide public acceptance of technological innovations. While many of these components might be beyond the reach of the scientific community, researchers can contribute to the advancement of the field by increasing transparency of their assessments (Sena and Hicks, 2018) and effectively communicating science to stakeholders and the public.

KEYWORDS

- **biological oxygen demand**
- **clean water act**
- **direct potable reuse**
- **environmental risk assessment**
- **life cycle assessment**
- **membrane bioreactors**

REFERENCES

Aarhus, V., (n.d.). *Marselisborg, Denmark-From Wastewater Plant to Power Plant*. State of Green. https://stateofgreen.com/en/profiles/aarhus-water-ltd/solutions/marselisborg-wwtp-energy-neutral-water-management (accessed on 10 June 2020).

Amann, A., Zoboli, O., Krampe, J., Rechberger, H., Zessner, M., &Egle, L., (2018). Environmental impacts of phosphorus recovery from municipal wastewater. *Resources, Conservation and Recycling*, *130*, 127–139.

Anastas, P. T., (2012). Fundamental changes to EPA's research enterprise: The path forward. *Environmental Science and Technology*, *46*(2), 580–586.

Angelakis, A. N., & Snyder, S. A., (2015). Wastewater treatment and reuse: Past, present, and future. *Water*, *7*, 4887–4895.

Aresta, M., Dibenedetto, A., &Barberio, G., (2005). Utilization of macro-algae for enhanced CO_2 fixation and bio-fuels production: Development of computing software for an LCA study. *Fuel Processing Technology*, *86*(14/15), 1679–1693.

Asano, T., Burton, F. L., Leverenz, H. L., Tsuchihashi, R., &Tchobanoglous, G., (2007). *Water Reuse: Issues, Technologies, and Applications*. New York, NY: McGraw-Hill.

Avery, L. M., Anchang, K. Y., Tumwesige, V., Strachan, N., &Goude, P. J., (2014). Potential for pathogen reduction in anaerobic digestion and biogas generation in Sub-Saharan Africa. *Biomass and Bioenergy*, *70*, 112–124.

Bachmann, N., (2015). *Sustainable Biogas Production in Municipal Wastewater Treatment Plants*. IEA Bio-energy. http://task37.ieabioenergy.com/files/daten-redaktion/download/Technical%20Brochures/Wastewater_biogas_grey_web-1.pdf (accessed on 10 June 2020).

Bakir, H. A., (2001). Sustainable wastewater management for small communities in the Middle East and North Africa. *Journal of Environmental Management*, *61*, 319–328.

Balkema, A. J., Preisig, H. A., Otterpohl, R., & Lambert, F. J., (2002). Indicators for the sustainability assessment of wastewater treatment systems. *Urban Water*, *4*(2), 153–161.

Baumann, D. D., (1983). Social acceptance of water reuse. *Applied Geography*, *3*(1), 79–84.

Beaver, D., (2014). *Wastewater Treatment Plant Process*. City of Beaver Dam, Wisconsin. http://209.43.45.22/egov/apps/document/center.egov?view=item&id=49 (accessed on 10 June 2020).

Blume, S., & Winker, M., (2011). Three years of operation of the urine diversion system at GTZ headquarters in Germany: User opinions and maintenance challenges. *Water Science and Technology*, *64*(3), 579–586.

Brown, R. R., &Farrelly, M. A., (2009). Delivering sustainable urban water management: A review of the hurdles we face. *Water Science and Technology*, *59*(5), 839–846.

Brown, S., Beecher, N., & Carpenter, A., (2010). Calculator tool for determining greenhouse gas emissions for biosolids processing and end use. *Environmental Science and Technology*, *44*(24), 9509–9515.

Buckwalter, P., Embaye, T., Gormly, S., & Trent, J. D., (2013). Dewatering microalgae by forward osmosis. *Desalination*, *312*, 19–22.

Cai, T., Park, S. Y., & Li, Y., (2013). Nutrient recovery from wastewater streams by microalgae: Status and prospects. *Renewable and Sustainable Energy Reviews*, *19*, 360–369.

Caporgno, M. P., Trobajo, R., Caiola, N., Ibanez, C., Fabregat, A., &Bengoa, C., (2015). Biogas production from sewage sludge and microalgae co-digestion under mesophilic and thermophilic conditions. *Renewable Energy*, *75*, 374–380.

Cartmell, E., Gostelow, P., Riddell-Black, D., Simms, N., Oakey, J., Morris, J., & Pollard, S. J., (2006). Biosolids a fuel or a waste? An integrated appraisal of five co-combustion scenarios with policy analysis. *Environmental Science and Technology*, *40*(3), 649–658.

Chae, K. J., & Kang, J., (2013). Estimating the energy independence of a municipal wastewater treatment plant incorporating green energy resources. *Energy Conversion and Management*, *75*, 664–672.

Chen, Z., Ngo, H. H., & Guo, W., (2012). A critical review on sustainability assessment of recycled water schemes. *Science of the Total Environment*, *426*, 13–31.

Ching, L., (2010). Eliminating 'yuck': A simple exposition of media and social change in water reuse policies. *International Journal of Water Resources Development*, *26*(1), 111–124.

CMU, (2008). *Economic Input-Output Life Cycle Assessment (EIO-LCA)*. Carnegie Mellon University (CMU) Green Design Institute. http://www.eiolca.net/ (accessed on 10 June 2020).

Copeland, C., &Tiemann, M., (2010). *Water Infrastructure Needs and Investment: Review and Analysis of Key Issues*. Washington, D.C: Congressional Research Services Report to Congress.

Cornejo, P. K., Zhang, Q., &Mihelcic, J. R., (2016). How does scale of implementation impact the environmental sustainability of wastewater treatment integrated with resource recovery? *Environmental Science and Technology*, *50*(13), 6680–6689.

Cornel, P., &Schaum, C., (2009). Phosphorus recovery from wastewater: Needs, technologies and costs. *Water Science and Technology*, *59*(6), 1069–1076.

Danfoss, (2014). *Case Story: Generating Surplus Power from Wastewater Treatment*. Danfoss VLT Drives, DKDD.PC.207.A1.02. http://drives.danfoss.nl/uploadedfiles/content/files/documents/pe/marselisborg-dkddpc207a102-casestory_nov2014.pdf (accessed on 10 June 2020).

De-Bashan, L. E., & Bashan, Y., (2004). Recent advances in removing phosphorus from wastewater and its future use as fertilizer (1997–2003). *Water Research*, *38*(19), 4222–4246.

Diaz-Elsayed, N., Rezaei, N., Guo, T., Mohebbi, S., & Zhang, Q., (2019). Wastewater-based resource recovery technologies across scale: A review. *Resources, Conservation, and Recycling*. In press.

Dolnicar, S., Hurlimann, A., & Nghiem, L. D., (2010). The effect of information on public acceptance-the case of water from alternative sources. *Journal of Environmental Management*, *91*(6), 1288–1293.

Egle, L., Rechberger, H., Krampe, J., &Zessner, M., (2016). Phosphorus recovery from municipal wastewater: An integrated comparative technological, environmental, and economic assessment of P recovery technologies. *Science of the Total Environment*, *571*, 522–542.

EPRI & WRF, (2013). *Electricity use and Management in the Municipal Water Supply and Wastewater Industries*. Electric Power Research Institute (EPRI) and Water Research Foundation (WRF).

Farrelly, M., & Brown, R., (2011). Rethinking urban water management: Experimentation as a way forward? *Global Environmental Change*, *21*(2), 721–732.

Fielding, K. S., &Roiko, A. H., (2014). Providing information promotes greater public support for potable recycled water. *Water Research, 61*, 86–96.

Gabor, T., Dan, V., Badila, I. N., Tiuc, A. E., & Sur, I. M., (2017). Improving the energy efficiency of residential buildings by using a drain water heat recovery system. *Environmental Engineering and Management Journal*, *16*(7), 1631–1636.

Garcia-Cuerva, L., Berglund, E. Z., & Binder, A. R., (2016). Public perceptions of water shortages, conservation behaviors, and support for water reuse in the US. *Resources, Conservation and Recycling*, *113*, 106–115.

Ge, Z., Li, J., Xiao, L., Tong, Y., & He, Z., (2013). Recovery of electrical energy in microbial fuel cells: Brief review. *Environmental Science and Technology Letters*, *1*(2), 137–141.

Gerrity, D., Pecson, B., Trussell, R. S., & Trussell, R. R., (2013). Potable reuse treatment trains throughout the world. *Journal of Water Supply: Research and Technology-AQUA*, *62*(6), 321–338.

GGTC, (2012). *Technology Assessment Report: Aqueous Sludge Gasification Technologies* (Report No.: EPA/600/R-12/540). Prepared by the Greenhouse Gas Technology Center. US Environmental Protection Agency, Washington, DC.

Gikas, P., (2014). Electrical energy production from bio-solids: A comparative study between anaerobic digestion and ultra-high-temperature gasification. *Environmental Technology*, *35*(17), 2140–2146.

Gleeson, T., Wada, Y., Bierkens, M. F., & Van, B. L. P., (2012). Water balance of global aquifers revealed by groundwater footprint. *Nature*, *488*(7410), 197–200.

Gu, Y., Li, Y., Li, X., Luo, P., Wang, H., Robinson, Z. P., & Li, F., (2017). The feasibility and challenges of energy self-sufficient wastewater treatment plants. *Applied Energy*, *204*, 1463–1475.

Guest, J. S., Skerlos, S. J., Barnard, J. L., Beck, M. B., Daigger, G. T., Hilger, H., &Mihelcic, J. R., (2009). A new planning and design paradigm to achieve sustainable resource recovery from wastewater. *Environmental Science and Technology*, *43*(16), 6126–6130.

Guo, T., &Englehardt, J. D., (2015). Principles for scaling of distributed direct potable water reuse systems: A modeling study. *Water Research*, *75*, 146–163.

Guo, T., Englehardt, J., & Wu, T., (2014). Review of cost versus scale: Water and wastewater treatment and reuse processes. *Water Science and Technology*, *69*(2), 223–234.

Haddaway, A., (2015). *Nutrient Recovery Technology Transforms World's Largest Wastewater Treatment Plant*. Water World. http://www.waterworld.com/articles/print/volume-31/issue-2/features/nutrient-recovery-technology-transforms-world-s-largest-wastewater-treatment-plant.html (accessed on 10 June 2020).

Harris-Lovett, S. R., Binz, C., Sedlak, D. L., Kiparsky, M., &Truffer, B., (2015). Beyond user acceptance: A legitimacy framework for potable water reuse in California. *Environmental Science and Technology*, *49*(13), 7552–7561.

Hartley, T. W., (2006). Public perception and participation in water reuse. *Desalination*, *187*(1–3), 115–126.

Hawthorne, M., (2016). Chicago turning river pollutants into fertilizer. *Chicago Tribune*. Retrieved from: http://www.chicagotribune.com (accessed on 10 June 2020).

Ho, G., (2005). Technology for sustainability: The role of on site, small and community scale technology. *Water Science and Technology*, *51*(10), 15–20.

Hong, J., Hong, J., Otaki, M., & Jolliet, O., (2009). Environmental and economic life cycle assessment for sewage sludge treatment processes in Japan. *Waste Management*, *29*(2), 696–703.

ILFI, (2017). *Living Building Challenge*. International Living Future Institute. https://living-future.org/lbc/ (accessed on 10 June 2020).

Ishii, S. K., & Boyer, T. H., (2015). Life cycle comparison of centralized wastewater treatment and urine source separation with struvite precipitation: Focus on urine nutrient management. *Water Research*, *79*, 88–103.

Ishii, S. K., & Boyer, T. H., (2016). Student support and perceptions of urine source separation in a university community. *Water Research*, *100*, 146–156.

ISO, (2006). *Environmental Management-Life Cycle Assessment-Principles and Framework (ISO 14040: 2006)*. International Organization for Standardization (ISO).

Jeffrey, P., (2002). Public attitudes to in-house water recycling in England and Wales. *Water and Environment Journal*, *16*(3), 214–217.

Karáth, K., (2016). World's first city to power its water needs with sewage energy. *New Scientist*. https://www.newscientist.com/article/2114761-worlds-first-city-to-power-its-water-needs-with-sewage-energy/ (accessed on 10 June 2020).

Kavvada, O., Horvath, A., Stokes-Draut, J. R., Hendrickson, T. P., Eisenstein, W. A., & Nelson, K. L., (2016). Assessing location and scale of urban nonpotable water reuse systems for life-cycle energy consumption and greenhouse gas emissions. *Environmental Science and Technology*, *50*(24), 13184–13194.

Kavvada, O., Tarpeh, W. A., Horvath, A., & Nelson, K. L., (2017). Life-cycle cost and environmental assessment of decentralized nitrogen recovery using ion exchange from source-separated urine through spatial modeling. *Environmental Science and Technology, 51*(21), 12061–12071.

Lamichhane, K. M., & Babcock, Jr. R. W., (2013). Survey of attitudes and perceptions of urine-diverting toilets and human waste recycling in Hawaii. *Science of the Total Environment, 443*, 749–756.

Larsen, T. A., Alder, A. C., Eggen, R. I., Maurer, M., &Lienert, J., (2009). Source separation: Will we see a paradigm shift in wastewater handling? *Environmental Science and Technology, 43*(16), 6121–6125.

Leverenz, H. L., Tchobanoglous, G., & Asano, T., (2011). Direct potable reuse: A future imperative. *Journal of Water Reuse and Desalination, 1*(1), 2–10.

Li, H., Jin, C., &Mundree, S., (2017). Hybrid environmental and economic assessment of four approaches recovering energy from sludge with variant organic contents. *Journal of Cleaner Production, 153*, 131–138.

Lienert, J., & Larsen, T. A., (2009). High acceptance of urine source separation in seven European countries: A review. *Environmental Science and Technology, 44*(2), 556–566.

Liu, B., Wei, Q., Zhang, B., & Bi, J., (2013). Life cycle GHG emissions of sewage sludge treatment and disposal options in Tai Lake Watershed, China. *Science of the Total Environment, 447*, 361–369.

Lundin, M., Olofsson, M., Pettersson, G. J., &Zetterlund, H., (2004). Environmental and economic assessment of sewage sludge handling options. *Resources, Conservation and Recycling, 41*(4), 255–278.

Marks, J. S., (2004). *Advancing Community Acceptance of Reclaimed Water*. Australian Water and Wastewater Association.

Marks, J., Martin, B., &Zadoroznyj, M., (2008). How Australians order acceptance of recycled water: National baseline data. *Journal of Sociology, 44*(1), 83–99.

Marlow, D. R., Moglia, M., Cook, S., & Beale, D. J., (2013). Towards sustainable urban water management: A critical reassessment. *Water Research, 47*(20), 7150–7161.

Maurer, M., Rothenberger, D., & Larsen, T. A., (2005). Decentralized wastewater treatment technologies from a national perspective: At what cost are they competitive? *Water Supply, 5*(6), 145–154.

McCarty, P. L., Bae, J., & Kim, J., (2011). Domestic wastewater treatment as a net energy producer-can this be achieved? *Environmental Science and Technology, 45*(17), 7100–7106.

Means, E., (2004). *Water and Wastewater Industry Energy Efficiency: A Research Roadmap*. AW Research Foundation. CEC-500-2004-901.

Meggers, F., &Leibundgut, H., (2011). The potential of wastewater heat and energy: Decentralized high-temperature recovery with a heat pump. *Energy and Buildings, 43*(4), 879–886.

Melin, T., Jefferson, B., Bixio, D., Thoeye, C., De Wilde, W., De Koning, J., &Wintgens, T., (2006). Membrane bioreactor technology for wastewater treatment and reuse. *Desalination, 187*(1–3), 271–282.

Meneses, M., Pasqualino, J. C., & Castells, F., (2010). Environmental assessment of urban wastewater reuse: Treatment alternatives and applications. *Chemosphere, 81*(2), 266–272.

Metcalf & Eddy, (2004). *Wastewater Engineering, Treatment, and Reuse*. New York, NY: McGraw Hill.

Mihelcic, J. R., Ren, Z. J., Cornejo, P. K., Fisher, A., Simon, A. J., Snyder, S. W., & Moeller, J., (2017). Accelerating innovation that enhances resource recovery in the wastewater sector: Advancing a national test bed network. *Environmental Science and Technology, 51*(14), 7749–7758.

Mikkonen, L., Rämö, J., Keiski, R. L., &Pongrácz, E., (2013). *Heat Recovery from Wastewater: Assessing the Potential in Northern Areas* (pp. 161–164). Water Research at Oulu University.

Mitchell, C., Fam, D., &Abeysuriya, K., (2013). *Transitioning to Sustainable Sanitation-A Transdisciplinary Project of Urine Diversion*. Institute for Sustainable Futures, University of Technology Sydney, Australia.

Mo, W., & Zhang, Q., (2012). Can municipal wastewater treatment systems be carbon neutral? *Journal of Environmental Management, 112*, 360–367.

Mo, W., & Zhang, Q., (2013). Energy-nutrients-water nexus: Integrated resource recovery in municipal wastewater treatment plants. *Journal of Environmental Management, 127*, 255–267.

Muñoz, I., Rodriguez, A., Rosal, R., & Fernandez-Alba, A. R., (2009). Life cycle assessment of urban wastewater reuse with ozonation as tertiary treatment: A focus on toxicity-related impacts. *Science of the Total Environment, 407*(4), 1245–1256.

Muralikrishna, I. V., & Manickam, V., (2017). Chapter eight-environmental risk assessment. In: *Environmental Management: Science and Engineering for Industry* (pp. 135–152). Kidlington, Oxford, UK: Butterworth-Heinemann.

Murray, A., Horvath, A., & Nelson, K. L., (2008). Hybrid life-cycle environmental and cost inventory of sewage sludge treatment and end-use scenarios: A case study from China. *Environmental Science and Technology, 42*(9), 3163–3169.

MWRDGC, (n.d.). *Stickney WRP*. Metropolitan Water Reclamation District of Greater Chicago (MWRDGC). https://www.mwrd.org/irj/portal/anonymous/stickney (accessed on 10 June 2020).

NEBRA, (2007). *A National Bio-Solids Regulation, Quality, End Use, and Disposal Survey: Final Report*. North East Bio-solids and Residuals Association (NEBRA). https://static1.squarespace.com/static/54806478e4b0dc44e1698e88/t/5488541fe4b03c0a9b8ee09b/1418220575693/NtlBiosolidsReport-20July07.pdf (accessed on 10 June 2020).

Nexus, (2015). *Nexus e Water*. http://www.nexusewater.com/ (accessed on 10 June 2020).

Ni, L., Lau, S., Li, H., Zhang, T., Stansbury, J., Shi, J., & Neal, J., (2012). Feasibility study of a localized residential grey water energy-recovery system. *Applied Thermal Engineering, 39*, 53–62.

Nourozi, B., Wang, Q., &Ploskić, A., (2019). Maximizing thermal performance of building ventilation using geothermal and wastewater heat. *Resources, Conservation and Recycling, 143*, 90–98.

Nowak, O., Kuehn, V., &Zessner, M., (2004). Sludge management of small water and wastewater treatment plants. *Water Science and Technology, 48*(11, 12), 33–41.

NRC, (2012). *Water Reuse: Potential for Expanding the Nation's Water Supply Through Reuse of Municipal Wastewater*. National Research Council, Washington, DC: National Academies Press.

NSF, (2017). *NSF/ANSI 350 Onsite Residential and Commercial Water Reuse Treatment*. http://info.nsf.org/Certified/Wastewater/Listings.asp?TradeName=&Standard=350 (accessed on 10 June 2020).

OCWD, (2019). *Green Acres Project*. Orange County Water District (OCWD). https://www.ocwd.com/what-we-do/water-reuse/green-acres-project/ (accessed February 18, 2019).

OCWD, (n.d.). *Groundwater Replenishment System*. Orange County Water District (OCWD). https://www.ocwd.com/media/4267/gwrs-technical-brochure-r.pdf (accessed on 10 June 2020).

Ostara, (2017). *Ostara: Crystal Green*. http://ostara.com/nutrients/ (accessed on 10 June 2020).

Pahl-Wostl, C., Schönborn, A., Willi, N., Muncke, J., & Larsen, T. A., (2003). Investigating consumer attitudes towards the new technology of urine separation. *Water Science and Technology, 48*(1), 57–65.

Pinellas County, (2013). *Potable Water Supply, Wastewater, and Reuse Element: Chapter 2, Wastewater*. The Pinellas County Planning Department. https://www.pinellascounty.org/Plan/comp_plan/9water/ch2.pdf (accessed on 10 June 2020).

Pinellas County, (2018). *South Cross Bayou Water Reclamation Facility: Environmental Commitment through Resource Recovery*. http://www.pinellascounty.org/utilities/south-cross.htm (accessed on 10 June 2020).

Pinellas County, (n.d.). *William E. Dunn Water Reclamation Facility: Facility Overview and Information*. https://www.pinellascounty.org/utilities/PDF/dunn-tour.pdf (accessed on 10 June 2020).

Porse, E., Mika, K. B., Litvak, E., Manago, K. F., Hogue, T. S., Gold, M., &Pincetl, S., (2018). The economic value of local water supplies in Los Angeles. *Nature Sustainability, 1*, 289–297.

Power, C., McNabola, A., & Coughlan, P., (2014). Development of an evaluation method for hydropower energy recovery in wastewater treatment plants: Case studies in Ireland and the UK. *Sustainable Energy Technologies and Assessments, 7*, 166–177.

PUB, (2019). *Four National Taps*. Singapore's Public Utilities Board (PUB). https://www.pub.gov.sg/watersupply/fournationaltaps (accessed February 18, 2019).

Quezada, G., Walton, A., & Sharma, A., (2016). Risks and tensions in water industry innovation: Understanding adoption of decentralized water systems from a socio-technical transitions perspective. *Journal of Cleaner Production, 113*, 263–273.

Radcliffe, J. C., (2006). Future directions for water recycling in Australia. *Desalination, 187*(1–3), 77–87.

Rahman, M. M., Salleh, M. A. M., Rashid, U., Ahsan, A., Hossain, M. M., & Ra, C. S., (2014). Production of slow release crystal fertilizer from wastewaters through struvite crystallization: A review. *Arabian Journal of Chemistry, 7*(1), 139–155.

Raucher, R. S., &Tchobanoglous, G., (2014). *The Opportunities and Economics of Direct Potable Reuse*. Alexandria, VA: WateReuse Research Foundation.

Remy, C., (2010). *Life Cycle Assessment of Conventional and Source Separation Systems for Urban Wastewater Management(Doctoral Dissertation)*. Retrieved from: https://d-nb.info/1001584597/34 (accessed February 18, 2019).

Rezaei, N., Diaz-Elsayed, N., Mohebbi, S., Xie, X., & Zhang, Q., (2019). A multi-criteria sustainability assessment of water reuse applications: A case study in Lakeland, Florida. *Environmental Science: Water Research and Technology, 5*(1), 102–118.

Ross, V. L., Fielding, K. S., & Louis, W. R., (2014). Social trust, risk perceptions and public acceptance of recycled water: Testing a social-psychological model. *Journal of Environmental Management, 137*, 61–68.

Sahlström, L., (2003). A review of survival of pathogenic bacteria in organic waste used in biogas plants. *Bioresource Technology*, *87*(2), 161–166.

San Diego, (2017). *Energy Efficiency Program*. City of San Diego. https://www.sandiego.gov/mwwd/environment/energy (accessed on 10 June 2020).

San Diego, (2018). *Metro Biosolids Center*. City of San Diego. https://www.sandiego.gov/mwwd/facilities/metrobiosolids (accessed on 10 June 2020).

San, B. G., Choulot, A., & Denis, V., (2010). *Energy Recovery in Existing Infrastructures with Small Hydropower Plants*. European Small Hydropower Association, Brussels, Belgium.

Scaglia, B., D'Imporzano, G., Garuti, G., Negri, M., & Adani, F., (2014). Sanitation ability of anaerobic digestion performed at different temperature on sewage sludge. *Science of the Total Environment*, *466*, 888–897.

Scherson, Y. D., Woo, S. G., &Criddle, C. S., (2014). Production of nitrous oxide from anaerobic digester centrate and its use as a co-oxidant of biogas to enhance energy recovery. *Environmental Science and Technology*, *48*(10), 5612–5619.

Sena, M., & Hicks, A., (2018). Life cycle assessment review of struvite precipitation in wastewater treatment. *Resources, Conservation and Recycling*, *139*, 194–204.

Sharma, A. K., Cook, S., Tjandraatmadja, G., & Gregory, A., (2012). Impediments and constraints in the uptake of water sensitive urban design measures in Greenfield and infill developments. *Water Science and Technology*, *65*(2), 340–352.

Sharma, A. K., Tjandraatmadja, G., Cook, S., & Gardner, T., (2013). Decentralized systems-definition and drivers in the current context. *Water Science and Technology*, *67*(9), 2091–2101.

Sharma, A., Burn, S., Gardner, T., & Gregory, A., (2010). Role of decentralized systems in the transition of urban water systems. *Water Supply, 10*(4), 577–583.

Sheikh, B., Cort, R. P., Kirkpatrick, W. R., Jaques, R. S., & Asano, T., (1990). Monterey wastewater reclamation study for agriculture. *Research Journal of the Water Pollution Control Federation, 62*(3), 216–226.

Słyś, D., &Kordana, S., (2014). Financial analysis of the implementation of a drain water heat recovery unit in residential housing. *Energy and Buildings*, *71*, 1–11.

Snyder, S. W., & Simon, A., (2017). *Concept of Operations: National Technology Test Bed Network for Water Resource Recovery*. INL/EXT-18-50231, LLNL-TR-743712.

Soda, S., Iwai, Y., Sei, K., Shimod, Y., & Ike, M., (2010). Model analysis of energy consumption and greenhouse gas emissions of sewage sludge treatment systems with different processes and scales. *Water Science and Technology*, *61*(2), 365–373.

Stillwell, A. S., Hoppock, D. C., & Webber, M. E., (2010). Energy recovery from wastewater treatment plants in the United States: A case study of the energy-water nexus. *Sustainability*, *2*(4), 945–962.

Stokes, J. R., & Horvath, A., (2010). Supply-chain environmental effects of wastewater utilities. *Environmental Research Letters, 5*(1), 014015.

Suh, Y. J., & Rousseaux, P., (2002). An LCA of alternative wastewater sludge treatment scenarios. *Resources, Conservation and Recycling*, *35*(3), 191–200.

Tangsubkul, N., Beavis, P., Moore, S. J., Lundie, S., & Waite, T. D., (2005). Life cycle assessment of water recycling technology. *Water Resources Management*, *19*(5), 521–537.

Tarallo, S., Shaw, A., Kohl, P., &Eschborn, R., (2015). *A Guide to Net-Zero Energy Solutions for Water Resource Recovery Facilities*. Alexandria, VA: Water Environment Research Foundation.

Tchobanoglous, G., &Leverenz, H., (2013). The rationale for decentralization of wastewater infrastructure. In: Larsent, T. A., Udert, K. M., &Lienert, J., (eds.), *Source Separation and Decentralization for Wastewater Management*. London, UK: IWA publishing.

Tchobanoglous, G., Leverenz, H., Nellor, M. H., & Crook, J., (2011). *Direct Potable Reuse: A Path Forward.* Alexandria, VA: WateReuse Research Foundation.

The Balmoral Group, LLC, (n.d.). *Florida Statewide Agricultural Irrigation Demand (FSAID).* https://www.freshfromflorida.com/content/download/61727/1412341/FSAID_II_Final_Report_The_Balmoral_Group_20150701.pdf (accessed on 10 June 2020).

Theregowda, R., Vidic, R., Dzombak, D. A., & Landis, A. E., (2015). Life cycle impact analysis of tertiary treatment alternatives to treat secondary municipal wastewater for reuse in cooling systems. *Environmental Progress and Sustainable Energy, 34*(1), 178–187.

Tillman, A. M., Svingby, M., & Lundström, H., (1998). Life cycle assessment of municipal waste water systems. *The International Journal of Life Cycle Assessment, 3*(3), 145–157.

Trussell, R. R., Trussell, R. S., Salveson, A., Steinle-Darling, E., He, Q., Snyder, S., & Gerrity, D., (2015). *Equivalency of Advanced Treatment Trains for Potable Reuse: User Manual for Treatment Train Toolbox*. Alexandria, VA: WateReuse Research Foundation.

University of Georgia, (2019). *Crop Code Sheets*. University of Georgia, College of Agricultural and Environmental Sciences. http://aesl.ces.uga.edu/publications/soil/CropSheets.pdf (accessed on 10 June 2020).

Urkiaga, A., De Las Fuentes, L., Bis, B., Chiru, E., Balasz, B., & Hernández, F., (2008). Development of analysis tools for social, economic, and ecological effects of water reuse. *Desalination, 218*(1–3), 81–91.

US EPA, (2000). *Bio-Solids Technology Factsheet: Land Application of Bio-Solids*. Report No.: 832-F-00-064. US Environmental Protection Agency, Washington, DC.

US EPA, (2003). *Bio-solids Technology Factsheet: Use of Incineration for Bio-solids Management.* Report No.: 832-F-03-013. US Environmental Protection Agency, Washington, DC.

US EPA, (2009). *Nutrient Control Design Manual: State of Technology Review Report.* Report No.: EPA/600/R-09/012, US Environmental Protection Agency, Washington, DC.

US EPA, (2012). *2012 Guidelines for Water Reuse*. Report No.: EPA/600/R-12/618, US Environmental Protection Agency, Washington, DC.

USDA, (2017). *What is Pyrolysis?* US Department of Agriculture, Agricultural Research Service. https://www.ars.usda.gov/northeast-area/wyndmoor-pa/eastern-regional-research-center/docs/biomass-pyrolysis-research-1/what-is-pyrolysis/ (accessed on 10 June 2020).

Van, D. H. J. P., De Fooij, H., &Struker, A., (2016). Wastewater as a res*ource:* Strategies to recover resources from Amsterdam's wastewater. *Resources, Conservation and Recycling, 113*, 53–64.

Vand, (n.d.). *Marselisborg, Denmark-From Wastewater Plant to Power Plant*. State of Green.https://stateofgreen.com/en/profiles/aarhus-water-ltd/solutions/marselisborg-wwtp-energy-neutral-water-management (accessed on 10 June 2020).

Viswanathan, S., (2010). From biomass to energy. *Water and Wastes Digest*, (pp. 20–21).

Water Issues Committee, (2016). *Green Acres Project Future Direction.* Orange County Water District. http://www.ocwd.com/media/4483/wic07bodpresentation.pptx (accessed on 10 June 2020).

Wilderer, P. A., &Schreff, D., (2000). Decentralized and centralized wastewater management: A challenge for developers. *Water Science and Technology, 41*(1), 1–8.

Wiley, P., Harris, L., Reinsch, S., Tozzi, S., Embaye, T., Clark, K., & Kagawa, H., (2013). Microalgae cultivation using offshore membrane enclosures for growing algae (OMEGA). *Journal of Sustainable Bio-Energy Systems*, *3*, 18–32.
Wood, A., Blackhurst, M., Hawkins, T., Xue, X., Ashbolt, N., & Garland, J., (2015). Cost-effectiveness of nitrogen mitigation by alternative household wastewater management technologies. *Journal of Environmental Management, 150*, 344–354.
Zimmerman, J. B., Mihelcic, J. R., & Smith, J., (2008). Global stressors on water quality and quantity. *Environmental Science and Technology*, *42*(12), 4247–4254.

CHAPTER 2

Evaluating Resource Recovery Options in Wastewater Treatment Plants Using Mathematical Models

PEDRAM RAMIN, ELHAM RAMIN, HANNAH FELDMAN, XAVIER FLORES-ALSINA, and KRIST V. GERNAEY

Process and Systems Engineering Center (PROSYS), Department of Chemical and Biochemical Engineering, Technical University of Denmark, Building – 228A, 2800 Kongens Lyngby, Denmark

ABSTRACT

Wastewater treatment plants (WWTPs) are traditionally designed and operated to ensure that a certain effluent quality is achieved. Attaining this goal has been conventionally possible via biological processes for COD and nutrient removal. For instance, nitrogen is typically removed via nitrification/denitrification, which requires energy (aeration) and in some cases additional carbon source. Phosphorus can also be removed biologically or chemically by the addition of metal salts (precipitation). Excess wastewater sludge, a byproduct of the traditional activated sludge process, is also used frequently to produce biogas through the anaerobic digestion process. However, recent studies show that only a fraction of the energy potential of the wastewater is captured through the above-mentioned conventional processes. Moreover, due to recent concerns about climate change (CC) and limited availability of natural resources, together with strong needs for recovering more water and energy from wastewater, more sustainable wastewater treatment solutions must be sought. In addition, uncertainty in energy prices as well as drivers at the national level, require the modification of WWTPs towards low energy alternatives. Nutrient recovery is a sustainable approach in wastewater treatment, through which the nutrients can be further reused (e.g., in the form of energy or fertilizer). The idea of transforming a wastewater treatment plant to a resource recovery facility has been introduced in the past decade, where municipal wastewater is no longer considered as a waste but as a source of water, energy, and nutrients for agriculture. In this perspective, different technologies have been suggested to concentrate the nutrients and remove them via separation processes. This chapter will review general platforms and modeling tools to build resource recovery facilities and will also discuss process alternatives and technologies available to increase the net energy yield from wastewater and to promote recovery of resources. Moreover, different simulation approaches are introduced in order to develop innovative configurations for future WWTPs and to predict the efficiency of such configurations in terms of maximizing energy production and recovery of valuable nutrients. Specifically, energy recovery in an anaerobic digestion process is discussed in more detail through a case study.

2.1 WHY DO WE NEED MODELS FOR RESOURCE RECOVERY FROM WASTEWATER?

While resource recovery from wastewater treatment is a sustainable way to produce energy, chemicals, and other products, it is important that the process is cost-effective, as wastewater treatment in itself is an additional cost for many industries and municipalities. Therefore, performing extensive laboratory and pilot-scale experiments is often not desired or prioritized, due to the high cost related to such experiments. Yet, process engineers are constantly looking for ways to optimize treatment processes to increase energy and nutrients recovery and decrease the consumption of chemicals. Here, mathematical models form a useful tool for process optimization that is not labor, material, and energy-intensive to use (Batstone and Keller, 2003).

In the past years, mathematical models have proven to be a valuable tool for design, for diagnosis and suggesting improved plant management strategies. Since the establishment of activated sludge and anaerobic digestion models (Henze et al., 2000; Batstone et al., 2002) the primarily goal was on COD and nutrient removal, mainly in the aqueous phase using biological, physico-chemical and mass transfer equations in series of completely stirred reactors. Later, more complex models were built including more detailed process descriptions, both in terms of kinetics as well as hydraulics. It is often argued that a wastewater treatment plant should be considered as an integrated system of different units, e.g., activated sludge, settling tanks, dewatering system, anaerobic digestion, as these units are all linked together and hence need to be simulated and controlled not as individual unit operations but as a system considering all the interactions between the components. This realization is also crucial when wastewater treatment plants (WWTPs) are to be optimally designed and operated, not only for effluent standard compliance, but also for maximizing resource recovery potentials. Mathematical tools can provide support for the decision making process, especially when trade-off situations are to be considered. The tools can be effectively used for process optimization, for studying control as well as for techno-economic analysis of process alternatives.

Plant-wide models can be categorized as steady-state models consisting of algebraic equations or as dynamic systems consisting of differential equations. As compared to large and complex plant-wide dynamic models, steady state models require much less input information and provide explicit solutions. This makes these models useful for designing and sizing the treatment plant for the specified capacity, and allows explicit calculation of unit operations performance criteria. For design purpose, once the overall scheme of the treatment plant is established, and the main parameters for the unit operations are estimated and the source of errors in data are eliminated, the dynamic model can be used to further refine the design of process units and to test their performance under dynamic conditions (Ekama, 2009).

Beside "modeling for design" and "modeling for optimization," the application of models in the field of wastewater treatment extends to teaching and learning (Gernaey et al., 2014). Such educational purpose can be for both students as well as operating personnel at a treatment facilities. While these groups can run laboratory experiments to increase their understanding of underlying concepts, mathematical models can be used as a complementary tool with relation to a specific learning objective, the level of education, time availability, and software to be used (Morgenroth et al., 2002).

2.2 PLANT-WIDE MODELING FOR RESOURCE RECOVERY, PLATFORMS, AND METHODOLOGIES

In practice, combining wastewater treatment with energy recovery was traditionally implemented by combining aerobic treatment and anaerobic digestion. This configuration only allowed capturing a portion of the energy potential of the wastewater as aerobic processes remove most easily degradable organic fractions. Although activated sludge treatment (AST) is superior to other alternative technologies to treat the organic matter, the process is very energy-intensive and requires high operation and maintenance (O&M) costs. Aeration can represent up to 75% of the total energy costs while sludge treatment and disposal can count for up to 60% of the total operation cost (Gude 2016). However, anaerobic digestion can also be used as a process to remove organic pollutants while recovering valuable bioenergy. Nevertheless, optimum energy and nutrient recovery in treatment plants often require a combination of treatment units. To allow exploring scenarios that maximize the nutrient recovery potential, a modeling framework is needed to put together a series of unit processes with optimized operating conditions.

Mathematical modeling for wastewater treatment processes started with the main focus on the activated sludge process (Ekama and Marais, 1977), which further led to the development of the famous activated sludge model no. 1 (ASM1) (Henze et al., 1987) for a more detailed description of carbon and nitrogen removal. The ASM1 considers two bacteria populations, heterotrophic biomass that consumes biodegradable organic matter and an autotrophic population, which is able to oxidize ammonia. The ASM1 set the stage for standardization of wastewater characterization and modeling practice in research and also in industry (Gernaey et al., 2014). Later on, the ASM2/ASM2d was developed to predict the phosphorus removal process performance in activated sludge processes, which also included a continuity equation for elemental mass and charge continuity evaluation (Henze et al., 2000). For anaerobic digestion, the first comprehensive model was the anaerobic digestion model no. 1 (ADM1) (Batstone et al., 2002), based on earlier studies in the field (Angelidaki et al., 1993; Pavlostathis and Gossett, 1986). ADM1 involves three main reactions: hydrolysis of carbohydrates, protein, and lipids to sugar, amino acids, and fatty acids followed by conversion to volatile fatty acids (VFAs), and short-chain fatty acids and hydrogen and finally the production of methane and CO_2.

2.2.1 *PLANT-WIDE MODELING APPROACHES*

With the aim of finding an optimum solution for improvement of the stability of a process, while also performing in terms of effluent quality and operational costs, mathematical models have been used to describe the steady-state and dynamic behavior of the system in an integrated way for all the process units in the water and sludge line. Steady-state plant-wide modeling has been used to draw links between COD and nutrient removal in activated sludge and anaerobic digestion systems. These models have shown internal consistency and externally compatible elements to create plant-wide dynamic simulations (Ekama et al., 2006). Developing plant-wide models, which ensures mass and charge continuity through the entire WWTP, is challenging. Most mathematical models such as ASMs and ADM1 are developed to represent the behavior of individual unit operations. In fact, these models have some inherent differences in terms of the central hypothesis on which they have been built, and in terms of the description of components and process transformations (COD fractionations, inert compounds, organic N, pH, and alkalinity, etc.) (Grau et al., 2009). Such discrepancies make these models to a large extent incompatible in a plant-wide perspective. Hence, there was a need to develop standardized methodologies and procedures for plant-wide modeling to understand the behavior of the system as a whole and to construct new process units in a compatible way. Several methodologies have been proposed, which can be summarized as following:

2.2.1.1 INTERFACE APPROACH

The model interface approach is an attempt to connect existing well-accepted process models for different process units by using appropriate conversions. In this method, the best way is sought to transform model components of one process unit to model components of another unit. A typical example is connecting an activated sludge model with settling models for secondary clarification, where all particulate fractions in the bioreactor are lumped into one fraction as total suspended solids (TSS), a variable which can be used for the settling model. Another example is connecting the ASM1 model (with 13 state variables) with the ADM1 model (with 26 state variables) (Copp et al., 2003), which is incorporated in the widely used WWTP benchmark simulation model no. 2 (BSM2) (Jeppsson et al., 2007). One of the challenges using the interface approach is maintaining elemental mass and charge continuity during conversion. The continuity-based interfacing method is developed to facilitate the coupling of any set of different process models using conservation principles (Volcke et al., 2006). Although the advantage of this method is preserving the use of well-established models, it is difficult to find an efficient interface for new process models without re-definition of the mass composition of the components.

2.2.1.2 SUPERMODEL APPROACH

In this approach, a unique model is used to describe all the relevant processes in all units. This means that all the transformations are present for the entire plant in all units that are modeled, regardless of the type of stream or the unit. For example, description of aerobic transformations is then present in the activated sludge model as well as the anaerobic digestion model. The rate of a process could be zero, but all the equations are simultaneously solved in an integrated way in this approach. This means that there is a unique and common vector which defines the state variables throughout the entire plant. Such approach was used to develop biological nutrient removal model no.1 (BNRM1) using the ASM2d wastewater characterization (Seco et al., 2004). The main advantage of the model is that no interface is needed to enable the connections between different units, and specific transformations can be switched on or off based on the conditions in a process unit. Moreover, estimated parameters during model calibration are not dependent on a specific treatment scheme and thus have less variability. However, the supermodel approach is not flexible in case of addition or removal of a unit operation to a process.

2.2.1.3 TRANSFORMATION-BASED APPROACH

This approach is similar to the supermodel approach but only includes the most relevant, biological, chemical, and physico-chemical processes for the plant under study. This method is based on selecting the necessary set of transformations that takes place in the entire plant from a library of all potential groups of reactions. The selection is depending on the process, the relevant bacteria population, and the transformations that would describe anabolic and catabolic activity of the population. This selection is done by the user, all at once, to model all the unit process elements for the entire plant. A general List of transformations (LT) for plant-wide modeling is provided (Figure 2.1), which after selection of appropriate transformations can follow a systematic procedure (Grau et al., 2007). The procedure consists of 4 steps. In the first step, based on the configuration of the plant and the purpose of the model, appropriate biological processes should be selected for the plant-wide model. The second step is the identification of active microorganisms involved in the selected processes, e.g., heterotrophs, and nitrifiers. In the third step, biochemical transformations for enzymatic hydrolysis, growth, and decay under aerobic, anoxic, and anaerobic conditions should be selected. In the last step, acid-base equilibria and liquid-gas transfer should finally be incorporated in the model. This step also involves addition of extra components (and their respective transformations) for corresponding acid/base-conjugates and gas-phase components. Following these steps, the transformations together with a vector of model components build the plant-wide model (Grau et al., 2007).

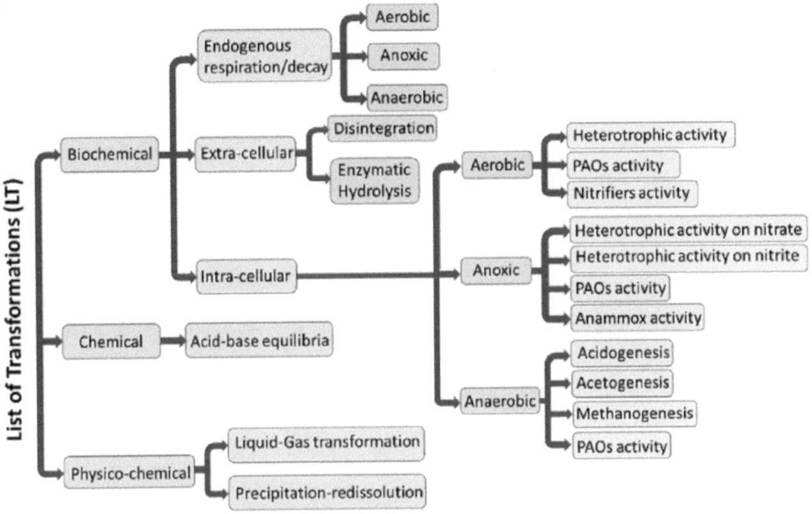

FIGURE 2.1 List of transformations (LT) for plant-wide modeling with the transformation-based approach as presented by Grau et al. (2007).

Source: Adapted with permission from Grau et al., 2007. © Elsevier.

2.2.2 PHYSICO-CHEMICAL MODELING

Plant-wide modeling of various chemical species requires a correct description of ion interactions and the formation of various complexes. For example, sulfate can be reduced to sulfide in the sewer and the primary treatment. Sulfide can further bind with iron leading to release of phosphate. However, in the activated sludge plant, sulfide is biologically oxidized to sulfur and sulfate, and also iron is oxidized. The released iron can then bind with phosphorus due to the higher stability of the resulting compound (phosphate recovery by precipitation). Unfortunately, the original ASMs and ADM1 models lack the physico-chemical components for nutrient recovery processes, and one has to incorporate various model extensions which have been developed more recently (Flores-Alsina et al., 2015; Wang et al., 2016; Huber et al., 2017). The extensions are mainly offering an accurate description of acid-base reactions and ion-pairing, liquid-gas mass transfer, slow precipitation kinetics, and sorption and desorption processes in a complex mixture (Batstone, 2012). Table 2.1 summarizes these transformations with examples.

TABLE 2.1 Transformations Involved in a Physico-Chemical Model as an Extension to the Original ASMs and ADM1 Models

Trans–formation	Description/ Mechanism	Example	Mathematical Expression	Type
Water chemistry/ speciation	Acid-base reactions	$SO_4^{2-} + H^+ \leftrightarrow HSO_4^-$	$Z_i = K_i \prod_{j=1}^{N} Z_j^{vij}; S_{i,tot} = \sum_{i=1}^{N} S_i$	Algebraic (fast)
	Ion-paring	$SO_4^{2-} + Fe^{2+} \leftrightarrow FeSO_4(aq)$		
	Redox trans–formations	$SO_4^{2-} + 9H^+ + 8e^+ \leftrightarrow HS^- + 4H_2O$		
Liquid-solid exchange	Precipitating-dissolution	$SO_4^{2-} + Ca^{2+} \leftrightarrow CaSO_4$	$kcXc\ (S - 1)^n$	Differential (slow)
Liquid-gas exchange	Stripping	H_2S (aq) $\leftrightarrow$ H_2S(g)	$K_{LG}(Sliq - H_T p_{gas})$	Differential (slow)

2.2.2.1 WATER CHEMISTRY/SPECIATION

The water chemistry/speciation description is targeted for chemical species participating in relevant processes such as mineral precipitation and gas exchange. They are described with two sets of equations, (i) non-linear algebraic equations expressed in terms of chemical activities (Z_i) for aqueous phase reaction *i* according to the law of mass-action (Kazadi et al., 2015). The formulation also includes equilibrium constants (*K*) and a stoichiometric coefficient of the reactant or product *j* ($v_{i,j}$); (ii) linear algebraic equations in which measurable total molar concertation of an aqueous phase ingredient (S_{tot}) is equal to the sum of the different compounds ($S = Z/\gamma$, γ is the activity coefficient) present in various chemical forms. The weak acid-base chemistry presented in Figure 2.2 demonstrates this for calcium.

The chemical speciation sub-model creates stiff equations due to its fast dynamics and instantaneous equilibrium as compared to the relatively slower physico-chemical (e.g., liquid-gas transfer), biological, and precipitation processes. Hence, solving such stiff systems of equations is often challenging and a compromise must be found between model accuracy and required simulation time. Examples of different fast and slow transformations between gas, liquid, and solid are presented in Table 2.1.

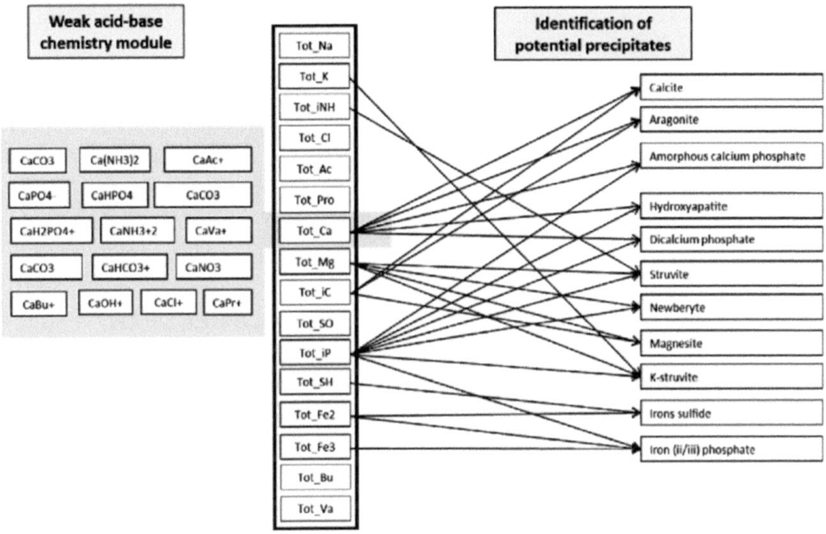

FIGURE 2.2 Physico-chemical description including weak acid-base chemistry module and potential precipitation reactions.

2.2.2.2 LIQUID-SOLID EXCHANGE

Reversible precipitation processes are triggered with the saturation index (SI) as a driving force. If SI>0 the aqueous phase is supersaturated and mineral precipitation can occur (Kazadi et al., 2015). The precipitation kinetic model consists of dynamic state equations for each mineral solid compound (X_c). S is the saturation ratio as a function of ion activity product (Vaneeckhaute et al., 2018), K_s is the kinetic rate coefficient which is normally estimated following a titration test (Kazadi et al., 2015).

The speciation model often includes a thermodynamic database consisting of hundreds of different reaction kinetics with corresponding acid-base and ion-pairing constants, solubility products, and other thermodynamics. MINTEQ and PHREEQC are the two commonly used databases involving water chemistry with homogeneous reactions (acid-base, ion-pairing, and redox transformations) (Vaneeckhaute et al., 2018). Figure 2.2 summarizes typical chemical components (equilibrium variables), which are involved in the formation of a number of common mineral precipitates in wastewater systems.

2.2.2.3 LIQUID-GAS EXCHANGE

In order to close the elemental mass balance around a process unit, liquid-gas exchange should also be described for both lowly/moderately soluble gases, e.g., CO_2, H_2S, and N_2 as well as highly soluble gases, e.g., NH_3. The exchange is driven by the concentration difference between the aqueous phase and the saturation concentration of the gaseous compound. The latter is calculated as the product of partial pressure in the gas (p_{gas}) and the Henry coefficient (H_T). The differential equation is hence established by addition of a liquid-gas transfer coefficient (K_{LG}) which needs to be adjusted manually or automatically (with a control system).

2.2.3 PLANT-WIDE MODELING FOR RESOURCE RECOVERY

There have been a number of studies proposing the transition from a WWTP to a Wastewater Resource Recovery Facility (WRRF's). Most of the new approaches for resource recovery are not primarily focused on aerobic biological treatment. In this regard, there are two main philosophies:(i) low energy mainline (LEM) focusing on anaerobic digestion with low strength followed by nutrient recovery processes (McCarty et al., 2011); (ii) partition-release-recover (PRR) through a multi-step approach (Batstone et al., 2015). The LEM layout includes a conventional primary settling tank together with an anaerobic membrane bioreactor (AnMBR) and a countercurrent air-stripping unit. As compared to the activated sludge process, LEM is shown to double the methane production, lower the quantity of digested sludge and in general, lower the costs and increase the energy production (McCarty et al., 2011). However, the disadvantage of the LEM approach is the weak separation processes and hence not achieving the full recovery potential of the resources present in the wastewater. As an alternative, dividing streams to major and minor (M&M) concentrated and dilute streams could achieve fractionation of components and recovery of energy and fertilizer. This method was demonstrated by up-concentration through UF/RO and anaerobic digestion (biogas) and a filter press in the downstream for recovery of nitrogen (liquid stream) and phosphorus (cake) (Verstraete et al., 2009). The PRR layout is based on accumulation of nutrients in the biosolids (partitioning), a digestion process (releasing) and a digestate treatment (recovering), where COD and nutrients are decoupled for better performance. However, the challenge of the PRR approach is the complex process and the loss of carbon to catabolic metabolism.

The plant-wide modeling methodology, as presented earlier, offers useful tools to perform a global techno-economical assessment in conventional treatment plants. It helps to understand key process variables that can be used in improving the overall economic outlook as well as for investigating the incorporation of new technologies to improve nutrient recovery and energy production (Batstone et al., 2015). Different recovery strategies can be evaluated by constructing or modifying plant-wide models through selection of different plant configurations, process models, and cost models. The great advantage of such methodology is the comparative assessment of already existing technologies and the ones currently under development. As an example, Fernández-Arévalo et al. (2017) proposed a WRRF layout based on the PRR concept and modeled it by modifying the BSM2 model (Figure 2.3). In this layout, the focus was to separate the treatment of nutrients (N/P) and COD, promote recovery, and maximize energy use via volarization of organic matter as biogas, phosphorus as struvite, and treating nitrogen with energy-efficient technologies (partial Nitritation, annamox). Through simulation analysis, it was shown that the proposed layout can reach self-sufficiency by an increase in electricity production, a decrease in iron dosing for P precipitation and a decrease in aeration cost. A low solids retention time (SRT) was considered to avoid excessive accumulation of inerts and nitrification, while at the same time keeping a high quantity of biodegradable organic matter in the anaerobic digestion. Moreover, a thermal hydrolysis technology is used to increase sludge degradability as well as promote ortho-P ions accumulation.

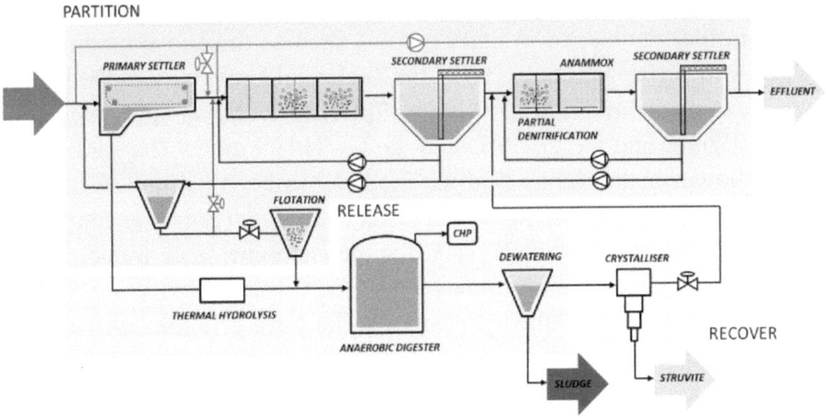

FIGURE 2.3 A WRRF configuration based on the PRR concept as presented by Fernández-Arévalo et al. (2017).

Source: Adapted with permission from Fernández-Arévalo et al., 2017. © Elsevier.

2.3 PLANT-WIDE MODELING OF PHOSPHORUS RECOVERY

Mathematical models are used as engineering tools to optimize the phosphorus removal processes by minimizing the impact of undesired effects such as increase in sludge production and alkalinity consumption, and to ensure compliance with effluent standards. In order to describe the full transformation of phosphorus in WWTPs, complex models are required to describe the interaction of phosphorus with other compounds in biological and chemical processes in the water and the sludge line (Damien et al., 2015):

1. **Water Line:** Phosphorus can be removed from wastewater through chemical or biological processes or a combination of both. The models for describing biological P removal processes in the water line are well developed and extensively applied. The standard activated sludge model no. 2d (ASM2d) is extensively used for describing enhanced biological phosphorus removal (EBPR) processes. It describes the Bio-P transformation with specific focus on two main mechanisms of phosphorus accumulating organisms (PAOs) in EBPR:

   - Uptake of carbon sources like VFAs in the form of carbon polymers (poly-β-hydroxyalkanoates or PHAs) under anaerobic conditions and simultaneous release of phosphate; and
   - Uptake of soluble P in phosphate forms using the stored PHAs as energy source under aerobic and anoxic conditions.

   The concept of Bio-P removal introduced by the ASM2d has been further developed in other models such as the Bio-P module of the ASM3 (Rieger et al., 2001), the Technical University of Delft (TUD) extension of ASM2d (Meijer, 2004), the Barker and Dold model (Barker and Dold, 1997) and the UCTPHO model (Hu et al., 2007). However, the focus of these models has been mainly on biological phosphorus (Bio-P) removal in the waterline (Batstone et al., 2012), and all these models lack the physico-chemical transformations essential to describe P recovery.

   ASM2d and other extensions so far have a simplified formulation for chemical precipitation of phosphorus. Recent extensions to ASM2d to take into account P transformations and interactions with iron and sulfur are as follows:

- A more general precipitation framework for a wide range of precipitation reactions (Flores-Alsina et al., 2015).
- Phosphorus content for all soluble and particulate components in the ASM2d through elemental balances (Takács et al., 2007).
- Sulfate transformations including sulfide oxidation and sulfate reduction by sulfate reducing bacteria (SRB) using acetate and fermentation products, as well as hydrogen sulfide inhibition (Solon et al., 2017).
- Hydrogen sulfide stripping.
- Iron transformations by adapting the hydrous ferric oxide model (Smith et al., 2008; Hauduc et al., 2015) focusing only on phosphates adsorption and co-precipitation.

2. **Sludge Line:** The biologically accumulated polyphosphate in the sludge is removed through the waste activated sludge and is extensively released from the sludge during anaerobic digestion together with ammonia, magnesium, calcium, and potassium. The released phosphorus binds with magnesium and calcium and forms struvite and other metal phosphates.

   Phosphorus binding with metals can significantly affect the physico-chemistry of the anaerobic digestion process. The default ADM1 contains the kinetic reactions most commonly used in anaerobic digestion systems. However, especially industrial wastewaters often have complex characteristics with high concentrations of compounds such as COD, sulfate, phosphate, and other minerals. The interaction of phosphorus with iron and sulfur cycles is depicted in Figure 2.4. The ADM1 model has been extended recently and there seems to be a general common agreement to model phosphorus and sulfur species and their interactions with Iron (Flores-Alsina et., 2016; Hauduc et., 2019; Ikumi and Ekama, 2019):

   - **Sulfur Species:** The presence of sulfate can lead to growth of SRB, which compete for substrate with the methanogenic biomass. Furthermore, SRBs produce hydrogen sulfide upon sulfate degradation, which is toxic to methanogenic biomass. These effects will need to be incorporated when modeling a system containing high concentrations of sulfate. Fedorovich et al. (2003) developed an extension to the ADM1, in which SRBs can take up VFAs or hydrogen, competing with acidogens, acetogens, and methanogens, and produce hydrogen sulfide. In addition, an inhibition term is added for hydrogen sulfide. Further model implementations have been developed since, including the reduction of iron with sulfides or hydrogen as electron donors (Hauduc et al., 2019).

- **Phosphorus Species:** Consistent with ASM2d, three additional state variables are included: PAOs (X_{PAO}), polyhydroxyalkanoates (X_{PHA}), and polyphosphates (X_{PP}). X_{PHA} are formed from uptake of VFAs by X_{PAO} under anaerobic conditions (P-release). Decay of X_{PAO} and lysis of X_{PHA} and X_{PP} are also considered in the model. Moreover, phosphorus is influencing the physico-chemistry of the wastewater via formation of multiple buffers and precipitation with various metals such as calcium and iron.
- **Iron:** It has a strong connection with the sulfur and phosphorus cycles. Iron is reduced from Fe(III) to Fe(II) by receiving electrons from hydrogen or hydrogen sulfide. Other electron donors can also contribute to reduction of Iron (e.g., VFAs, and nitrate). Fe(II) can remove sulfide via ferrous sulfide precipitation and can also form iron phosphate. In anaerobic digestion, Fe in iron phosphate (previously formed for example in the activated sludge process) can rebind with sulfide due to the higher stability of the resulting compound, which also results in more phosphorus release.

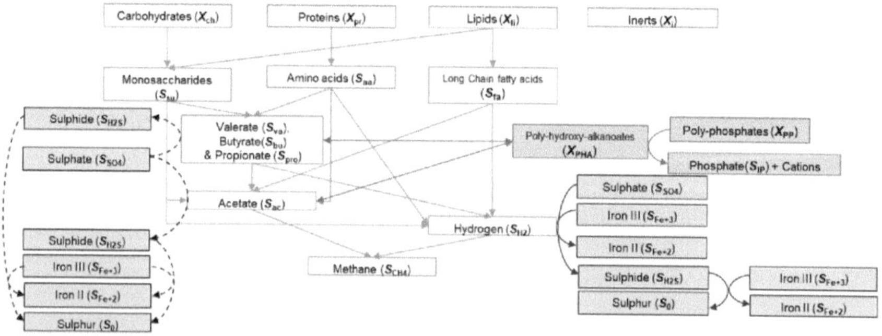

FIGURE 2.4 ADM1 model extensions with phosphorus, iron, and sulfur cycles (Flores-Alsina et al., 2016).

Source: Reprinted with permission from Flores-Alsina et al., 2016. © Elsevier.

- **Water Chemistry/Speciation:** To describe the interactions between different ions, a speciation model is added to the ADM1 implementation. The pH of the influent, reactor, and effluent is dependent on the concentration of the multiple species present in the wastewater.

This extension takes into account the different species of anions and cations, and subsequently calculates pH. Moreover, a high mineral content in the wastewater could potentially lead to precipitation, which can cause scaling inside the reactor and pipes or even loss of activity inside the reactor. As mentioned earlier, ADM1 can be extended with a multiple-mineral precipitation model such as the one developed by Kazadi-Mbamba et al. (2015).

2.3.1 PLANT-WIDE MODELING

A plant-wide modeling approach is necessary to capture the full transformation of P through the entire plant including water and sludge line. Therefore, it is required to include biological and physico-chemical transformations in the standard ASM and ADM models (Batstone et al., 2018). Transformation of phosphorus, iron, and sulfur in the BSM2 layout is presented in Figure 2.5 with connection to Table 2.2. Solon et al. (2017) presented an extended version of BSM2 with description of the main P transformations in ASM2d and ADM1 models and concluded their work with the following important findings:

- Potential uptake of organics by X_{PAO} to form X_{PHA} could impact overall biogas production.
- SRB plays an important role in treating sludge with high S content. The SRBs lead to H_2S formation, which can potentially reduce methane formation due to inhibition and loss of electrons.

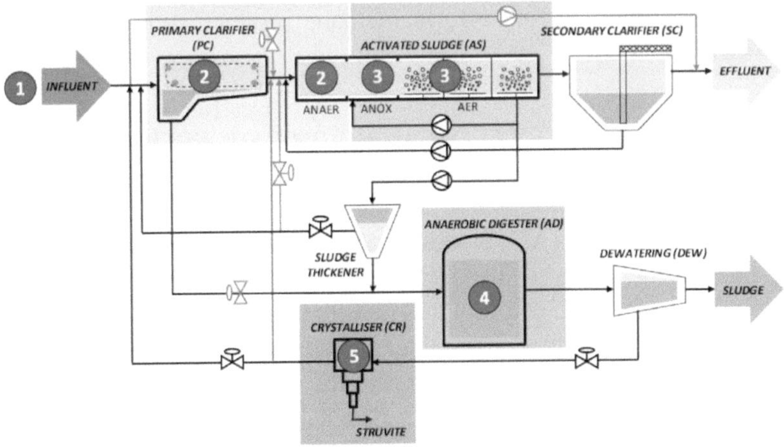

FIGURE 2.5 Transformation of phosphorus, iron, and sulfur in the wastewater treatment plant-adopted from (Solon et al., 2017)-see Table 2.2 for associated transformations in each unit. *Source:* Reprinted with permission from Solon et al., 2017. © Elsevier.

TABLE 2.2 Dominant transformation of Phosphorus, Iron, and Sulfur in the Wastewater Treatment Plant

	PC/ANAER	ANOX/AER	CR
P	PHA storage Phosphate release Phosphate absorption to HFO Precipitation with Ca and Mg	Poly-P storage PAO growth Phosphate absorption to HFO Precipitation with Ca and Mg	Precipitation with Ca and Mg
S	Sulfur and sulfate reduction to sulfide on acetate and fermentation products	Sulfide oxidation to sulfate Hydrogen sulfide stripping	Sulfur and sulfate reduction to sulfide on hydrogen, acetate and fermentation products
Fe	Fe-S formation HFO reduction to Fe (II) on acetate and fermentation products Precipitation of P with Fe(II) HFO formation	Fe (II) oxidation to Fe (III) HFO formation	Precipitation of P with Fe(II) HFO formation

PHA: poly-β-hydroxyalkanoates; PAO: phosphorus-accumulating organisms; SRB: sulfate reducing bacteria; HFO: Hydrous ferric oxide; PC = primary clarifier; ANAER = anaerobic; ANOX = anoxic; AER = aerobic; AD = anaerobic digestion; CR = crystallizer.

- Iron sulfide precipitation reduces hydrogen sulfide gas production and results in an increase in methane production in the anaerobic digester (AD).

2.4 ENERGY RECOVERY FROM ANAEROBIC DIGESTION: A CASE STUDY

The anaerobic digestion (AD) process produces biogas in the form of methane and carbon dioxide. The methane can be utilized as energy source (heat and electricity) using a combined heat and power (CHP) unit. The process can be applied to a wide variety of wastewater sources, which adds complexity to a model capable of describing the AD process with any influent. For this purpose, in the ADM1 (Batstone et al., 2002) fractionated COD components are present in the model as presented in Figure 2.4. The highly fractionated influent states of the ADM1 can be a disadvantage, as either the states need to be measured separately, or derived from algorithms (Kleerebezem and Van Loosdrecht, 2006). Identification of COD fractions as well as associated kinetic parameters can be achieved based on anaerobic respirometry during which the methane production rate is recorded in a batch experiment (Girault et al., 2012).

The goal of implementing a model of an anaerobic digestion process is often motivated by optimizing the energy recovery and decreasing the costs of operating the reactor. ADM1 applications have been reported for a wide variety of wastewaters, simulating the treatment of municipal wastewaters to agro-cultural and industrial wastewaters. Nevertheless, applications on full-scale industrial reactors are scarce, as many studies take industrial wastewater and operate a reactor on either lab- or pilot-scale, where operational conditions are easier to control. Treatment of industrial wastewaters can be a challenge to model, due to the high rates, the high concentrations of compounds present, such as the previously mentioned COD, sulfate, and phosphate, and the high variations of flow rate and wastewater composition due to variation in industrial production schedules. Furthermore, many industrial reactors are a type of upflow anaerobic sludge bed (UASB) reactors, which contain granular biomass. This adds an extra layer of complexity to the model.

The plant in the case study is an industrial expanded granular sludge bed (EGSB) reactor (Figure 2.6) with a volume of 2200 m^3. This reactor treats the wastewater from a fermentation factory and is high in COD loading. The reactor, under mesophilic conditions, is comprised of four sections, mixing (where most COD is converted to biogas), expanded sludge bed, polishing, and gas-liquid separator. The inlet to the reactor is at the mixing section where the influent water is mixed with the treated wastewater that is recirculated from the top of the reactor. The rising gas automatically controls the internal recirculation during the anaerobic process. Biomass is present as granules with higher concentration in the bottom of the reactor. Two modeling approaches are considered to simulate anaerobic activity and gas production. While the biological models are identical, the hydraulic configurations are different in both approaches.

- **Model I:** Artificial recirculation of particulate matter through an ideal separator (Feldman et al., 2018). In this configuration, the hydraulic residence time (HRT) is decoupled from solids residence time (SRT) by retaining the solids in the reactors. The reactor is simplified to two continuously stirred tank reactors (CSTRs) allowing for different solid loading. In addition, the model includes a combiner and a separator. The effluent of each CSTR is separated into particulate and soluble matter, where the particulate matter is recirculated back to the reactor. The solids retention time (SRT) of the system is set at 100 days (Feldman et al., 2018).

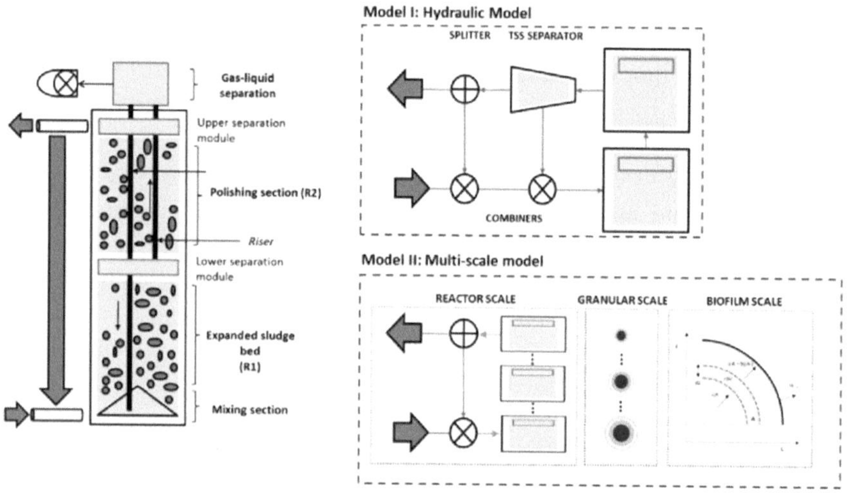

FIGURE 2.6 Schematic representation of the industrial anaerobic digester in the case study (left), and the two models tested as presented by Feldman et al. (2017).

Source: Reprinted with permission from Feldman et al., 2017. © Elsevier.

- **Model II:** This model is formulated at various scales, i.e., reactor, granule, and biofilm. The reactor is described as two or three CSTRs, increasing the solid loading and the size of the granules from bottom to top (Feldman et al., 2017). Biofilm granules are modeled in a 1-D biofilm model (gradients in the direction of the biofilm radius). The transport of soluble compounds is solely described by diffusion and for particulate compounds by advection. Advection is governed by biofilm growth and the growth velocity is described as net growth (growth minus decay). The biofilm thickness is also modeled as a combination of both growth and detachment phenomena. The latter is modeled as the maximum radius of the granule, set as a fixed parameter. Biomass is assumed to be spherical and have a fixed density that is not changing as a function of radius (Vangsgaard et al., 2013). The number of granules inside the reactor is dependent on reactor volume, granule size, density, and volatile suspended solids (VSS).

The model prediction accuracy was tested during a 18-day monitoring period. The average deviations between measurements and simulations for both models lies below 15% for each operational period. The simulation results of Model II are shown in Figure 2.7. The predictions for CH_4, CO_2, H_2S, and SO_4 profiles reflect the adequate description of methanogenesis, liquid-gas mass transfer, sulfidogenesis as well as the competition between methanogens and SRBs. Details of the model calibration and further model adjustments can be reviewed in Feldman et al. (2017).

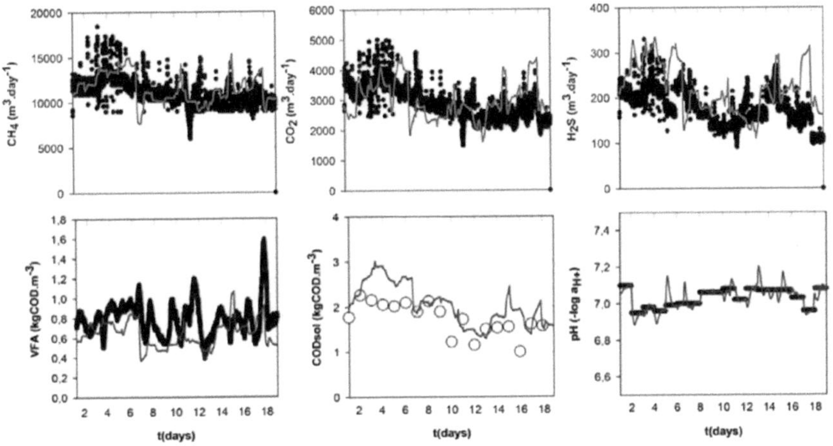

FIGURE 2.7 Continuous monitoring of the industrial anaerobic EGSB reactor with simulation results using Model II (Feldman et al., 2017).

Source: Reprinted with permission from Feldman et al., 2017. © Elsevier.

Similar results were produced using Model I (Feldman et al., 2018). The calibrated Model I is further used for reactor optimization and economic analysis. Simulation scenarios were specifically created to investigate the role of pH and the S (sulfate and sulfide) fraction on the reactor performance with the purpose of increasing energy recovery and decreasing operational cost (chemical dosing). The performance was then calculated as reactor performance indicator (RPI) (Gernaey et al., 2014) applying a unit of M€/year. Considering the base case operation of the reactor (pH=7.2 and S=1) and the main economic factors (e.g., sodium hydroxide consumption, price of electricity and heat), the base RPI leads to around 2 M€/year (Feldman et al., 2018). In the simulated scenario, a lower S fraction would lead to higher methane formation mainly due to lowered competition between SRBs and methanogens favoring the second (Hao et al., 2014), and lower sulfide toxicity. Moreover, with lower sulfide levels, less sodium hydroxide is required to control pH. This leads to an increase of RPI by 0.48 M€/year (+25%). However, it is argued that the potential cost of S removal may not balance the financial gain (Feldman et al., 2018). On the other hand, decreasing the pH would also lower the need for more sodium hydroxide and leads to less precipitation problems (not active in Model I). Considering the optimal pH range for methanogens and the achieved reactor performance, the pH of 6.8 increases the RPI by 0.88 M€/year (+45.2%). Simultaneous S removal and pH decrease would increase the RPI by 1.35 M€/year (+69.2%). A more detailed scenario analysis can be found in Feldman et al. (2018).

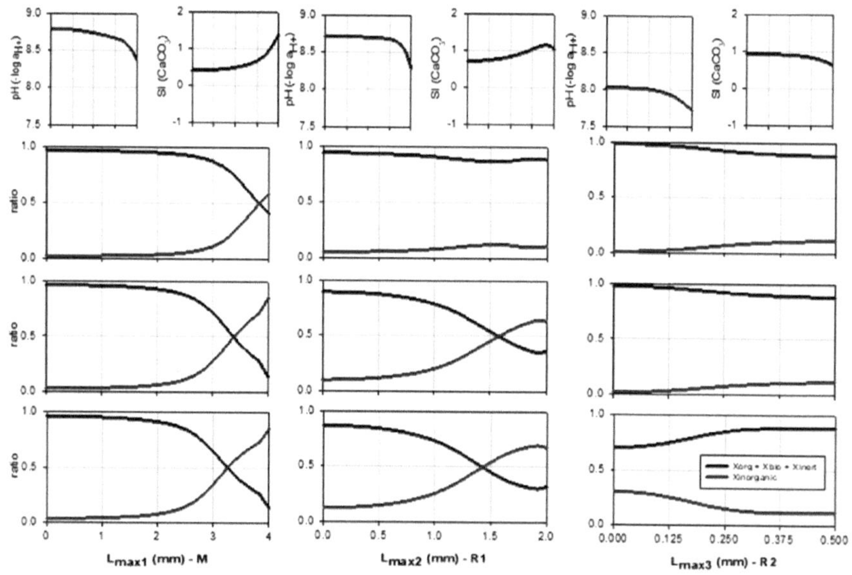

FIGURE 2.8 pH, saturation index (SI), and distribution of organics and inorganics inside the granules for the industrial EGSB reactor. The simulations are made for the mixing section (M) lower (R1) and upper (R2) section of the reactor and represent the evolution over time (0, 50, 100 days) when moving from bottom to top in the presented ratio graphs (Feldman et al., 2019). X-axis is the granule radius where 0 is the center of the granule.

Source: Reprinted with permission from Feldman et al., 2019. IWA Publishing. Open access.

In addition, in Model II the distribution of different biomass fractions can be simulated inside the granules. For example, methanogens tend to accumulate in the inner layers of the granule, while acidogens and acetogens are more distributed in the outer layers. The distribution is strongly linked to the solid loading rate, the mass transfer limitation, and the relative affinity of microorganisms for substrate uptake. Importantly, the model can be used to simulate the distribution of organic and inorganic fractions, including precipitation of minerals. These distributions together with the SI values are shown in Figure 2.8. The simulations, over time (0, 50 and 100 days), are made for the mixing section (M), expanded sludge bed (R1) and the polishing section (R2) (see Figure 2.6), assuming that the granules in these parts of the system have a diameter of 4, 2 mm and 0.5 mm respectively. Simulation results show that the formation of precipitates accumulates at the bottom of the reactor and changes depending on the granule size. Thus, in larger granules, precipitates are expected to form in the outer layers of the granule. In smaller granules, precipitates will tend to spread all the way to the center of the granules. This eventually leads to a lower methanogenic capacity of the granules and a decreased production of biogas.

While the models produce similar results in the outputs (biogas, nutrients, sulfur compounds) (Figure 2.7), the biofilm model (Model II) describes the occurrence of precipitation within the granule. This can significantly impact the reactor performance, due to competition for space in the granule between biomass and precipitates (see Figure 2.9). When precipitation is not an issue, the biofilm model may not be necessary, as the time it takes to run such a model is significantly longer due to the calculation of all kinetics and chemical reactions in each biofilm layer. In that case, model I is more practical.

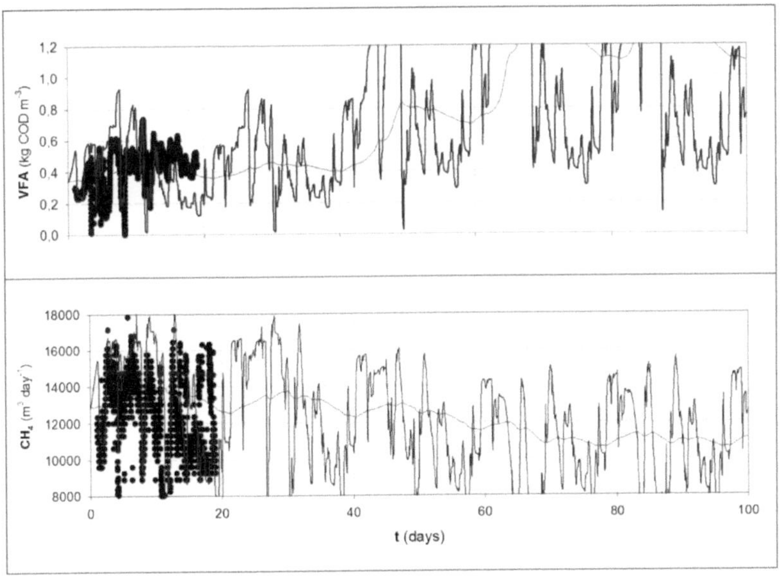

FIGURE 2.9 Scenario analysis results assuming a long-term (LT) period assuming intra-granule. Top, VFA production, bottom CH_4 generated. Markers indicate measurements and red lines are simulated results. exponential moving average filter is included to better visualize trends (Feldman et al., 2019).

Source: Reprinted with permission from Feldman et al., 2019. IWA Publishing. Open access.

Overall, in this case study, it was demonstrated that mathematical models can be developed at various complexity levels (i.e., Model I and Model II) to serve specific purposes. Model simulations cannot only be used to make predictions, but also as a tool for process optimization and ultimately as a decision support system, when different process options have to be considered. Hence, the models can also be helpful for techno-economical evaluation and cost-profit analysis.

KEYWORDS

- **activated sludge model no. 1**
- **anaerobic digestion**
- **biological nutrient removal model no. 1**
- **continuously stirred tank reactors**
- **enhanced biological phosphorus removal**
- **expanded granular sludge bed**

REFERENCES

Angelidaki, I., Ellegaard, L., & Ahring, B. K., (1993). A mathematical model for dynamic simulation of anaerobic digestion of complex substrates: Focusing on ammonia inhibition. *Biotechnol. Bioeng. 42*, 159–166. doi: 10.1002/bit.260420203.

Barker, P. S., & Dold, P. L., (1997). General model for biological nutrient removal activated-sludge systems: Model presentation. *Water Environ. Res., 69*, 969–984. doi: 10.2175/106143097X125669.

Batstone, D. J., Amerlinck, Y., Ekama, G., Goel, R., Grau, P., Johnson, B., Kaya, I., et al., (2012). Towards a generalized physicochemical framework. *Water Sci. Technol., 66*, 1147–1161. doi: 10.2166/wst.2012.300.

Batstone, D. J., Flores-Alsina, X., & Hauduc, H., (2018). *Modeling the Phosphorus Cycle in the Wastewater Treatment Process*. IWA Publishing.

Batstone, D. J., Hülsen, T., Mehta, C. M., & Keller, J., (2015). Platforms for energy and nutrient recovery from domestic wastewater: A review. *Chemosphere, 140*, 2–11. doi: 10.1016/j.chemosphere.2014.10.021.

Batstone, D. J., Keller, J., Angelidaki, I., Kalyuzhnyi, S. V., Pavlostathis, S. G., Rozzi, A., Sanders, W. T. M., et al., (2002). Anaerobic digestion model no. 1. *Water Sci. Technol., 45*, 65–73.

Batstone, D. J., Puyol, D., Flores, A. X., & Rodríguez, J., (2015). Mathematical modeling of anaerobic digestion processes: Applications and future needs. *Rev. Environ. Sci. Biotechnol., 14*, 595–613. doi: 10.1007/s11157-015-9376-4.

Copp, J. B., Jeppsson, U., & Rosen, C., (2003). Towards an asm1-adm1 state variable interface for plant-wide wastewater treatment modeling. *Proc. Water Environ. Fed.,* 498–510. doi: 10.2175/193864703784641207.

Ekama, G. A., & Marais, G. V. R., (1977). The activated sludge process (Part II)-dynamic behavior. *Water SA, 3,* 18–50.

Ekama, G. A., (2009). Using bioprocess stoichiometry to build a plant-wide mass balance based steady-state WWTP model. *Water Res., 43,* 2101–2120. doi: 10.1016/j.watres.2009.01.036.

Ekama, G. A., Sötemann, S. W., & Wentzel, M. C., (2006). Mass balance-based plant-wide wastewater treatment plant models-Part 3: Biodegradability of activated sludge organics under anaerobic conditions. *Water SA, 32,* 287–296. doi: 10.4314/wsa.v32i3.5273.

Federovich, V., Lens, P., & Kalyuzhnyi, L., (2003). Extension of anaerobic digestion model no 1 with processes of sulfate reduction. *Appl. Biochem. Biotechnol., 109,* 33–45. doi: 10.1385/ABAB:109:1-3:33.

Feldman, H., (2018). *PhD Thesis, Optimization of the Energy Recovery in an Industrial Wastewater Treatment Plant.* doi: 10.1111/tbed.12529.

Feldman, H., Flores-Alsina, X., Kjellberg, K., Jeppsson, U., Batstone, D. J., & Gernaey, K. V., (2018). Model-based analysis and optimization of a full-scale industrial high-rate anaerobic bioreactor. *Biotechnol. Bioeng., 115,* 2726–2739. doi: 10.1002/bit.26807.

Feldman, H., Flores-Alsina, X., Ramin, P., Kjellberg, K., Batstone, D. J., Jeppsson, U., & Gernaey, K. V., (2019). Assessing intra-granule precipitation in industrial anaerobic granular sludge reactors. *Wat. Sci. and Technol., 79*(7), 1327–1337. doi.org/10.2166/wst.2019.129.

Feldman, H., Flores-Alsina, X., Ramin, P., Kjellberg, K., Jeppsson, U., Batstone, D. J., & Gernaey, K. V., (2017). Modeling an industrial anaerobic granular reactor using a multi-scale approach. *Water Res., 126,* 488–500. doi: https://doi.org/10.1016/j.watres.2017.09.033.

Fernández-Arévalo, T., Lizarralde, I., Fdz-Polanco, F., Pérez-Elvira, S. I., Garrido, J. M., Puig, S., Poch, M., et al., (2017). Quantitative assessment of energy and resource recovery in wastewater treatment plants based on plant-wide simulations. *Water Res., 118,* 272–288. doi: 10.1016/j.watres.2017.04.001.

Flores-Alsina, X., Kazadi, M. C., Solon, K., Vrecko, D., Tait, S., Batstone, D. J., Jeppsson, U., & Gernaey, K. V., (2015). A plant-wide aqueous phase chemistry module describing pH variations and ion speciation/pairing in wastewater treatment process models. *Water Res., 85,* 255–265. doi: 10.1016/j.watres.2015.07.014.

Flores-Alsina, X., Solon, K., Kazadi, M. C., Tait, S., Gernaey, K. V., Jeppsson, U., & Batstone, D. J., (2016). Modeling phosphorus (P), sulfur (S), and iron (Fe) interactions for dynamic simulations of anaerobic digestion processes. *Water Res., 95,* 370–382. doi: 10.1016/j.watres.2016.03.012.

Gernaey, K. V., Jeppsson, U., P. A. V., & Copp, J. B., (2014). *Benchmarking of Control Strategies for Wastewater Treatment Plants.* IWA Scientific and Technical Report No. 23. IWA Publishing, London, UK. doi: 10.2166/9781780401171.

Girault, R., Bridoux, G., Nauleau, F., Poullain, C., Buffet, J., Steyer, J. P., Sadowski, A. G., & Béline, F., (2012). A waste characterization procedure for ADM1 implementation based on degradation kinetics. *Water Res., 46,* 4099–4110. doi: 10.1016/j.watres.2012.04.028.

Grau, P., Copp, J., Vanrolleghem, P. A., Takács, I., & Ayesa, E., (2009). A comparative analysis of different approaches for integrated WWTP modeling. *Water Sci. Technol., 59,* 141–147. doi: 10.2166/wst.2009.589.

Grau, P., De Gracia, M., Vanrolleghem, P. A., & Ayesa, E., (2007). A new plant-wide modeling methodology for WWTPs. *Water Res., 41,* 4357–4372. doi: 10.1016/j.watres.2007.06.019.

Gude, V.G., (2016). Wastewater treatment in microbial fuel cells - An overview. *J. Clean. Prod.* 122. doi:10.1016/j.jclepro.2016.02.022.

Hao, T., Xiang, P., Mackey, H. R., Chi, K., Lu, H., Chui, H., Van, L. M. C. M., & Chen, G. H., (2014). A review of biological sulfate conversions in wastewater treatment. *Water Res., 65*, 1–21. doi: 10.1016/j.watres.2014.06.043.

Hauduc, H., Takács, I., Smith, S., Szabo, A., Murthy, S., Daigger, G.T., Spérandio, M., (2015). A dynamic physicochemical model for chemical phosphorus removal. *Water Res. 73*, 157–170. doi:10.1016/j.watres.2014.12.053.

Hauduc, H., Wadhawan, T., Takács, I., Johnson, B., Bott, C., Ward, M., (2019). Incorporating sulfur reactions and interactions with iron and phosphorus into a general plant-wide model. *Water Sci. Technol. 79*, 26–34. doi:10.2166/wst.2018.482.

Henze, M. Jr., C. P. L. G., Gujer, W., Marais, G. V. R., & Matsuo, T., (1987). *Activated Sludge Model No. 1*. Scientific and Technical Report No. 1. IWA Publishing, London, UK.

Henze, M., Gujer, W., Mino, T., & Van, L. M. C. M., (2000). *Activated Sludge Models ASM1, ASM2, ASM2d AND ASM3, Scientific and Technical Report No 9*. IWA Publishing, London.

Hu, Z. R., Wentzel, M. C., & Ekama, G. A., (2007). A general kinetic model for biological nutrient removal activated sludge systems: Model evaluation. *Biotechnol. Bioeng., 98*, 1259–1275. doi: 10.1002/bit.21507.

Huber, P., Neyret, C., Fourest, E., (2017). Implementation of the anaerobic digestion model (ADM1) in the PHREEQC chemistry engine. *Water Sci. Technol. 76*, 1090–1103. doi:10.2166/wst.2017.282.

Ikumi, D.S., Ekama, G.A., (2019). Plantwide modelling – Anaerobic digestion of waste sludge from parent nutrient (N and p) removal systems. *Water SA 45*, 305–316. doi:10.17159/wsa/2019.v45.i3.6698.

Jeppsson, U., Pons, M. N., Nopens, I., Alex, J., Copp, J. B., Gernaey, K. V., Rosen, C., Steyer, J. P., & Vanrolleghem, P. A., (2007). Benchmark simulation models no 2: General protocol and exploratory case studies. *Water Sci. Technol., 56*, 67. doi: 10.2166/wst.2007.604.

Kazadi, M. C., Batstone, D. J., Flores-Alsina, X., & Tait, S., (2015). A generalized chemical precipitation modeling approach in wastewater treatment applied to calcite. *Water Res., 68*, 342–353. doi: 10.1016/j.watres.2014.10.011.

Kleerebezem, R., & Van, L. M. C. M., (2006). Waste characterization for implementation in ADM1. *Water Sci. Technol., 54*, 167–174. doi: 10.2166/wst.2006.538.

McCarty, P. L., Bae, J., & Kim, J., (2011). Domestic wastewater treatment as a net energy producer-can this be achieved? *Environ. Sci. Technol., 45*, 7100–7106. doi: 10.1021/es2014264.

Morgenroth, E., Arvin, E., & Vanrolleghem, P., (2002). The use of mathematical models in teaching wastewater treatment engineering. *Water Sci. Technol., 45*, 229–233.

Pavlostathis, S. G., & Gossett, J. M., (1986). A kinetic model for anaerobic digestion of biological sludge. *Biotechnol. Bioeng., 28*, 1519–1530. doi: 10.1002/bit.260281010.

Rieger, L., Koch, G., Kühni, M., Gujer, W., & Siegrist, H., (2001). The EAWAG Bio-P module for activated sludge model No. 3. *Water Sci. Technol., 35*, 3887–3903. doi: 10.1016/S0043-1354(01)00110-5.

Seco, A., Ribes, J., Serralta, J., & Ferrer, J., (2004). Biological nutrient removal model No. 1 (BNRM1). *Water Sci. Technol., 50*, 69–78.

Smith, S., Takács, I., Murthy, S., Daigger, G. T., & Szabo, A., (2008). Phosphate complexation model and its implications for chemical phosphorus removal. *Water Environ. Res., 80*, 428–438. doi: 10.2175/106143008X268443.

Solon, K., Flores-Alsina, X., Kazadi, M. C., Ikumi, D., Volcke, E. I. P., Vaneeckhaute, C., Ekama, G., et al., (2017). Plant-wide modeling of phosphorus transformations in wastewater treatment systems: Impacts of control and operational strategies. *Water Res., 113*, 97–110. doi: 10.1016/j.watres.2017.02.007.

Takács, I., Vanrolleghem, P. A., Wett, B., & Murthy, S., (2007). Elemental balance based methodology to establish reaction stoichiometry in environmental modeling. *Water Sci. Technol., 56,* 37–41. doi: 10.2166/wst.2007.606.

Vaneeckhaute, C., Claeys, F. H. A., Tack, F. M. G., Meers, E., Belia, E., & Vanrolleghem, P. A., (2018). Environmental modeling and software development, implementation, and validation of a generic nutrient recovery model (NRM) library. *Environ. Model. Softw., 99*, 170–209. doi: 10.1016/j.envsoft.2017.09.002.

Vangsgaard, A. K., Mutlu, A. G., Gernaey, K. V., & Smets, B. F., (2013). *Calibration and Validation of a Model Describing Complete Autotrophic Nitrogen Removal in a Granular SBR System*. doi: 10.1002/jctb.4060.

Verstraete, W., Van, D. C. P., & Diamantis, V., (2009). Maximum use of resources present in domestic "used water." *Bioresour. Technol., 100*, 5537–5545. doi: 10.1016/j.biortech.2009.05.047.

Volcke, E. I. P., "van Loosdrecht, M. C. M., & Vanrolleghem, P. A., (2006). Continuity-based model interfacing for plant-wide simulation: A general approach. *Water Res., 40,* 2817–2828. doi: 10.1016/j.watres.2006.05.011.

Wang, R., Li, Y., Chen, W., Zou, J., Chen, Y., (2016). Phosphate release involving PAOs activity during anaerobic fermentation of EBPR sludge and the extension of ADM1. *Chem. Eng. J. 287*, 436–447. doi:10.1016/j.cej.2015.10.110

CHAPTER 3

Osmotic Membrane Bioreactor and Its Hybrid Systems for Resource Recovery from Wastewater

WENHAI LUO,[1] ZHICHENG XU,[1] and XINHUA WANG,[2]

[1]*Beijing Key Laboratory of Farmland Soil Pollution Prevention and Remediation, College of Resources and Environmental Sciences, China Agricultural University, Beijing – 100193, China*

[2]*Jiangsu Key Laboratory of Anaerobic Biotechnology, School of Environmental and Civil Engineering, Jiangnan University, Wuxi – 214122, China*

ABSTRACT

Osmotic membrane bioreactor (OMBR), which integrates forward osmosis (FO) with biological treatment process, has been recently developed to advance wastewater treatment and reuse. During OMBR operation, biologically treated water transports from the mixed liquor into a highly concentrated draw solution, through a semi-permeable FO membrane and driven by osmotic pressure gradient. Compared to conventional MBR, OMBR has several advantages, including better product water quality, lower fouling propensity, and higher fouling reversibility. OMBR can be operated in the osmotic dilution mode when the draw solution, such as liquid fertilizers or seawater, can be reused or discharged directly. In most cases, OMBR is integrated with an additional process, commonly including nanofiltration, reverse osmosis, membrane distillation, and electrodialysis, to form hybrid systems for sustainably reconcentrating draw solutions and producing clean water for reuse. In addition, several membrane processes, such as microfiltration, ultrafiltration, microbial fuel cell, and electrodialysis, are combined with OMBR to address its inherent issue, salinity build-up in the bioreactor, and achieve resource (e.g., nutrients and energy) recovery. Despite the promise of OMBR and its hybrid systems, several key factors should be optimized and/or addressed for their further development to recover resources from wastewater. These include salinity build-up in the bioreactor, membrane properties, draw solution types and concentration, and contaminant accumulation in the draw solution.

3.1 INTRODUCTION

Membrane bioreactor (MBR), which combines conventional activated sludge (CAS) treatment with a membrane separation process, typically microfiltration (MF) and ultrafiltration (UF), has been widely used for wastewater treatment and reuse (Xiao et al., 2019). Compared to CAS treatment, MBR is more robust and versatile to produce higher quality reused water, but with lower sludge production and smaller physical footprint (Judd, 2008). For instance, MBR can increase the removal of trace organic contaminants (TrOCs) of significant concern from wastewater, due to its longer operating sludge retention time (SRT) than CAS technology (Besha et al., 2017; Dharupaneedi et al., 2019).

A novel variation of MBR, called osmotic membrane bioreactor (OMBR), has been recently developed to advance wastewater treatment and reuse (Achilli et al., 2009). OMBR integrates biological treatment with forward osmosis (FO), which is an osmotically driven process. In the OMBR process, water is transported from the mixed liquor into a highly concentrated draw solution (e.g., sodium chloride and seawater) through a selective, semi-permeable FO membrane. Compared to conventional MBR, OMBR exhibits higher contaminant removal (Luo et al., 2015b), lower fouling propensity (Achilli et al., 2009; Luo et al., 2015c), and higher fouling reversibility (Mi and Elimelech, 2010).

OMBR can be applied as a stand-alone system or integrated with additional processes for wastewater treatment and reuse. When draw solutions, such as liquid fertilizers (Kim et al., 2016) and seawater (Ansari et al., 2016), can be directly used or discharged, OMBR can be operated in the osmotic dilution mode. In most cases, a desalination process, such as nanofiltration (NF), reverse osmosis (RO), membrane distillation (MD), and electrodialysis (ED), is integrated with OMBR to extract clean water and concentrated raw solutions. Moreover, several additional processes, such as MF, UF, and microbial fuel cells (MFCs) have been combined with OMBR to extract salts and nutrients accumulated in the bioreactor during OMBR operation (Lu and He, 2015; Holloway et al., 2015c; Luo et al., 2016b).

This chapter aims to fundamentally introduce the performance of OMBR and its hybrid systems for resource recovery from wastewater. Key factors affecting the practical applications of these systems were systematically delineated. These include salinity build-up in the bioreactor, membrane properties, membrane fouling and stability, draw solution types and concentration, and contaminant accumulation in the draw solution.

3.2 OSMOTIC MEMBRANE BIOREACTOR (OMBR) AND ITS HYBRID SYSTEMS

3.2.1 *OMBR AND ITS CONFIGURATIONS*

FO is an osmotically driven membrane process, in which water moves from a low saline feed solution into a highly concentrated draw solution across a semi-permeable FO membrane (Figure 3.1A). Given its low fouling propensity (Cath et al., 2006), FO can be a good candidate to integrate with biological treatment to form OMBR (Figure 3.1B). Recent years have witnessed the successful combination of FO with several biological processes, particularly the activated sludge treatment (AST) (Holloway et al., 2015a; Song et al., 2018). Anaerobic processes are also attractive to combine with FO for energy-efficient wastewater treatment (Wang et al., 2016a). During anaerobic OMBR operation, biogas can be produced by biodegrading organic substances in wastewater (Table 3.1) and further converted into electricity and heat and/or circulated to the bioreactor to control membrane fouling and mix the mixed liquid (Chen et al., 2014). In addition, FO has also be coupled with other biological processes, such as sequential batch reactor (Linares et al., 2016), moving bed biofilm reactor (Nguyen et al., 2015), attached growth biofilm reactor (Nguyen et al., 2016), and aerobic/anoxic reactor (Morrow et al., 2018).

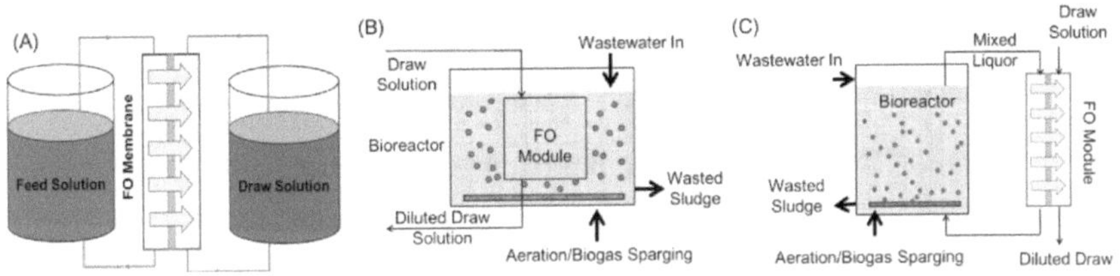

FIGURE 3.1 Schematic diagram of (A) forward osmosis and (B) submerged and (C) side-stream osmotic membrane bioreactor.

TABLE 3.1 The Potential of OMBR and its Hybrid Systems for Resource Recovery from Wastewater

Configuration	Resources Recovered From Wastewater
OMBR	Clean water from the FO membrane
	Biogas from organic matters in wastewater under anaerobic condition
OMBR-RO/MD/ED	Clean water from the dual high retention membrane barriers
	Biogas from organic matters in wastewater under anaerobic condition
ED-OMBR-RO/MD/ED	Clean water from the dual high retention membrane barriers
	Biogas from organic matters in wastewater under anaerobic condition
	Inorganic salts from OMBR mixed liquor
MF/UF-OMBR	Clean water from the FO membrane and the MF/UF membrane
	Biogas from organic matters in wastewater under anaerobic condition
	Nutrients (e.g., phosphorus, and nitrogen) and inorganic salts from the MF/UF permeate
MFC/MRC-OMBR	Clean water from the FO membrane and the MF/UF membrane
	Biogas or bioelectricity from organic matters in wastewater under anaerobic condition
	Nutrients (e.g., phosphorus, and nitrogen) recovered from OMBR mixed liquor

OMBR can be configurated as submerged or side-stream depending on how FO is connected with the bioreactor (Figure 3.1C). In the submerged OMBR, the plate-and-frame cassette (Luo et al., 2017) or hollow fiber membrane module (Zhang et al., 2012) is typically immersed into the bioreactor. Only draw solution needs to be circulated through the membrane module during submerged OMBR operation. By contrast, the cross-flow membrane cell is externally combined with the bioreactor for side-stream OMBR. Thus, both draw solution and mixed liquid in the bioreactor are circulated through the membrane cell. It is noteworthy that the flow channels of the membrane cell should be narrow enough to ensure a relatively high cross-flow velocity and thus reduce concentration polarization (Wang et al., 2016a). Compared with the submerged OMBR, severe blockage and membrane fouling may occur in the membrane cell in the side-stream configuration, particularly at relatively high biomass concentrations (Luo et al., 2015b). Moreover, the high cross-flow velocity in the side-stream OMBR would considerably increase the pump energy demand and potentially destroy the biomass floc structure. As a result, recent efforts have been dedicated on the development of submerged OMBR for practical application.

3.2.2 *OSMOTIC MEMBRANE BIOREACTOR (OMBR) HYBRID SYSTEMS*

A desalination process is usually integrated with OMBR to produce clean water and regenerate the draw solution (Figure 3.2A). Recent studies have demonstrated the compatibility of RO (Luo et al., 2017), MD (Nguyen et al., 2017; Luo et al., 2018a), and ED (Lu and He, 2015) with OMBR. ED can also be integrated with OMBR to extract salts from the bioreactor for reusing as the draw solute (Figure 3.2B). In these hybrid systems, OMBR can effectively remove contaminants and thus prevent subsequent membrane processes from severe fouling (Luo et al., 2017). As a result, there can be a reduction in operational cost related to membrane cleaning and replacement in wastewater treatment. Indeed, it has been reported that OMBR-RO could reduce the operational cost of wastewater treatment and reuse by 5–25% compared to conventional MBR-RO using the UF membrane (Cornelissen et al., 2011). On the other hand, the desalination process downstream to OMBR not only reconcentrates the draw solution, but also enhances the produced water quality via providing another barrier to further remove contaminants permeating through the FO membrane.

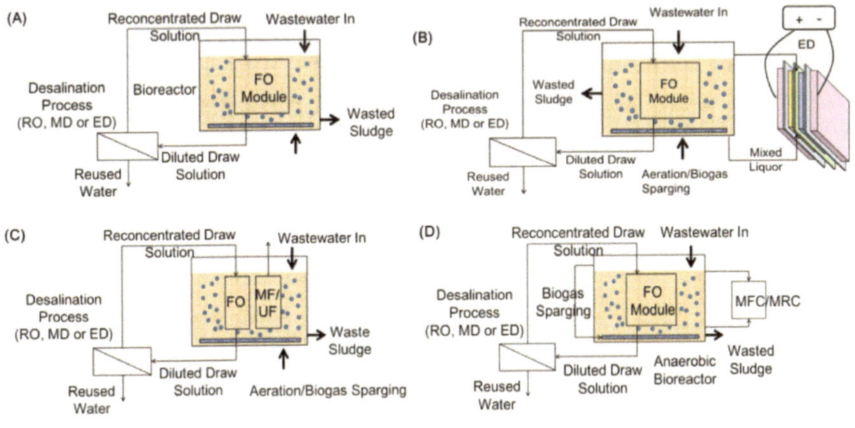

FIGURE 3.2 Schematic diagram of submerged OMBR hybrid systems: (A) OMBR-RO/MD/ED, (B) ED-OMBR-RO/MD/ED, (C) MF/UF-OMBR, and (D) MFC/MRC-OMBR.

Porous membranes (e.g., MF and UF) can be coupled with OMBR to mitigate salinity build-up in the bioreactor and recover nutrients (Figure 3.2C). Salinity build-up in the bioreactor is an inherent issue associated with OMBR due to the high salt rejection by the FO membrane and the reverse salt transport from the draw solution. Since dissolved salts can permeate through the porous MF or UF membranes, salinity build-up in bioreactor can be effectively controlled during OMBR operation (Wang et al., 2014b; Holloway et al., 2015b). Nutrients accumulated in the bioreactor during OMBR operation can also be discharged by the MF/UF membrane and subsequently recovered by chemical precipitation (Figure 3.3). Moreover, the MF/UF permeate can be potentially used for non-potable household, landscape, and agronomic activities.

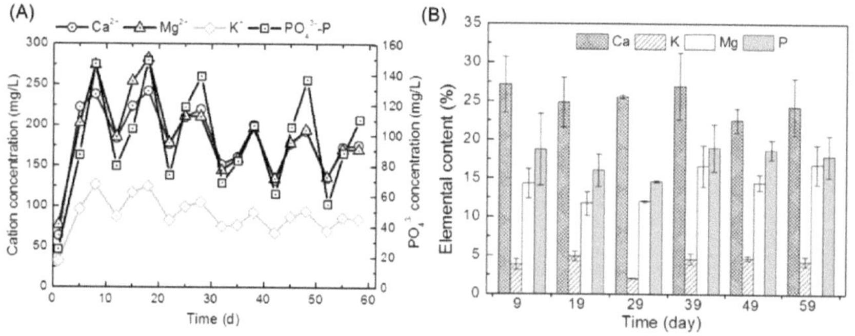

FIGURE 3.3 (A) Nutrient concentrations in the mixed liquor during MF-OMBR operation and (B) their contents in the solids recovered periodically from the MF permeate. Results are reproduced from Luo et al. (2016b). In that study, OMBR-RO was run in a cycle of 8 d on and 2 d off. MF was operated for 24 h to extract 3 L water from the bioreactor for phosphorus recovery during OMBR-RO off time. Nutrients were recovery from the MF permeate by precipitation at pH 10.

MFCs can be integrated with anaerobic OMBR to develop a MFC-OMBR hybrid system (Figure 3.2D) for simultaneous wastewater treatment and energy production (Table 3.1). MFCs use electrochemically active microorganisms to produce bioelectricity from waste streams (Hou et al., 2016). In this hybrid system, solute accumulation in the bioreactor during anaerobic OMBR operation can increase the mixed liquor conductivity and buffer capacity, thereby improving ion transfer for power production by MFCs (Liu et al., 2017). On the other hand, MFCs are potentially beneficial to anaerobic OMBR by alleviating membrane fouling and reducing sludge production. In addition to energy production, high quality water can be produced from these

MFC-OMBR hybrid systems, due to the complementarity of FO rejection and biological treatment.

A microbial recovery cell (MRC) has been recently developed by Hou et al. (2017) to incorporate with anaerobic OMBR for nutrient recovery (Figure 3.2C). The MRC was modified from microbial ED cells by adding two recovery chambers between anodes and cathodes with hydraulic connection. The recovery chamber was formed by a cation exchange membrane and an anion exchange membrane facing anode and cathode, respectively. In this hybrid system, nutrients are recovered by the MRC chambers and clean water is produced from the FO unit.

3.3 RESOURCE RECOVERY

3.3.1 *CLEAN WATER PRODUCTION*

Whilst the development of OMBR and its hybrid systems is on-going, its potential for effective contaminant removal has already been demonstrated (Figure 3.4). For instance, organic matter and phosphorus can be effectively removed by OMBR with removal rate higher than 98%. Such high removal can be attributed to the complementarity between membrane rejection and biological treatment. The high rejection capacity of the FO membrane increase the retention time of organic matter in the bioreactor and therefore facilitate their biodegradation. On the other hand, biodegradation reduces the accumulation of rejected organic matter in the bioreactor. Since biological phosphorus, removal under either aerobic or anaerobic conditions is limited and largely depends on microbial assimilation by phosphate accumulating bacteria (Oehmen et al., 2007). Thus, the high phosphorus removal by OMBR is mainly attributed to its effective rejection by the FO membrane. Holloway et al. (2007) reported that the FO membrane could almost completely reject phosphate (PO_4^{3-}) ions due to their large hydraulic radius and negative charge.

Moderate removal of total nitrogen (TN) can be achieved by OMBR (Figure 3.4). This result is caused by the relatively low rejection (60–70%) of nitrogen species, including ammonia (NH_4^+), nitrite (NO_2^-), and nitrate (NO_3^-) by the FO membrane (Irvine et al., 2013). No matter whether OMBR is operated under aerobic or anaerobic conditions, TN removal by biological treatment is limited and largely depends on microbial assimilation (Gerrity et al., 2013). For example, during aerobic OMBR operation, NH_4^+ can be effectively removed by nitrification (Figure 3.4), while its products (NO_x^-) accumulate considerably in the bioreactor without denitrification and subsequently pass through the FO membrane to reduce the TN removal (Luo et al., 2017). Nevertheless, the TN removal can be enhanced by several novel OMBR systems, which utilize sequential batch reactor (Linares et al., 2016), moving bed biofilm reactor (Nguyen et al., 2015), and attached growth biofilm reactor (Nguyen et al., 2016) to provide aerobic and anoxic conditions for simultaneous nitrification and denitrification. In addition, contaminants that escape from the FO membrane can be further removed by the draw solution recovery process (e.g., RO, and MD), which provides an additional barrier for product water purification.

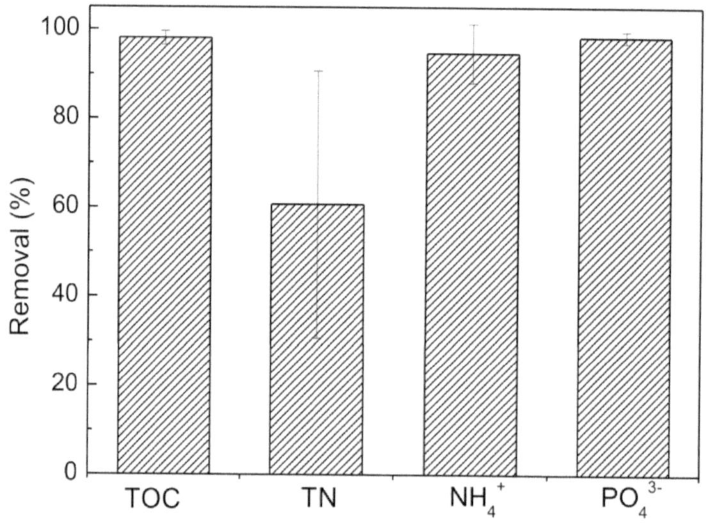

FIGURE 3.4 Removal of organic matter and nutrients by aerobic OMBR. Results are reproduced based previous studies (Achilli et al., 2009; Qiu and Ting, 2013; Chen et al., 2014; Wang et al., 2014; Tan et al., 2015; Nguyen et al., 2015, 2016; Luo et al., 2018).

TrOCs occur ubiquitously in wastewater at trace concentrations (from several nanograms per liter to a few micrograms per liter) and challenge water reuse applications (Luo et al., 2014). By integrating highly selective FO membrane with biological treatment, TrOCs can be effectively removed by OMBR (Figure 3.5). Compared to conventional MBR, OMBR can enhance the removal of TrOCs, particularly these biologically recalcitrant and hydrophilic compounds. Further removal can be achieved when an additional process (e.g., RO and MD) is integrated with OMBR for draw solution recovery and clean water production.

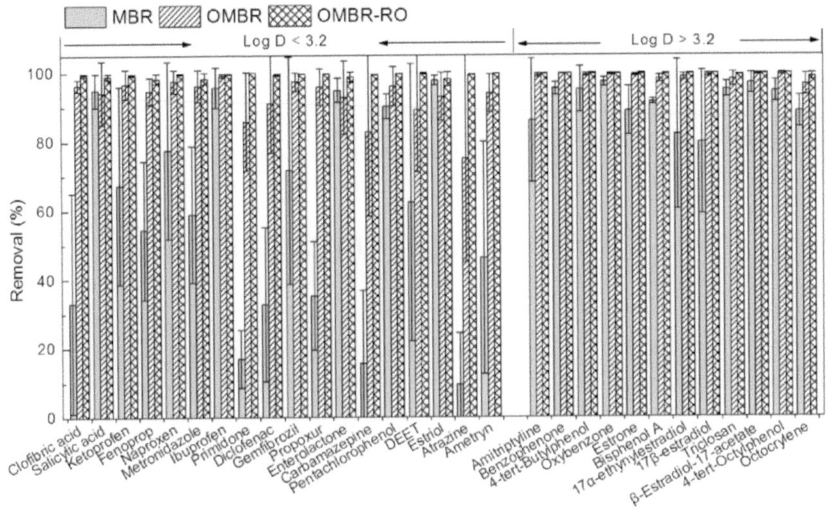

FIGURE 3.5 Removal of TrOCs by OMBR and its hybrid system. TrOCs are ordered based on their hydrophobicity as indicated by their effective octanol-water partition coefficients (*Log D*) at a solution pH of 8, which are obtained from SciFinder Scholar (ACS) database. Removal results are reproduced based previous studies: MBR (Tadkaew et al., 2011; Wijekoon et al., 2013), OMBR (Lay et al., 2012; Alturki et al., 2012; Luo et al., 2018b), and OMBR-RO (Luo et al., 2017; Zhang et al., 2017).

3.3.2 NUTRIENT RECOVERY

During OMBR operation, nutrients can be rejected by the FO membrane and accumulate in the bioreactor for subsequent recovery when cannot be removed by biological treatment. Qiu and Ting (2014) periodically settled the mixed liquor to recover phosphorus from the supernatant during OMBR operation. By chemical precipitation, phosphorus could be recovered in the form of amorphous calcium phosphate and accounted for 11% in the recovered solids. They also reported that phosphorus could be recovered with the content of 11.1–13.3% by operating MF simultaneously with OMBR (Qiu et al., 2015). Luo et al. (2016b) demonstrated a higher phosphorus recovery from wastewater when MF was operated intermittently with OMBR and thus developed a hybrid system for simultaneous phosphorus and water recovery from wastewater.

Anaerobic OMBR is promising for nutrient recovery in comparison to its aerobic counterparts. During anaerobic OMBR operation, both NH_4^+ and PO_4^{3-} can be enriched in the bioreactor, providing an opportunity for recovery (Table 3.1). For example, Hou et al. (2017) reported that the MRC-OMBR hybrid system could recover up to 45% NH_4^+ and 65% PO_4^{3-} from wastewater. Nutrient removal from the bioreactor in turn benefit anaerobic OMBR operation by reducing membrane scaling and NH_4^+ toxicity to anaerobic digesters (ADs) (Siles et al., 2010).

3.3.3 BIOGAS AND BIOELECTRICITY PRODUCTION

Energy can be recovered from wastewater in the form of biogas and bioelectricity by operating OMBR under anaerobic conditions (Table 3.1). For instance, methane (CH_4) yield from anaerobic OMBR has been reported to range from 0.21 to 0.3L CH_4/g chemical oxygen demand (COD) (Chen et al., 2014; Gu et al., 2015; Wang et al., 2017b). This production is comparable to conventional anaerobic MBR under similar operational conditions (Lin et al., 2013).

Bioelectricity can be produced from wastewater by integrating MFC with anaerobic OMBR. Hou et al. (2016) observed an increase in the power density of MFCs from 3 up to 11.5 W/m^3 when integrated with anaerobic OMBR. Liu et al. (2017) subsequently optimized the MFC-OMBR hybrid system by incorporating MF to control salinity build-up in the bioreactor and anaerobic acidification to fasten the conversion of organic matter into short chain fatty acids and alcohols to enhance bioelectricity production. In their study, the hybrid system achieved a continuous and relatively stable power generation with maximum power density of 4.38 W/m^3 (Liu et al., 2017).

3.4 AFFECTING FACTORS

3.4.1 SALINITY BUILD-UP IN THE BIOREACTOR

Salinity build-up in the bioreactor is an inherent challenge to OMBR, which is caused by the high rejection capacity of the FO membrane and reverse draw solute diffusion (Song et al., 2018). Salinity build-up in the bioreactor can negatively affect OMBR performance by reducing the effective driving force for water transport (Qiu and Ting, 2013), inhibiting microbial activity (Wang et al., 2014a), altering microbial community structure and sludge characteristics (Luo et al., 2016c), and aggravating membrane fouling (Lay et al., 2010; Luo et al., 2015a).

Salinity build-up in the bioreactor can be effectively alleviated by integrating porous membranes (e.g., MF, and UF) with OMBR. Luo et al. (2015b) demonstrated that the mixed liquor conductivity stabilized at approximately 700 μS/cm during MF-OMBR operation in comparison to an increase from 268 to 8270 μS/cm within a week under similar conditions but without MF as reported by Alturki et al. (2012). Similar results have also been reported by Holloway et al. (2015c) who demonstrated a stable operation of a pilot-scale UF-OMBR hybrid system over 125 days. Dissolved salts in the bioreactor can also be extracted by ED. Lu and He (2015) reported that ED at 3 Voltage could maintain a relatively low mixed liquor conductivity of 8 mS/cm, which prolonged OMBR operation by 6 times. Salts recovered by ED could be reused as the draw solute for OMBR.

3.4.2 DRAW SOLUTION TYPES

Draw solution provides the osmotic driving force in OMBR and thus determines its performance and viability. An ideal draw solution should contribute a high waterflux, low fouling potential, low reverse solute diffusion, low or no toxicity to microorganisms, low cost and ease of regeneration (Achilli et al., 2010; Bowden et al., 2012). Draw solutions with small molecular weight and charged ions are highly diffusive and thus can provide high-water flux for OMBR by reducing the internal concentration polarization (Ge et al., 2013; Johnson et al., 2017). Thus, inorganic salts, such as sodium chloride (NaCl) and magnesium chloride ($MgCl_2$), are commonly used in OMBR. It is noteworthy that inorganic draw solutes with low molecular weight have relatively high reverse solute flux and thus result in severe salinity build-up in the bioreactor during OMBR operation (Achilli et al., 2010). Qiu and Ting (2014) reported that the mixed liquor conductivity increased from 0.1 to 15 mS/cm within 63 days when 0.5 M $MgCl_2$ was used as the draw solution; and a more dramatic increase up to 33 mS/cm within 20 days was observed when 0.83 M NaCl draw was subsequently used to contribute the same osmotic pressure of 4 MPa.

Organic-based draw solutions, which are the combination of organic acids (anions) with organic or inorganic bases (cations), have been proposed to control salinity build-up in OMBR. Given their large molecular weight and thus low diffusivity, organic-based draw solutes have lower reverse solute flux than their inorganic counterparts (Bowden et al., 2012; Ansari et al., 2015). Although the reverse diffusion of organic solutes occurs, they are readily biodegraded and do not cause long-term salinity build-up in the bioreactor. Nguyen et al. (2015) reported that the mixed liquor conductivity only increased from 0.26 to 0.54 mS/cm within 68 days when ethylene-diamine-tetra acetic acid disodium salt (EDTA-2Na) coupled with Triton X-100 were used as the draw solution for OMBR. Such a small increase in the bioreactor salinity could be attributed to the adsorption of Triton X-114 surfactant onto the membrane supporting layer by hydrophobic interaction, thereby constricting membrane pores to reduce the reverse diffusion of EDTA-2Na (Nguyen et al., 2016). Luo et al. (2016a) also observed a significant reduction in the bioreactor salinity when sodium acetate (NaOAc) and EDTA-2Na were used as the draw solutions in comparison with NaCl during OMBR operation. It is noteworthy that the reverse diffusion of NaOAc exacerbated the FO membrane fouling by providing extra carbon sources for biofilm development andEDTA-2Na transported reversely inhibited microbial growth and activity (Luo et al., 2016a). Similar results have also been demonstrated by Siddique et al. (2017). Thus, further research is necessary to determine comprehensively the compatibility of organic-based draw solutions with OMBR.

Liquid fertilizers can be potentially used as the draw solutions for OMBR. In this case, draw solution recovery is not needed, since the diluted liquid fertilizer can be used directly for fertigation. Kim et al. (2016) developed a protocol to select suitable fertilizer draw solutes for anaerobic OMBR and concluded that mono-ammonium phosphate was appropriate given its contribution to small salinity build-up in the bioreactor, excellent water flux, high dilution capacity, and indiscernible toxicity to anaerobic microbes. Wang et al. (2017a) compared the performance of three liquid fertilizers in aerobic OMBR and highlighted the potential of ammonium sulfate with a relatively high water flux (6.85 L/m^2h) and low reverse salt flux (3.02 g/m^2h).

3.4.3 DRAW SOLUTION CONCENTRATION

Draw solution concentration determines the osmotic pressure gradient across the FO membrane for water transport during OMBR operation. Wang et al. (2019) investigated the effects of draw solution concentration on anaerobic OMBR performance, and reported that the water flux of FO increased linearly when the draw solution concentration increased in the range of 0.25–1.0 M NaCl, while the flux decline was more remarkable and the water flux deteriorated when the draw solution concentration was elevated to 2 M. This result indicates that a critical draw solution concentration or driving force exists, below which an increase in the draw solution concentration can increase the FO water flux, whereas above which the flux performance of FO membrane becomes worse due to an aggravated membrane fouling induced by a higher initial flux. It is noteworthy that high draw solution concentration exacerbates salinity build-up in the bioreactor and increases energy consumption of subsequent recovery processes, particularly RO, which is significantly affected by the feed salinity.

3.4.4 MEMBRANE PROPERTIES

Currently commercial FO membranes are almost exclusively fabricated on either cellulose or polyamide. In specific, cellulose triacetate (CTA) and polyamide thin film composite (TFC) FO membranes are commonly used in current OMBR studies. The CTA membrane consists of an active layer made of CTA and a polyester mesh for mechanical support (Cath et al., 2013). The TFC FO membrane comprises a thin, selective polyamide active layer on a porous supporting layer, typically made of polysulfone. The polyamide TFC FO membrane has higher water permeability, lower salt permeability, and better solute rejection in comparison with its CTA counterpart (Cath et al., 2013; Xie et al., 2014).

OMBR performance varies significantly with different FO membranes. Zhang et al. (2017) reported that the TFC membrane exhibited a higher initial water flux but more dramatic flux decline compared to the CTA membrane during OMBR operation. This result could be attributed to the fact that polyamide membranes are more susceptible to fouling than cellulose-based membranes (Mazlan et al., 2016). Wang et al. (2016b) further identified that biofouling was more significant for the CTA membrane; while inorganic fouling was more severe to the TFC membrane during OMBR operation. Compared to the CTA membrane, the TFC membrane has higher selectivity and thus increases the removal of TrOCs, particularly the hydrophilic and biologically persistent compounds in OMBR (Zhang et al., 2017). Since the TFC membrane has smaller reverse salt flux and lower stable water flux, salinity build-up in the bioreactor was less significant in comparison to the CTA membrane for OMBR (Hou et al., 2016).

Biomimetic membranes, based on aquaporins, have been recently developed to advance the OMBR performance (Tang et al., 2015; Fane et al., 2015). Aquaporins are water-channel proteins in biological cell membrane, with each channel being able to transport up to 10^9 water molecules per second and completely reject all other solutes (Jensen and Mouritsen, 2006). Indeed, Luo et al. (2018b) reported that aquaporin FO membrane had better transport properties in comparison with conventional CTA and TFC FO membranes. In particular, the aquaporin FO membrane exhibited much lower salt permeability and thus smaller reverse salt flux, resulting in a less severe salinity build-up in the bioreactor during OMBR operation.

3.4.5 CONTAMINANT ACCUMULATION IN THE DRAW SOLUTION

Contaminants that pass through the FO membrane, but are rejected by the downstream membrane can accumulate in the draw solution, and thereby deteriorating the productivity and sustainability of OMBR hybrid system. Luo et al. (2017) observed a notable build-up of organic matters, such as protein- and humic-like substances, in the draw solution during OMBR-RO operation. Contaminant accumulation in the draw solution is more severe for OMBR-MD, given the complete rejection of non-volatile solutes by the MD membrane (Alkhudhiri et al., 2012; Tijing et al., 2015). As a result, membrane fouling occurs to the subsequent draw solution recovery process (Luo et al., 2018a). More importantly, contaminant accumulation in the draw solution would deteriorate the product water quality of OMBR hybrid systems.

Contaminant accumulation in the draw solution can be alleviated by additional processes. For example, Xie et al. (2013) demonstrated that granular activated carbon (GAC) adsorption and ultraviolet (UV) oxidation could effectively reduce the accumulation of organic substances in the draw solution during the operation of FO-MD for sewer mining. Nevertheless, the long-term operation concerns of either GAC adsorption or UV oxidation should be considered when controlling contaminant accumulation in the draw solution of OMBR hybrid systems. For instance, periodic regeneration or replacement of GAC is necessary to avoid contaminant breakthrough. Moreover, there remain uncertainties regarding the impact of UV oxidation by-products or residuals on product water quality and membrane polymer.

3.5 CONCLUDING REMARKS

By integrating highly selective FO membranes with biological processes, typically CAS treatment, OMBR can potentially advance wastewater treatment and resource recovery. OMBR analogs can be further developed by combining the FO membrane with varying biological processes, such as ADs, moving bed biofilm reactor, attached growth biofilm reactor, and sequential batch reactor. In particular, energy can be recovered from wastewater in the form of biogas and bioelectricity when OMBR is operated under anaerobic conditions. OMBR and its analogs can be operated in osmotic dilution mode when the draw solution is readily available or discharged directly. Alternatively, a desalination process, such as RO, MD, and ED, can be integrated with OMBR to regenerate the draw solution and provide an additional barrier to produce clean water for reuse.

Additional processes are required to integrate with OMBR to control salinity build-up in the bioreactor and recover resources (nutrients and energy) from wastewater. These processes mainly include MF, UF, ED, MFC, and MRC. Nutrients (e.g., phosphorus and ammonia) can be effectively recovered from MF and UF permeates as well as MRC solutions by chemical precipitation without the addition of external salt sources. ED can extract dissolved salts from the bioreactor, which can be directly used as the draw solutes for OMBR. MFC can be combined with anaerobic OMBR to enhance bioelectricity production as salinity build-up in the bioreactor can increase the mixed liquor conductivity and buffer capacity to increase ion transfer for power production.

Despite the promising of OMBR and its hybrid systems in wastewater reuse and resource recovery, their development is still in the infant stage and affected by several key factors. These include salinity build-up in the bioreactor, membrane properties, draw solution types, and concentration, and contaminant accumulation in the draw solution. Thus, optimizing these factors would accelerate the practical application of OMBR and its hybrid systems for resource recovery from wastewater.

KEYWORDS

- **conventional activated sludge**
- **forward osmosis**
- **osmotic membrane bioreactor**
- **resource recovery**
- **salinity build-up**
- **wastewater reuse**

REFERENCES

Achilli, A., Cath, T. Y., & Childress, A. E., (2010). Selection of inorganic-based draw solutions for forward osmosis applications. *J. Membr. Sci., 364*(1/2), 233–241.

Achilli, A., Cath, T. Y., Marchand, E. A., & Childress, A. E., (2009). The forward osmosis membrane bioreactor: A low fouling alternative to MBR processes. *Desalination, 239*, 10–21.

Alkhudhiri, A., Darwish, N., & Hilal, N., (2012). Membrane distillation: A comprehensive review. *Desalination, 287*, 2–18.

Alturki, A., McDonald, J., Khan, S. J., Hai, F. I., Price, W. E., & Nghiem, L. D., (2012). Performance of a novel osmotic membrane bioreactor (OMBR) system: Flux stability and removal of trace organics. *Bioresour. Technol., 113*, 201–206.

Ansari, A. J., Hai, F. I., Guo, W., Ngo, H. H., Price, W. E., & Nghiem, L. D., (2015). Selection of forward osmosis draws solutes for subsequent integration with anaerobic treatment to facilitate resource recovery from wastewater. *Bioresour. Technol., 191*, 30–36.

Ansari, A. J., Hai, F. I., Price, W. E., & Nghiem, L. D., (2016). Phosphorus recovery from digested sludge centrate using seawater-driven forward osmosis. *Sep. Purif. Technol., 163*, 1–7.

Besha, A. T., Gebreyohannes, A. Y., Tufa, R. A., Bekele, D. N., Curcio, E., & Giorno, L., (2017). Removal of emerging micro pollutants by activated sludge process and membrane bioreactors and the effects of micro pollutants on membrane fouling: A review. *J. Environ. Chem. Eng., 5*(3), 2395–2414.

Bowden, K. S., Achilli, A., & Childress, A. E., (2012). Organic ionic salt draw solutions for osmotic membrane bioreactors. *Bioresour. Technol., 122*, 207–216.

Cath, T. Y., Childress, A. E., & Elimelech, M., (2006). Forward osmosis: Principles, applications, and recent developments. *J. Membr. Sci., 281*(1), 70–87.

Cath, T. Y., Elimelech, M., McCutcheon, J. R., McGinnis, R. L., Achilli, A., Anastasio, D., Brady, A. R., et al., (2013). Standard methodology for evaluating membrane performance in osmotically driven membrane processes. *Desalination, 312*, 31–38.

Chen, L., Gu, Y., Cao, C., Zhang, J., Ng, J. W., & Tang, C., (2014). Performance of a submerged anaerobic membrane bioreactor with forward osmosis membrane for low-strength wastewater treatment. *Water Res., 50*, 114–123.

Cornelissen, E. R., Harmsen, D., Beerendonk, E. F., Qin, J. J., Oo, H., De Korte, K. F., & Kappelhof, J. W. M. N., (2011). The innovative osmotic membrane bioreactor (OMBR) for reuse of wastewater. *Water Sci. Technol., 63*(8), 1557–1565.

Dharupaneedi, S. P., Nataraj, S. K., Nadagouda, M., Reddy, K. R., Shukla, S. S., & Aminabhavi, T. M., (2019). Membrane-based separation of potential emerging pollutants. *Separation and Purification Technology, 210*, 850–866.

Fane, A. G., Wang, R., & Hu, M. X., (2015). Synthetic membranes for water purification: Status and future. *Angew. Chem. Int. Ed Engl., 54*(11), 3368–3386.

Ge, Q., Ling, M., & Chung, T. S., (2013). Draw solutions for forward osmosis processes: Developments, challenges, and prospects for the future. *J. Membr. Sci., 442*, 225–237.

Gerrity, D., Pecson, B., Trussell, R. S., & Trussell, R. R., (2013). Potable reuse treatment trains throughout the world. *J. Water Supply: Res. Technol. AQUA, 62*(6), 321–338.

Gu, Y., Chen, L., Ng, J. W., Lee, C., Chang, V. W. C., & Tang, C. Y., (2015). Development of anaerobic osmotic membrane bioreactor for low-strength wastewater treatment at mesophilic condition. *J. Membr. Sci., 490*, 197–208.

Holloway, R. W., Achilli, A., & Cath, T. Y., (2015a). The osmotic membrane bioreactor: A critical review. *Environ. Sci.: Water Res. Technol., 1*(5), 581–605.

Holloway, R. W., Childress, A. E., Dennett, K. E., & Cath, T. Y., (2007). Forward osmosis for concentration of anaerobic digester cent rate. *Water Res., 41*(17), 4005–4014.

Holloway, R. W., Wait, A. S., Da Silva, A. F., Herron, J., Schutter, M. D., Lampi, K., & Cath, T. Y., (2015b). Long-term pilot scale investigation of novel hybrid ultrafiltration-osmotic membrane bioreactors. *Desalination, 363*, 64.

Holloway, R. W., Wait, A. S., Fernandes, D. S. A., Herron, J., Schutter, M. D., Lampi, K., & Cath, T. Y., (2015c). Long-term pilot scale investigation of novel hybrid ultra filtration-osmotic membrane bioreactors. *Desalination, 363*, 64–74.

Hou, D., Lu, L., & Ren, Z. J., (2016). Microbial fuel cells and osmotic membrane bioreactors have mutual benefits for wastewater treatment and energy production. *Water Res., 98*, 183–189.

Hou, D., Lu, L., Sun, D., Ge, Z., Huang, X., Cath, T. Y., & Ren, Z. J., (2017). Microbial electrochemical nutrient recovery in anaerobic osmotic membrane bioreactors. *Water Res., 114*, 181–188.

Irvine, G. J., Rajesh, S., Georgiadis, M., & Phillip, W. A., (2013). Ion selective permeation through cellulose acetate membranes in forward osmosis. *Environ. Sci. Technol., 47*(23), 13745–13753.

Jensen, M. O., & Mouritsen, O. G., (2006). Single-channel water permeability's of *Escherichia coli* aquaporins AqpZ and GlpF. *Biophys. J., 90*(7), 2270–2284.

Johnson, D. J., Suwaileh, W. A., Mohammed, A. W., & Hilal, N., (2017). Osmotic's potential: An overview of draw solutes for forward osmosis. *Desalination*. doi: 10.1016/j.desal.2017.09.017.

Judd, S., (2008). The status of membrane bioreactor technology. *Trends Biotechnol., 26*(2), 109–116.

Kim, Y., Chekli, L., Shim, W. G., Phuntsho, S., Li, S., Ghaffour, N., Leiknes, T., & Shon, H. K., (2016). Selection of suitable fertilizer draw solute for a novel fertilizer-drawn forward osmosis-anaerobic membrane bioreactor hybrid system. *Bioresour. Technol., 210*, 26–34.

Lay, W. C. L., Zhang, Q., Zhang, J., McDougald, D., Tang, C., Wang, R., Liu, Y., & Fane, A. G., (2012). Effect of pharmaceuticals on the performance of a novel osmotic membrane bioreactor (OMBR). *Sep. Sci. Technol., 47*(4), 543–554.

Lay, W. C., Liu, Y., & Fane, A. G., (2010). Impacts of salinity on the performance of high retention membrane bioreactors for water reclamation: A review. *Water Res., 44*(1), 21–40.

Lin, H., Peng, W., Zhang, M., Chen, J., Hong, H., & Zhang, Y., (2013). A review on anaerobic membrane bioreactors: Applications, membrane fouling and future perspectives. *Desalination, 314*, 169–188.

Linares, R. V., Li, Z., Yangali-Quintanilla, V., Li, Q., Vrouwenvelder, J. S., Amy, G. L., & Ghaffour, N., (2016). Hybrid SBR-FO system for wastewater treatment and reuse: Operation, fouling and cleaning. *Desalination, 393*, 31–38.

Liu, J., Wang, X., Wang, Z., Lu, Y., Li, X., & Ren, Y., (2017). Integrating microbial fuel cells with anaerobic acidification and forward osmosis membrane for enhancing bio-electricity and water recovery from low-strength wastewater. *Water Res., 110*, 74–82.

Lu, Y., & He, Z., (2015). Mitigation of salinity buildup and recovery of wasted salts in a hybrid osmotic membrane bioreactor-electro dialysis system. *Environ. Sci. Technol., 49*(17), 10529–10535.

Luo, W., Hai, F. I., Kang, J., Price, W. E., Guo, W., Ngo, H. H., Yamamoto, K., & Nghiem, L. D., (2015a). Effects of salinity build-up on biomass characteristics and trace organic chemical removal: Implications on the development of high retention membrane bioreactors. *Bioresour. Technol., 177*, 274–281.

Luo, W., Hai, F. I., Kang, J., Price, W. E., Nghiem, L. D., & Elimelech, M., (2015b). The role of forward osmosis and microfiltration in an integrated osmotic-microfiltration membrane bioreactor system. *Chemosphere, 136*, 125–132.

Luo, W., Hai, F. I., Price, W. E., & Nghiem, L. D., (2015c). Water extraction from mixed liquor of an aerobic bioreactor by forward osmosis: Membrane fouling and biomass characteristics assessment. *Sep. Purif. Technol., 145*, 56–62.

Luo, W., Hai, F. I., Price, W. E., Elimelech, M., & Nghiem, L. D., (2016a). Evaluating ionic organic draw solutes in osmotic membrane bioreactors for water reuse. *J. Membr. Sci., 514*, 636–645.

Luo, W., Hai, F. I., Price, W. E., Guo, W., Ngo, H. H., Yamamoto, K., & Nghiem, L. D., (2016b). Phosphorus and water recovery by a novel osmotic membrane bioreactor-reverse osmosis system. *Bioresour. Technol., 200*, 297–304.

Luo, W., Phan, H. V., Hai, F. I., Price, W. E., Guo, W., Ngo, H. H., Yamamoto, K., & Nghiem, L. D., (2016c). Effects of salinity build-up on the performance and bacterial community structure of a membrane bioreactor. *Bioresour. Technol., 200*, 305–310.

Luo, W., Phan, H. V., Li, G., Hai, F. I., Price, W. E., Elimelech, M., & Nghiem, L. D., (2018a). An osmotic membrane bioreactor-membrane distillation system for simultaneous wastewater reuse and seawater desalination: Performance and implications. *Environ. Sci. Technol., 51*, 14311–14320.

Luo, W., Phan, H. V., Xie, M., Hai, F. I., Price, W. E., Elimelech, M., & Nghiem, L. D., (2017). Osmotic versus conventional membrane bioreactors integrated with reverse osmosis for water reuse: Biological stability, membrane fouling, and contaminant removal. *Water Res., 109*, 122–134.

Luo, W., Xie, M., Song, X., Guo, W., Ngo, H. H., Zhou, J., & Nghiem, L. D., (2018b). Biomimetic aquaporin membranes for osmotic membrane bioreactors: Membrane performance and contaminant removal. *Bioresour. Technol., 249*, 62–68.

Luo, Y., Guo, W., Ngo, H. H., Nghiem, L. D., Hai, F. I., Zhang, J., Liang, S., & Wang, X. C., (2014). A review on the occurrence of micro pollutants in the aquatic environment and their fate and removal during wastewater treatment. *Sci. Total Environ., 473, 474*, 619–641.

Mazlan, N. M., Marchetti, P., Maples, H. A., Gu, B., Karan, S., Bismarck, A., & Livingston, A. G., (2016). Organic fouling behavior of structurally and chemically different forward osmosis membranes: A study of cellulose triacetate and thin film composite membranes. *J. Membr. Sci., 520*, 247–261.

Mi, B., & Elimelech, M., (2010). Organic fouling of forward osmosis membranes: Fouling reversibility and cleaning without chemical reagents. *Journal of Membrane Science, 348*(1/2), 337–345.

Morrow, C. P., Furtaw, N. M., Murphy, J. R., Achilli, A., Marchand, E. A., Hiibel, S. R., & Childress, A. E., (2018). Integrating an aerobic/anoxic osmotic membrane bioreactor with membrane distillation for potable reuse. *Desalination, 432*, 46–54.

Nguyen, N. C., Chen, S. S., Nguyen, H. T., Chen, Y. H., Ngo, H. H., Guo, W., Ray, S. S., et al., (2017). *Applicability of an Integrated Moving Sponge Biocarrier-Osmotic Membrane Bioreactor MD System for Saline Wastewater Treatment Using Highly Salt-Tolerant Microorganisms*. Separation and Purification Technology.

Nguyen, N. C., Chen, S. S., Nguyen, H. T., Ngo, H. H., Guo, W., Hao, C. W., & Lin, P. H., (2015). Applicability of a novel osmotic membrane bioreactor using a specific draw solution in wastewater treatment. *Sci. Total Environ., 518, 519*, 586–594.

Nguyen, N. C., Nguyen, H. T., Chen, S. S., Ngo, H. H., Guo, W., Chan, W. H., Ray, S. S., Li, C. W., & Hsu, H. T., (2016). A novel osmosis membrane bioreactor-membrane distillation hybrid system for wastewater treatment and reuse. *Bioresour. Technol., 209*, 8–15.

Oehmen, A., Lemos, P. C., Carvalho, G., Yuan, Z., Keller, J., Blackall, L. L., & Reis, M. A., (2007). Advances in enhanced biological phosphorus removal: From micro to macro scale. *Water Res., 41*(11), 2271–2300.

Qiu, G. L., & Ting, Y. P., (2013). Osmotic membrane bioreactor for wastewater treatment and the effect of salt accumulation on system performance and microbial community dynamics. *Bioresour. Technol., 150*, 287–297.

Qiu, G., & Ting, Y. P., (2014). Direct phosphorus recovery from municipal wastewater via osmotic membrane bioreactor (OMBR) for wastewater treatment. *Bioresour. Technol., 170*, 221–229.

Qiu, G., Law, Y., Das, S., & Ting, Y., (2015). Direct and complete phosphorus recovery from municipal wastewater using a hybrid microfiltration-forward osmosis membrane bioreactor process with seawater brine as draw solution. *Environ. Sci. Technol., 49*, 6156–6163.

Siddique, M. S., Khan, S. J., Shahzad, M. A., Nawaz, M. S., & Hankins, N. P., (2017). Insight into the effect of organic and inorganic draw solutes on the flux stability and sludge characteristics in the osmotic membrane bioreactor. *Bioresour. Technol., 249*, 758–766.

Siles, J. A., Brekelmans, J., Martin, M. A., Chica, A. F., & Martin, A., (2010). Impact of ammonia and sulphate concentration on thermophilic anaerobic digestion. *Bioresour. Technol., 101*(23), 9040–9048.

Song, X., Xie, M., Li, Y., Li, G., & Luo, W., (2018). Salinity build-up in osmotic membrane bioreactors: Causes, impacts, and potential cures. *Bioresource Technology, 257*, 301–310.

Tadkaew, N., Hai, F. I., McDonald, J. A., Khan, S. J., & Nghiem, L. D., (2011). Removal of trace organics by MBR treatment: The role of molecular properties. *Water Res., 45*(8), 2439–2451.

Tang, C., Wang, Z., Petrinić, I., Fane, A. G., & Hélix-Nielsen, C., (2015). Biomimetic aquaporin membranes coming of age. *Desalination, 368*, 89–105.

Tijing, L. D., Woo, Y. C., Choi, J. S., Lee, S., Kim, S. H., & Shon, H. K., (2015). Fouling and its control in membrane distillation: A review. *J. Membr. Sci., 475*, 215–244.

Wang, H., Wang, X., Meng, F., Li, X., Ren, Y., & She, Q., (2019). Effect of driving force on the performance of anaerobic osmotic membrane bioreactors: New insight into enhancing water flux of FO membrane via controlling driving force in a two-stage pattern. *J. Membr. Sci., 569*, 41–47.

Wang, J., Pathak, N., Chekli, L., Phuntsho, S., Kim, Y., Li, D., & Shon, H., (2017a). Performance of a novel fertilizer-drawn forward osmosis aerobic membrane bioreactor (FDFO-MBR): Mitigating salinity build-up by integrating microfiltration. *Water, 9*(1), 21–34.

Wang, X. H., Yuan, B., Chen, Y., Li, X. F., & Ren, Y. P., (2014b). Integration of micro-filtration into osmotic membrane bioreactors to prevent salinity build-up. *Bioresour. Technol., 167*, 116–123.

Wang, X., Chang, V. W. C., & Tang, C. Y., (2016a). Osmotic membrane bioreactor (OMBR) technology for wastewater treatment and reclamation: Advances, challenges, and prospects for the future. *J. Membr. Sci., 504*, 113–132.

Wang, X., Chen, Y., Yuan, B., Li, X., & Ren, Y., (2014a). Impacts of sludge retention time on sludge characteristics and membrane fouling in a submerged osmotic membrane bioreactor. *Bioresour. Technol., 161*, 340–347.

Wang, X., Wang, C., Tang, C. Y., Hu, T., Li, X., & Ren, Y., (2017b). Development of a novel anaerobic membrane bioreactor simultaneously integrating microfiltration and forward osmosis membranes for low-strength wastewater treatment. *J. Membr. Sci., 527*, 1–7.

Wang, X., Zhao, Y., Yuan, B., Wang, Z., Li, X., & Ren, Y., (2016b). Comparison of biofouling mechanisms between cellulose triacetate (CTA) and thin-film composite (TFC) polyamide forward osmosis membranes in osmotic membrane bioreactors. *Bioresour. Technol., 202*, 50–58.

Wijekoon, K. C., Hai, F. I., Kang, J., Price, W. E., Guo, W., Ngo, H. H., & Nghiem, L. D., (2013). The fate of pharmaceuticals, steroid hormones, phytoestrogens, UV-filters, and pesticides during MBR treatment. *Bioresour. Technol., 144*, 247–254.

Xiao, K., Liang, S., Wang, X., Chen, C., & Huang, X., (2019). Current state and challenges of full-scale membrane bioreactor applications: A critical review. *Bioresource Technology, 271*, 473–481.

Xie, M., Nghiem, L. D., Price, W. E., & Elimelech, M., (2013). A forward osmosis-membrane distillation hybrid process for direct sewer mining: System performance and limitations. *Environment Science and Technology, 47*(23), 13486–13493.

Xie, M., Nghiem, L. D., Price, W. E., & Elimelech, M., (2014). Relating rejection of trace organic contaminants to membrane properties in forward osmosis: Measurements, modeling and implications. *Water Res., 49*, 265–274.

Zhang, B., Song, X., Nghiem, L. D., Li, G., & Luo, W., (2017). Osmotic membrane bioreactors for wastewater reuse: Performance comparison between cellulose triacetate and polyamide thin film composite membranes. *J. Membr. Sci., 539*, 383–391.

Zhang, J., Loong, W. L. C., Chou, S., Tang, C., Wang, R., & Fane, A. G., (2012). Membrane biofouling and scaling in forward osmosis membrane bioreactor. *J. Membr. Sci., 403–404*, 8–14.

CHAPTER 4

Energy Consumption and Recovery in Wastewater Treatment Systems

GIDEON SARPONG and VEERA GNANESWAR GUDE

Department of Civil and Environmental Engineering, Mississippi State University, Mississippi State, MS – 39762, USA

ABSTRACT

This chapter discusses the energy consumption and recovery trends in wastewater treatment systems. A comprehensive update on the present status of the wastewater treatment systems was presented. Various energy recovery options along with energy consumption minimization alternatives were discussed. Based on the feasibility levels, three energy-positive treatment configurations or alternatives were proposed considering both energy production and energy conservation potentials in current wastewater treatment systems. Basic, moderate, and advanced configurations were proposed based on the energy recovery targets and ease of implementation. Co-digestion of the primary and secondary sludge was identified as a feasible mechanism to increase the economic benefits of anaerobic digestion for revenue generation in the basic configuration. Equipment upgrades, process modifications, and co-digestion were considered in the moderate configuration to develop an energy-positive wastewater treatment scheme. The advanced configuration bypasses several conventional methods of treating the wastewater and focuses on advanced solids separation and biofilm and phototrophic technologies to enhance bioenergy production from wastewater treatment. Case studies were presented to highlight some of the key modifications that result in enhanced energy recovery. Mass and energy flow balances were also discussed for the proposed basic, moderate, and advanced configurations. Finally, research and practical considerations required to address the limitations associated with the proposed approaches were presented in detail.

4.1 INTRODUCTION

The worldwide population growth has increased the burden of managing higher volumes of waste. Domestic wastewater is the most common waste stream, which has important and adverse impacts on the environment. About 78% of the United States (US) population receives collection and treatment services from over 15,000 municipal wastewater treatment plants (WWTPs) consuming nearly ~4% of the entire US's electrical demand and treating an average wastewater flow of about 32,345 million gallons per day (MGD) (Moand Zhang, 2013; Yanwen et al., 2015).

WWTPs contribute to anthropogenic greenhouse gas (GHG) emissions, which is a major cause of global warming. Carbon dioxide (CO_2), methane (CH_4), and nitrous oxide (N_2O) are the main constituents of GHG emissions that are emitted from wastewater treatment processes. CO_2 is formed under aerobic condition during microbial degradation and through combustion of organic matter. CH_4 is generated through the degradation of organics under anaerobic conditions, while N_2O is produced as the result of the biological removal of nitrogen (N) through enhanced nitrification and denitrification processes (Hiroko et al., 2014). The need to reduce these emissions and to identify the factors controlling the GHG emissions from WWTPs is on ascendency (Kampschreur et al., 2009). CH_4 emissions from WWTPs occurs mostly during anaerobic decomposition (and anaerobic digestion of sludge) whereby methanogens are activated. CH_4 can be collected and used as an energy source, indirectly reducing CO_2 emissions (Oshita et al., 2014). Besides the anaerobic decomposition or digestion process, the sludge thickener produces the highest amount of CH_4 (2.1 gCH_4/kg BOD_5) whereas the aerobic reactor produces the highest amount of N_2O (1.26 gN_2O/kg TN) (Hwang et al., 2016). Therefore, WWTPs are recognized as one of the major sources of GHG emissions (Yan et al., 2014).

There is an increasing number of literature contributions focused on low energy or even on energy positive wastewater treatment processes (WERF, 2009; Elías-Maxil et al., 2014; McCarty et al., 2011; Funamizu et al., 2001; Gude, 2015a; Chae and Kang, 2013; Nowak et al., 2011; Frijns et al., 2013). The current technologies used in most of the WWTP were designed years ago when GHG emissions and energy production were not a topic of concern. According to the US EPA, there are over 14,700 municipal WWTPs and 48% of the plants use anaerobic digester (AD) for sludge stabilization and less than 10% actually use the biogas produced from the AD for heat or electricity production. There are about five utilities in the US and four in Europe that have achieved 100%

or excess energy production. This chapter discusses energy consumption and recovery trends in wastewater treatment systems and presents three different classifications for energy positive wastewater treatment configurations. Case studies are presented in detail with mass and energy balances.

4.2 ENERGY CONSUMPTION IN WASTEWATER TREATMENT SYSTEMS

In the U.S., approximately 3% to 4% of national electricity consumption is used for conveyance and treatment of water and wastewater (Goldstein and Smith, 2002; Galbraith, 2011). Typically, about 30% of the operational cost is assigned for energy use (Tchobanoglous et al., 2003). Energy costs represent a large portion of operating costs for utilities since it is normally required in all stages of the treatment process, from influent pumping to discharge of treated effluent. Figure 4.1 (A–D) shows the variability of energy requirements within different wastewater treatment technologies (A); in various countries (B); at different treatment capacities (C); and in different unit processes (D).

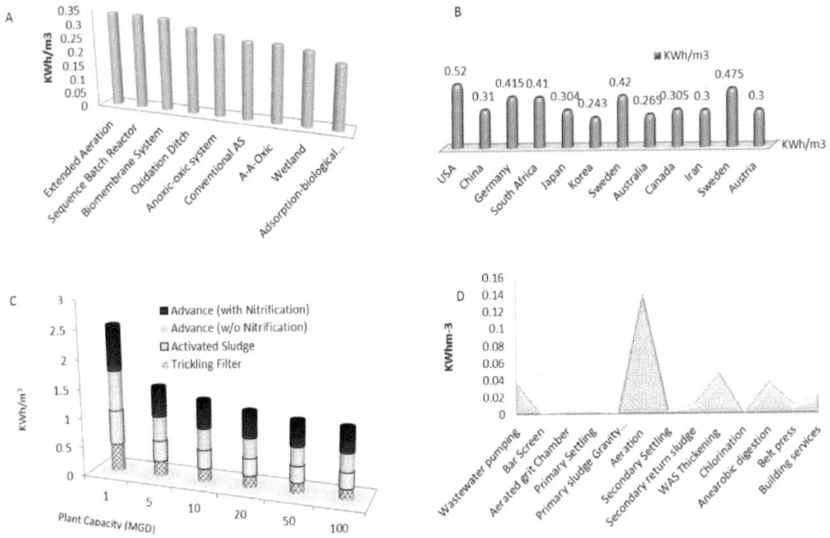

FIGURE 4.1 Specific energy consumption in wastewater treatment; (A) Energy requirement for different biological treatment technologies; (B) Energy consumed by wastewater treatment processes across the globe; (C) Energy consumption intensity for different treatment technologies; (D)-wastewater treatment plants energy intensity across the globe; (D) Specific electricity consumption in individual unit processes (Renan et al., 2017; Goldstein and Smith, 2002).

Specific energy consumption for wastewater treatment depends on the process technology, configuration and treatment goals (Figure 4.1A and Table 4.1). According to Table 4.1, the specific electrical energy consumption for different conventional treatment technologies range from 0.3 to 0.6 kWh/m^3. Adaptation of lagoon or pond type technology yields less energy consumption (0.07–0.3 kWh/m^3); whereas treatment technologies based on mechanical aeration such as oxidation ditch or high purity oxygen and activated sludge processes consume the highest energy (> 1 kWh/m^3).

Advanced technologies such as membrane reactors and extended aeration (EA) systems further increase the specific energy consumption. For instance, the addition of a reverse osmosis (RO) for water reuse will triple or quadruple the plant's energy consumption. Figure 4.1C shows that energy consumption varies depending on the treatment technology and plant capacity. Specific energy consumption is inversely proportional to the plant capacity for plants with capacities under 10 MGD and it does not change significantly beyond that capacity.

Among the few selected countries shown in Figure 4.1B, US plants consumed an estimate of 0.52 kWh of electrical energy for every cubic meter of wastewater treated; this is probably due to the aging infrastructure. The European countries and South Africa have the second highest (> 0.4 kWh/m^3) energy consumption; Australia, Iran, and the Asian countries record the lowest energy requirement (< 0.31 kWh) for treating wastewater (Renan et al., 2017).

WWTP capacity has a significant impact on the specific energy consumption. Figure 4.1C shows the effect of plant capacity on four different treatment technologies. The specific energy consumption for all the different technologies decreases as plant capacity or size increases. In recent years, design, and operation of WWTP has increasingly focused on improving or minimizing energy consumption and reducing cost of operation, without compromising on the treated water quality. Figure 4.1D shows a typical distribution of energy use in a conventional activated sludge (AST) process with treatment capacity of 10 MGD (Goldstein and Smith, 2002). About 44% (0.14 kWh/m^3) of the energy consumed by the wastewater operation is used for biological process such as the aeration tank, followed by waste activated sludge thickening process (~15%), anaerobic digestion and pumping both at 12%.

The application of high efficiency equipment and improvement of design and operation schemes can potentially lower energy consumption and maximize energy recovery. However, if additional energy present in wastewater were captured for use and even less were used for wastewater treatment, then wastewater treatment could become a net energy producer rather than a consumer (Logan, 2005).

TABLE 4.1 Wastewater Treatment Technologies and Their Specific Energy Consumption

Technology Type	Plant Capacity (MGD)	Treatment Technology	Consumption (kWh/m^3)	Source
	4	Activated Sludge Process	0.353	Schwarzenbeck et al., 2008
Conventional	5	CAS with Nitrification	0.509	Goldstein and Smith, 2002
	5	Activated Sludge Process	0.434	Wolfgangsee-Ischl WWTP
	5	Activated Sludge	0.362	Goldstein and Smith, 2002
	15	Activated Sludge Process	0.308	Willis, 2012; Dorr, 2011; Theiszen, 2013
	12	Activated Sludge Process	0.341	USDOE, 2012; Proctor, 2011
	67	Activated Sludge Process	0.447	Joss et al., 2010; Cao, 2011
		Activated Sludge Process	0.33–0.60	Gude, 2015a
Pond		Microalgae Stabilization Pond	0.079–0.28	Wang et al., 2016
		Aeration Ditch	0.48–1.03	Wang et al., 2016
		Lagoon	0.09–0.29	Gude, 2015a
Filter	5	Trickling Filter	0.258	Goldstein and Smith, 2002
	20	Trickling Filter	0.198	Goldstein and Smith, 2002
		Thickening filter	0.19–0.41	Wang et al., 2016
	10.1	Biotower/Activated Sludge	0.392	PG&E, 2003
		Trickling Filter	0.18–0.42	Gude, 2015a
Advanced		Immersed biological membrane reactors	0.8	Ortiz et al., 2007
	0.1	Primary filtration and Trickling Filter	0.087	Gikas, 2016
	5.5	High Purity Oxygen Activated Sludge	1.06	PG&E, 2003
	528.3	Anoxide-anaerobic-oxide	0.13	Kang and Chae, 2013

4.3 ENERGY RECOVERY TRENDS IN WASTEWATER TREATMENT SYSTEMS

Wastewater solids contain approximately 60% (dry basis) of organic compounds; which is 50–55% carbon and mostly biodegradable (in the form of bCOD), 10–15% is nitrogen (as N) and 1–3% is phosphorus (as P) (Gude, 2015b). The energy in the nutritional components of the wastewater such as N and P is approximately 0.7 kWh/m^3 (Chae and Kang, 2013). The energy contained in wastewater solids is 3.2 kJ/g of total solids (TS) (Nowak et al., 2011). The sludge from the primary treatment is reported to contain 15–22.8 kJ/g; secondary treatment is 12.4–16.1 kJ/g; digested sludge contains about 11 kJ/g on a dry mass basis (Figure 4.2) (Zanoni et al., 1982; Gude, 2015b; Shizas and Bagley, 2004). Shizas and Bagley (2004) reported that about 66% of energy content entering the WWTP is captured in the primary sludge, 42% of the remaining energy is retained in the secondary sludge, and the biogas contains 47% of the energy entering the digester. It is apparent that there is enough energy in wastewater (in the form of biogas from the AD) which represents a renewable fuel source that could be converted into electricity and heat. The available thermal heat for heat-pump extraction is about 7 kWh/m^3. According to Figure 4.3, it requires ~1.5 kWh/m^3 to treat 1 kg of COD which contains ~3.9 kWh/m^3 (Chae and Kang, 2013). Similarly, the energy required (and contained) to remove nitrogen and phosphorus are ~13 kWh/m^3 (~19 kWh/m^3) and ~6.44 kWh/m^3 (~2 kWh/m^3) respectively (McCarty et al., 2011).

Energy contained in wastewater can be harvested using various physical chemical and biological processes such as thermal treatment (gasification, incineration, liquefaction, and pyrolysis); composting to produce various valuable biofuels and nutrient-rich biosolids and finally anaerobic digestion (AD). Energy can be recovered from influent organic matter and nutrients, kinetic energy from wastewater flow, and residual heat in treated wastewater (Mo and Zhang, 2013). The most common practice is that, resources recovered in the form of "energy" are used directly by the WWTPs and other facilities reducing environmental loads by WWTPs (Goldstein and Smith, 2002; Wilkinson, 2000). Sometimes, onsite energy generation helps not only reduce energy cost, but also remove the hazardous contaminants in the wastewater and improve treated water quality (Goldstein and Smith, 2002). Some of the technologies used for wastewater resource recovery are combined heat and power (CHP) (EPA, 2007; Stillwell et al., 2010), biosolids incineration, effluent hydropower, onsite wind and solar power, and bioelectrochemical systems (BES).

According to the USEPA, the CHP units produce electricity at a cost below retail price, displace purchase fuel for thermal needs, qualify as a renewal source, reduce carbon footprint, and are reliable for onsite heat and power generation (EPA, 2007). However, CHP unit requires a high capital cost from \$2,000–\$7,500/kW. EPA also reported that the CHPs are cost-effective only for WWTPs with a flow rate above 5 MGD. Stillwell et al. reported that, WWTPs could achieve a reduction of 26% in electricity consumption if CHP is adopted (Stillwell et al., 2010). Anaerobic Digestion (AD) is the key component for energy (biogas) production with a CHP. AD has been tested and demonstrated to be the best option for recovering the maximum energy from primary and secondary sludge of a municipal wastewater treatment plant through energy-rich biogas production. During the anaerobic digestion process, the organic waste is decomposed to CH_4 (60% by volume) and CO_2 (30% by volume) (Tchobanoglous et al., 2003). Hence, electricity is generated by using biogas as a fuel source. Most municipal wastewater treatment utilities incinerate dewatered biosolids as a means of disposal, which requires dewatering prior to incineration. Other biosolids management methods include use as fertilizers or soil stabilizers or disposal in a landfill (Tchobanoglous et al., 2003). According to the US EPA and US DoE, the heating value of biogas is approximately 37.3 kJ/m^3 (550 BTU/ft^3), which is about 60% of the heat value of natural gas. An estimated 628–4,940 million kWh could be saved annually in the US by AD if all WWTPs could use the biogas produced (Stillwell et al., 2010). The use of biogas by individual utilities can result in significant energy savings if done properly. Biogas can be used on-site in different ways, such as generating heat for the process; generating heat for space heating and cooling; powering engines used to drive equipment directly; powering engines used with generators to drive remote equipment; and powering engines used with generators to produce general purpose electrical power (EPA and USDE, 1995).

Biosolids incineration is another technology that is widely employed in most utilities, however, it comes with some major disadvantages which include the release of persistent environmental pollutants, quality inconsistency, and the relatively high capital investment (\$66/dry Mg) and energy cost for dewatering the biosolids (EPA, 2007; Cartmell et al., 2006; Mahmood and Elliott, 2006; Wang et al., 2008).

One of the attractive approaches is onsite application of renewable energy resources such as wind and solar power. This application produces electricity from wind and/or solar energy by taking advantage of the large available land of the WWTPs. Table 4.2 shows a few state-of-the-art WWTPs with onsite wind and/or solar power generation. Location, climate condition, and large capital investments are some of the drawbacks for solar and wind onsite electricity generation.

TABLE 4.2 WWTPs with Solar and Wind Electricity Generation

Technology Integration	Location	Utility Name	Energy Production Potential	Application	References
Solar	CA, USA	Oroville Wastewater Treatment Plant	520 kW	Provide 80% of facility needs	SPG Solar
Solar	CO, USA	Boulder Wastewater Treatment Plant	1000 kW	Provide 15% of facility needs	Boulder, 2012
Solar	NJ, USA	Atlantic County Utilities Authority	500 kW	Provide 660,000 kWh of energy to the facility per year	ACUA, 2011
Wind			7500 kW	Provide 70% of facility needs	ACUA, 2011
Wind	MT, USA	Browning Wastewater Treatment Plant	40 kW	Displace grid electricity used at facility	Browning, 2001

Electricity or energy production via heat pump has been reported to produce 597×10^3 MWh low-temperature heat energy using 199×10^3 MWh electrical energy for a treatment capacity of 119 MGD. Heat pumps are mainly useful when there is a need for onsite heating and cooling within a short-range.

Microbial fuel cell (MFC), a type of bio-electrochemical systems, is another promising technology that has been widely studied over the last 15 years for resource recovery. MFC directly converts microbial metabolic or enzyme catalytic energy into electricity by using conventional electrochemical technology. The technology has the potential for harvesting the energy contained in wastewater; however, it has only been applied on pilot scales for wastewater treatment so far (Allen and Bennetto, 1993; Park and Zeikus, 2000; Roller et al., 1984; Foley et al., 2010; Kim, 2009). Beyond energy generation, another key advantage of the MFC is, it can also reduce the sludge by 20% when compared with the conventional treatment, thereby reducing the sludge disposal costs. However, there are some drawbacks prohibiting the large-scale use of MFC, which include energy loss during the electricity generation process, low organic utilization rates and high capital costs (around 800 times of an anaerobic system) (McCarty et al., 2011; Lui et al., 2004).

Phototrophic technology is another promising technology for onsite or offsite energy generation (Mo and Zhang, 2013). Inorganic or organic carbon and nutrients from wastewater are utilized for microalgae cultivation and microalgae are reported utilize carbon dioxide much faster than conventional biofuel crops (ESMAP, 2008). Currently, integrating the phototrophic technology in WWTPs is still in research phase. The main challenges of this integration include: (a) algal cultivation cost reduction; (b) harvesting, dewatering, and lipid extraction phase energy reduction; and (c) microalgae species selection for optimal performance (ESMAP, 2008).

4.4 CONSIDERATIONS FOR ENERGY-EFFICIENT WASTEWATER TREATMENT

Energy recovered in a plant can directly offset the energy costs of the plant; however, there are several limitations and uncertainties, such as capital costs, lack of reliability and specific requirements for climate and local conditions (Moand Zhang, 2013). In the case of biogas production for a CHP, the major challenges are economic and political factors, which often prevent the direct sale of digester gas. Given that over 90% of WWTPs in the U.S. are small plants, the major challenge is to improve/innovate technologies that have low capital costs, are simple and affordable to operate, and are easy to integrate into the existing small plants (Mo and Zhang, 2013). Life cycle energy benefits associated with reducing and reusing organic and nutrient loadings from wastewater and waste volume for downstream handling are infrequently studied (Mo and Zhang, 2013). Lack of life cycle analysis and studies examining the integration and tradeoffs of energy and resource recovery is another challenge. Studies are needed to evaluate the maximum amount of energy that can be generated onsite with consideration of such integrations and tradeoffs (Mo and Zhang, 2013).

Reducing electricity consumption of WWTP can be approached through improvement of both the hardware (mechanical equipment) and soft technology (process and operation). Among the hardware, the biological process (i.e., aeration facility) is the main electricity consumer and minimizing energy consumption of the aeration process is the key. On the other hand, current sludge regulations on biosolids disposal have become the driving force for municipal wastewater plants to focus on energy recovery. The solids treatment process is another challenge; it significantly affects the cost of buildings and operating a WWTP, which accounts for about 50% of a wastewater plant's capital cost (Joss et al., 2010).

By employing the best available technologies, long term planning for energy self-sufficiency is achievable. The energy intensity of a conventional wastewater treatment plant with nutrient removal and tertiary treatment is assumed to be 0.47 kWh/m^3 for a 10 MGD plant capacity (Goldstein and Smith, 2002). For example, up to 30% of energy savings can be achieved through improvement in the aeration system by selection of higher efficiency equipment and optimal process control. This will result in reduction of specific energy consumption from 0.473 to 0.331 kWh/m^3. Some of the recommended options to achieve energy-positive status are; enhancing primary settling tank performance by harvesting more bCOD to AD; incorporating sludge pretreatment to increase VSS destruction; using high efficiency electrical generators; and co-digestion.

4.5 ENHANCING ENERGY RECOVERY IN WASTEWATER TREATMENT SYSTEMS

Figure 4.2 shows possible ways of energy reduction and production in WWTPs and classifies them as "basic," "moderate" and "advanced" configurations. The

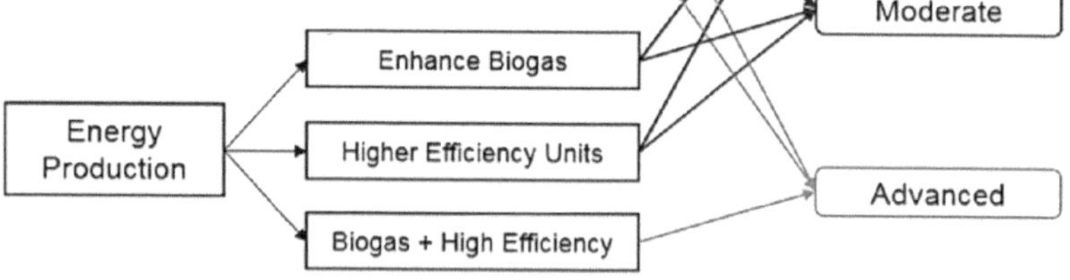

FIGURE 4.2 WRRF Classification-*basic* (possible with process upgrades) configuration consists of traditional wastewater treatment process with no upgrades; *moderate* (possible in near future with upgrades in equipment, process configuration, and treatment scheme) configuration is a modification of the basic process configuration to include the "Anammox" process which focuses on nitrogen removal by using nitrite as electron acceptor and CO_2 as energy source; *advanced* (possible in future, more preliminary work is required) configuration incorporates major process modifications to replace the energy intensive biological process of the basic configuration with a less energy consuming treatment technology such as microalgae systems, trickling filter, etc. This configuration also adopts an advanced primary treatment filtration, which focuses on higher biodegradable solids removal for enhanced energy production).

basic category is mainly focused on improving energy recovery with supplemental biogas production via co-digestion and minor upgrades to minimize energy consumption. The moderate category WRRF employs higher energy efficiency processes and process components to significantly reduce the energy consumption; and the advanced group consists of hypothetical designs that reduce energy consumption and enhance energy production. Some of the energy positive wastewater treatment systems are listed in Table 4.3.

4.5.1 BASIC TECHNOLOGY

As mentioned above, WRRFs in this category are mainly focused on increasing energy production from the AD by mixing wastewater sludge and highly biodegradable waste such as food waste, fat, oil, grease, diary effluents as co-substrates for biogas production. However, they also adopt minor plant modifications to minimize energy consumption as shown in Figure 4.3. Minor changes such as replacing energy-inefficient pumps with higher efficiency (low hp) pumps, changing aeration diffusors to fine bubble diffusors, on-line sensor monitoring or dynamic process control such as DO to minimize excess energy wasting are some of the basic upgrades that can save energy when implemented. Co-digestion of wastewater sludge and other highly biodegradable wastes is another common practice among this group.

The basic configuration quantitative assessment shown in Figure 4.4 assumes an 11% of total energy reduction from equipment upgrade. Wastewater energy content analysis follows literature as discussed in section 3 (Zanoni et al., 1982; Shizas and Bagley, 2004). The total specific energy consumed for the configuration is 0.378 kWh/m^3 (Goldstein and Smith, 2002). The influent conditions assumed were 10 MGD plant capacity, and a typical domestic wastewater concentration of 500 mg/L of COD.

As shown in Figure 4.5, 40% (7571 kg/day) of total COD entering the treatment plant was removed from the primary treatment tank (Tchobanoglous et al., 2003; Rossle and Pretorius, 2001). Out of the 40%, 26% (1968 kg/day) of primary sludge and 7% (48 kg/day) of secondary sludge was converted to biogas (Parkin and Owen, 1986; Miron et al., 2000). The published theoretical chemical energy obtained from converting 1 gram of COD to methane is 13.9 kJ (Heidrich et al., 2010). Thus, the recoverable energy (to electricity) from both the primary and secondary sludge was estimated as the sum of (13.9 kJ/gCOD/3600 kJ/kWh) × 52 g/m^3 from primary treatment sludge and (13.9 kJ/gCOD/3600 kJ/kWh) × 1.26 g/m^3 from secondary sedimentation sludge,

TABLE 4.3 Net Positive (100% Plus) Energy Wastewater Resource Recovery Facility

Location	Plant Name	Plant Capacity (MGD)	Energy Production (Biogas GWh)	Energy Produced-Biogas (kWh/y)	Energy Produced-Biogas (KWh/d)	Energy Produced-Biogas (kWh/m^3)	Anaerobic Digester Feedstock	References
NY, USA	Gloversville-Johnstown Joint WWTP	11	28	28×10^6	76.7×10^3	1.842	PS+WAS+HSW	Ostapczuk, 2011
WI, USA	Sheboygan Regional WWTP	11	32	32×10^6	87.7×10^3	2.105	PS+WAS+HSW+FOG	USDOE-Oregon, 2012; Doerr, 2011; Thieszen, 2013
OR, USA	Gresham WWTP	13	17.2	17.2×10^6	47.1×10^3	0.958	PS+WAS (~0.06 MGD)+FOG	Proctor, 2011
CA, USA	East Bay Municipal Utility District WWTP (EBMUD)	70	90	90×10^6	246.6×10^3	0.931	PS+WAS+HSW+FOG+FW	Williams, 2012; EBMUD
CA, USA	Point Loma WWTP	175	193	19.3×10^6	52.8×10^3	0.798	PS+WAS (~1 MGD)	Wiser et al., 2012; Boranyak, 2012; Greer, 2011; Mazanec, 2013
Germany	Grevesmuhlen WWTP	4	1.95	1.95×10^6	5.3×10^3	0.353	PS (10%)+WAS (60%)+GSIS (30%)	Schwarzenbeck, 2008

Location	**Plant Name**	**Plant Capacity (MGD)**	**Energy Production (Biogas GWh)**	**Energy Produced-Biogas (kWh/y)**	**Energy Produced-Biogas (KWh/d)**	**Energy Produced-Biogas (kWh/m^3)**	**Anaerobic Digester Feedstock**	**References**
Austria	Wolfgangsee-Ischl WWTP	5	3	3×10^6	8.2×10^3	0.434	PS+WAS	Nowak et al., 2011; Nowak et al., 2015
Austria	Strass imZillertal WWTP	6	10	10×10^6	27.4×10^3	1.206	PS+WAS+FOG (0.009 MGD)	Crawford, 2010; Wett, 2007a
Switzerland	Zurich Werdholzli WWTP	67	41.6	41.6×10^6	113.9×10^3	0.449	PS+WAS+FOG	Cao, 2011; Williams, 2012

Note: PS - Primary Sludge; WAS - Waste Activated Sludge; HSW - High Strength Waste; FOG - Fat, Oil and Grease; GIS - Grease Interceptor Sludge; FW - Food Waste

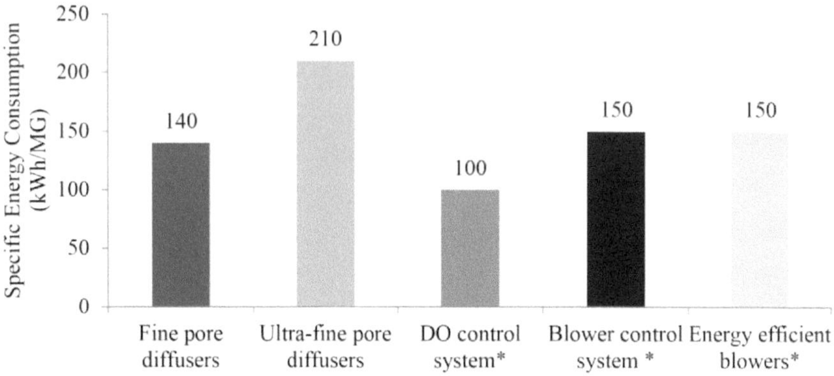

FIGURE 4.3 Potential energy savings on upgrading old process equipment to energy efficient units in a traditional wastewater treatment plant.

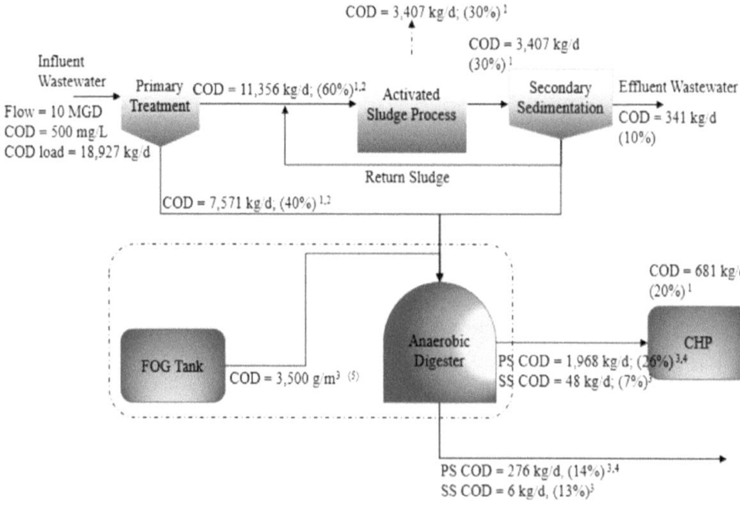

FIGURE 4.4 Basic configuration analysis; Main focus for this configuration is focusing on equipment upgrade and addition of supplemental waste.

which is 0.21 kWh/m^3. This energy production represents almost 48% energy efficiency without equipment upgrade (total specific energy without equipment upgrade is 0.448 kWh/m^3) and approximately 56% with equipment upgrades. The only way for a utility of this kind to meet or even exceed the energy demand is to include supplemental waste for co-digestion. According to John et al., (2009), co-digestion of fat-oil-grease (FOG) with primary and secondary sludge will increase energy production by a factor 2.95 (this represents a

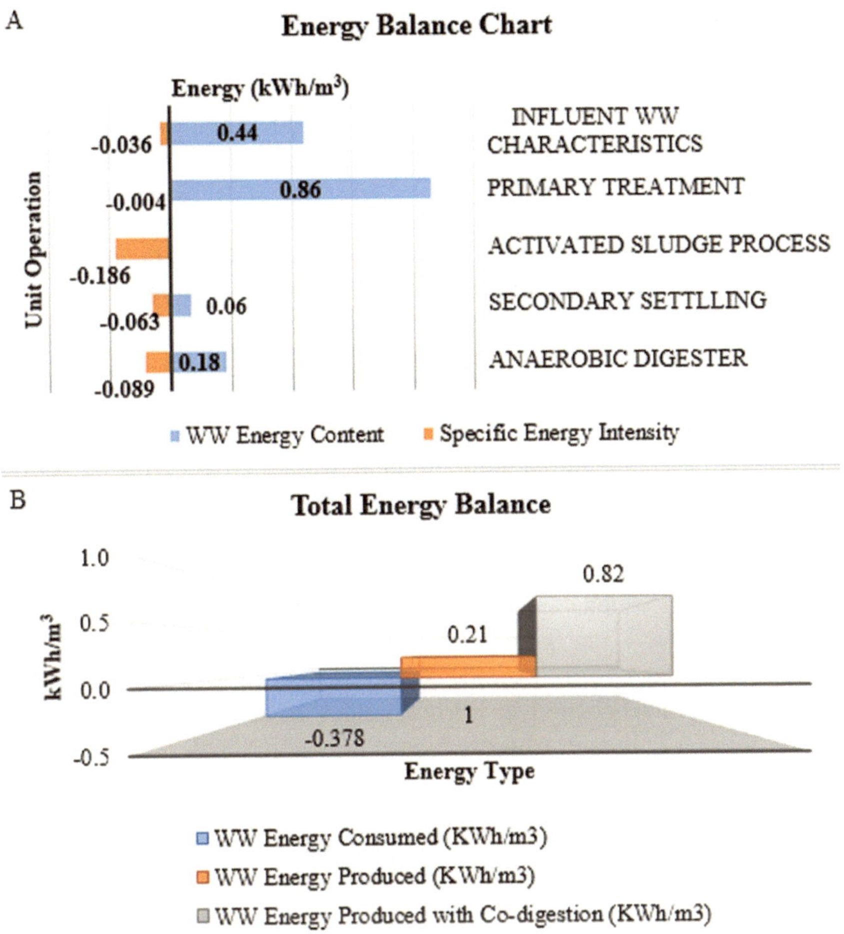

FIGURE 4.5 (A) Energy content and consumption in individual process streams and units, and (B) energy balance of the basic configuration.

soluble COD concentration of 3,500 mg/L). A combined energy production with co-digestion was estimated to be 0.82 kWh/m^3. This puts the plant above it energy by 216% (0.442 kWh/m^3 excess energy to the grid).

4.5.1.1 *CO-DIGESTION AS SUPPLEMENTAL BIOGAS PRODUCTION*

In the U.S., over 251 million metric tons of municipal solid waste are generated annually, among which only 87 million tons are composted or recycled, and the remaining are disposed to the landfill (USDA, USEPA, USDOE,

2014). Landfills are known to be a large source of GHG emissions. Most of the waste disposed in landfills has high organics which could be properly managed by harvesting the energy content of the biodegradable waste, thereby reducing the negative impact landfills have on the environment. Anaerobic treatment process is the alternative for handling the organic fraction collected separately from the municipal solid waste. The anaerobic digestion is the most cost-effective technology that can potentially maximize recycling and recovery of waste components compared to landfills due to the high energy recovery linked to the process and its limited environmental impact.

Co-digestion improves the yield of anaerobic digestion process by enhancing biogas production due to the positive synergism established in the digestion medium and the supply of missing nutrients by the co-substrates (Mata et al., 2000). Co-digestion of wastewater sludge with other organic wastes such as FOG and/or high strength wastes is receiving increasing attention in recent years (Edelmann et al., 2000; Mata et al., 2000, 2011, 2014).

Under mesophilic conditions, fats, oil, and grease (FOG) have high VSS destruction ratio which ranges from 70 to 80% and high reported biogas generation of up to 1.3 m^3/kg VSS destroyed, compared to a normal biosolids gas generation rate of 1 m^3/kg VSS destroyed (Johnson et al., 2009). Biogas production via co-digestion in Europe ranges from 2.5 to 4.0 m^3 biogas/day-m^3 digester tank age, whereas biogas production with only wastewater sludge in the U.S. ranges from 0.9 to 1.1 m^3 biogas/day-m^3 digester tank age (Schafer et al., 2013). The main cause of high VSS/TSS ratio and destruction percentage is due to the high-energy content and COD values of fat and grease as shown in Table 4.4.

TABLE 4.4 Energy Content and COD Values of fats, carbohydrates and proteins

Fats-Oil-Grease	kJ/kg	kg COD/kgVSS
Fat	16.7	3
Carbohydrate	16.7	1.32
Protein	37.7	1.32

The inclusion of FOG or other high strength waste has a synergetic effect on the digestion process as mentioned above, with higher biogas yield than that would be expected by the sum of separate biogas yields from mixture of biosolids and FOG (Cao, 2011). Existing co-digestion utilities have different feedstock ratio; however, digestion operation appears to remain stable with FOG to wastewater sludge ratio of 0.3 of the total digester feed volatile solids.

It has been reported that gas production due to co-digestion can increase by up to 15–30%, which clearly makes a significant contribution to electricity and heat recovery (Johnson et al., 2009). Hence, co-digestion provides advantages such as high biogas yield (Johnson et al., 2009; Edelmann et al., 2000; Mata et al., 2000, 2011, 2014); extra revenue for WWTPs with a tipping fee varying from $50 to $170 per ton (Parry, 2013); and finally, co-digestion enhances anaerobic digestion performance with an integrated waste-to-energy process using mixed feedstock sourced locally (Yanwen et al., 2015). However, there are a few challenges that hinder the co-digestion application such as upgrading existing facilities to incorporate the various waste feedstock; high variability of co-digestion feedstock composition and volume; and digester overloading rates and potential toxicity of other wastes (Yanwen et al., 2015; Bond et al., 2012; Long et al., 2012; Zhang et al., 2014; USEPA, 2006; Chapman and Krugel, 2011; Ganidi et al., 2009; Kougias et al., 2014).

4.5.1.2 CASE STUDIES

The East Bay Municipal Utility District (EBMUD) located in the San Francisco Bay area, covers a service area of 83 square miles with more than 600,000 residential and 20,000 commercial customers (EBMUD, 2012, 2014). Activated sludge process with a design capacity of 415 MGD and daily average flow of 70 MGD is employed. The EBMUD has implemented energy-efficient measures since 1980's by upgrading equipment and modifying process methods to maintain high-quality standards while reducing energy costs (EBMUD, 2012, 2014). As shown in Figure 4.6, the Utility upgraded their influent pumps (five 700 Hp) and effluent pumps (four 1000 Hp) to high efficiency pumps equipped with variable-frequency drives which lowered their influent and effluent pumps energy requirement by at least 50% which amount to $273,000 in savings (EBMUD, 2012, 2014). Addition of plastic balls to the oxygen production vaporizer pit has saved the plant over $2 million annually by preventing heat and evaporation losses (EBMUD, 2012, 2014).

The modifications shown in Figure 4.6 were improved by using a distributed control system to pace influent pump flow, a control water storage unit, and selective pumping patterns to take advantage of lower utility rates. These improvements reduced the total plant energy costs by about 60%. The estimated annual cost of electricity is $1.9 million (~ 30 kWh) but the plant yields a combined annual savings of approximately $2.8 million from $2 million in revenue from FOG/food waste tipping fee (EBMUD, 2012, 2014; Horenstein, 2013).

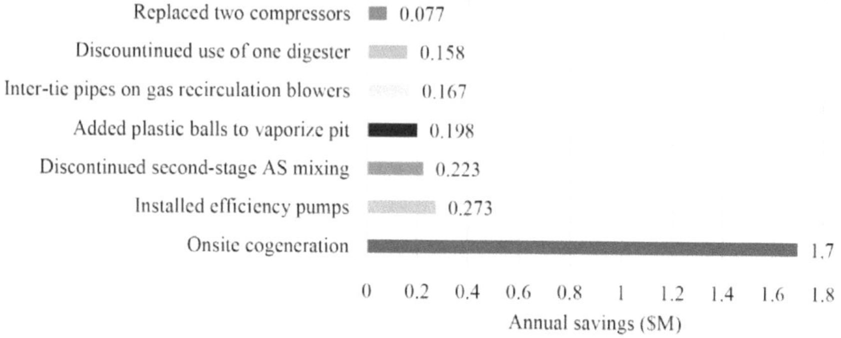

FIGURE 4.6 Summary of upgrades implemented at the EBMUD WWTP.

The Gloversville-Johnstown Joint Wastewater Treatment Plant (GJJWWTP) serves the Cities of Gloversville and Johnstown in New York. The utility upgraded its treatment process by adding a second stage secondary treatment and solids handling facility in 1988 (ASERTTI, 2009; Ostapczuk, 2011). The plant serves a combined population of about 25,000 including 24 industries. GJJWWTP has a design capacity of 13 MGD with a daily average flow of 11 MGD. In 1990, the New York State Energy Research and Development Authority (NYSERDA) sponsored an energy audit to identify energy consuming components and recommend solutions for reduction (ASERTTI, 2009; Ostapczuk, 2011). Upgrading aeration equipment saved the utility nearly $0.2 million energy costs annually. Increasing biogas production by co-digestion saved the plant roughly $0.3 million annually in energy costs. In addition, other minor upgrades such as changing high energy consuming mechanical equipment such as motors and pumps with high efficiency equipment reduced the overall energy demand (ASERTTI, 2009; Ostapczuk, 2011).

The Sheboygan regional wastewater treatment plant (SRWWTP) in Sheboygan, Wisconsin, provides wastewater treatment services to almost 68,000 people in the surrounding area. SRWWTP treats a daily wastewater flow average of approximately 10 MGD with peak design capacity of 58.6 MGD (Willis et al., 2012; Wiser et al., 2012). In 2002, the city of Sheboygan carried out a plant-wide audit to identify opportunities to minimize the WWTP's energy consumption. After a thorough evaluation, several upgrades were made to minimize the energy consumption. The sludge boilers were replaced with higher efficiency boilers cutting natural gas consumption by 78%, which saved an average of $0.155 million of operation costs annually. High efficiency pumps with variable frequency drives (VFD) were installed which saved the utility about 30% in energy consumption. Modified aeration units, centrifugal air compressors saved about 20% of aeration energy. Overall, SRWWTP was able to save roughly 20% (Willis et al., 2012; Wiser et al., 2012).

Gresham WWTP (Oregon) serves a population of about 114,000 with an average daily flow of 13 MGD. About 8% of Gresham's electric power consumption demand is generated from a set of 420 kW solar panels (Steve, 2016; USDOE-Oregon, 2012). In 2012, the plant increased their biogas production by incorporating FOG as co-substrate for digestion and made some upgrades such as adding a new high efficiency neuros blowers and aeration diffusers. A 80 hp gas mixer with a 5 hp linear motion mixers (5 hp each) reduced the plant's energy consumption by over 6.5% (Steve, 2016; USDOE-Oregon, 2012).

4.5.2 MODERATE CATEGORY

This category combines high-energy production and highly energy-efficient processes. The advantage of this technology is that, more energy can be produced while at the same time consuming less energy. Reducing oxygen demand will inherently reduce the amount of energy. Replacing aeration diffusers, applying an anoxic-aerobic configuration and higher COD removal from the primary settling tank are a few examples of reducing the amount of oxygen required for aeration. However, the anaerobic ammonium oxidation process, popularly known as ANAMMOX or DEMON, has been proven to be the most effective way of reducing the oxygen demand and energy requirements. A moderate configuration is shown in Figure 4.7.

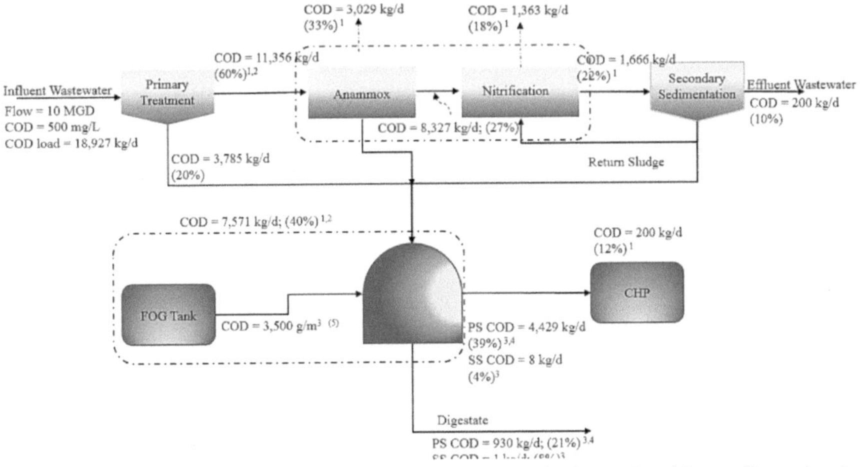

FIGURE 4.7 Moderate flow configuration analysis; main focus for this configuration is modification to the basic configuration to include the Anammox technology.

The anammox process uses nitrite as an electron acceptor and CO_2 as the energy source and was first reported by Mulder et al. in 1995 (WEF, Wett, 2006, 2007a, b). This technology was developed in Delft University of Technology and can reduce energy up to 63% (Siegrist et al., 2008; Lackner et al., 2014). Strass WWTP in Austria was the first plant to implement the anammox process on a full-scale. Chesapeake Bay watershed, Alexandria Sanitation Authority WRF, and HRSD's James River WWTP, were also upgraded to implement DEMON (Jin et al., 2012). However, Hampton Roads Sanitation District (HRSD)'s utility is the first full-scale anammox-based deammonification process in the U.S. for side stream nitrogen removal (Nifong et al., 2013; Daigger, 2011).

A *moderate* configuration including anammox-nitrification was analyzed (Figure 4.8). The assessment assumes 11% of total energy reduction for equipment upgrade, and a plant capacity of 10 MGD with an influent typical wastewater COD concentration of 500 mg/L. The specific energy consumption was estimated to be 0.34 kWh/m^3. As less COD load is available for cell synthesis, less aeration energy is required (0.145 kWh/m^3 compared to 0.182 kWh/m^3 for the basic configuration). Figure 4.8 shows a 40% (7571 kg/day) COD removal from the primary treatment tank; the anammox process reduces the COD entering the biological unit by 33% (equivalent to 3029 kg/day). A combined 43% of the AD sludge is converted to biogas. As mentioned above, the theoretical chemical energy obtained from converting 1 gram of COD to methane is 13.9 kJ. Hence, the estimated energy production is 0.45 kWh/m^3. This alone makes the plant energy-positive (133%) without co-digestion. The inclusion of the anammox process makes it even more suitable for the implementation of co-digestion; because of the nutrient recycle from the AD. Thus, by adding FOG as a supplemental waste to co-digest with sludge from primary and secondary settling units, a total energy production of 1.81 kWh/m^3 was estimated. This represents roughly 532% of energy efficiency.

4.5.2.1 CHALLENGES

Anammox process presents several key challenges. First, the process is susceptible to process instabilities that can occur during startup or even after extended periods of stable operation (Joss et al., 2011; Wells et al., 2013; Jin et al., 2012; Rosenthal et al., 2009). In addition, anammox bacteria have low growth rates, low cellular yields, and are sensible to adverse environmental conditions (Tang et al., 2013). Second, factors such as dissolved oxygen, several heavy metals, sulfide, salt, and toxic organic matter (antibiotics, phenol) are inhibitory to anammox. Substrates of the anammox process such as NO_2^- and NH_4^+, can act as inhibitors. In addition, studies show that free ammonium and free nitrous acid have negative effects on anammox bacteria (Cema et al., 2013). Additional research is required to address these concerns including the need for infrastructural modification, and lack of skilled operators (Han et al., 2014).

4.5.2.2 CASE STUDY

The Strass Im Zillertal WWTP is located in Austria, with a treatment capacity of 10 MGD. It serves a population that ranges from approximately 60,000

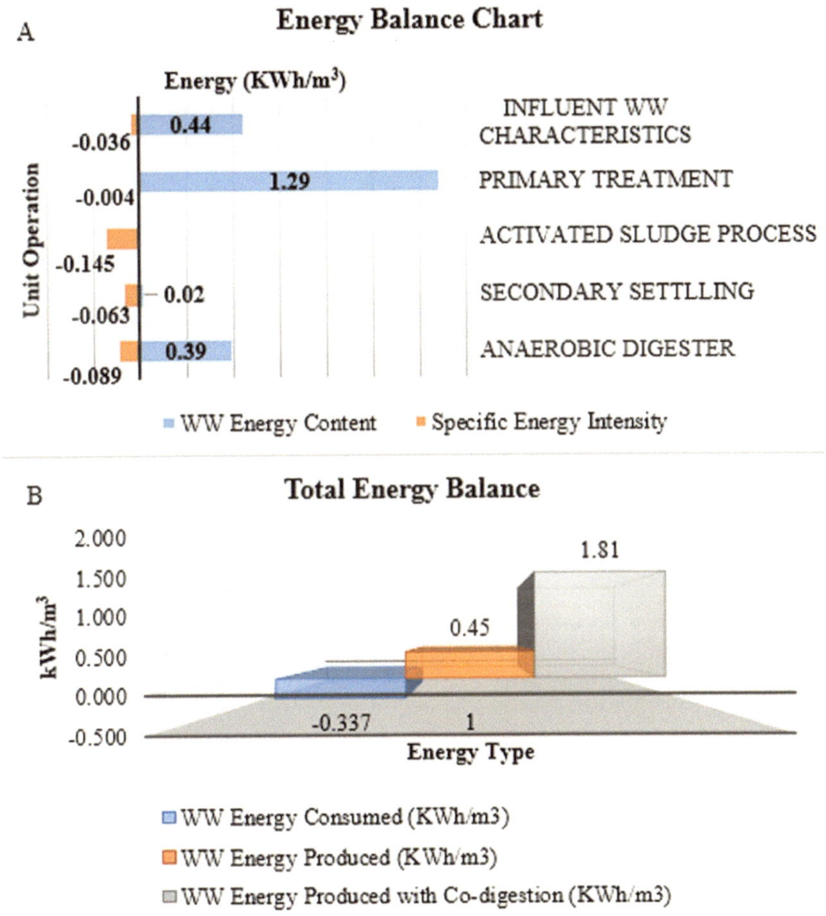

FIGURE 4.8 (A) Energy content and consumption in individual process streams and units, and (B) energy balance of the advanced configuration.

in the summer to 250,000 during the winter tourist season. The population constitutes roughly 31 communities in the Achental and Zillertal, Austria. The Strass plant has mastered the skills of maximizing energy recovery and energy efficiency and conservation. Since the inception of the utility in 1999, the plant has progressively implemented high-level technological upgrades as shown in Figure 4.9 (Wett, 2007a). The plant was able to minimize the energy costs from approximately 6.5 €/kg NH_4-N removed to 2.9 €/kg NH_4-N in the span of 5 years. This was achieved by optimizing the dissolved oxygen levels and by implementing ultra-high efficiency strip aeration. In addition, by implementing the novel side stream DEMON process, the plant saved 44% of energy required for nitrification/denitrification (Wett, 2007b). Energy self-sufficiency improved steadily from 49% in 1996 to 108% in 2005 without the need for implementation of co-digestion. The upgrades at the Strass plant did not only make the plant highly efficient in terms of energy consumption and production, but also reduced the total carbon footprint by up to 40% (Wett, 2007b).

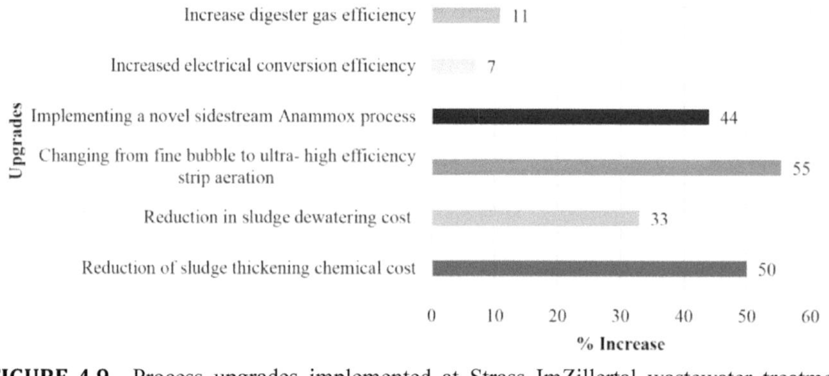

FIGURE 4.9 Process upgrades implemented at Strass ImZillertal wastewater treatment plant to increase energy production.

4.5.3 *ADVANCED CATEGORY*

Aeration accounts for 40–60% of the total energy consumption and presents as the limiting factor for reducing energy consumption in a WWTP (Cao, 2011). The *advanced* configuration seeks to eliminate the aeration needs and becomes a decisive step to bypass conventional aerobic treatment (Gikas, 2016). The *advanced* configuration consists of a hypothetical process scheme leading to energy-positive wastewater treatment. Many existing plants will consider equipment upgrades and retrofitting rather than complete replacement of the process which can be achieved through *basic* and *moderate* configurations. In the *advanced* configuration, the focus is shifted towards maximizing the energy content of wastewater sludge streams and minimizing energy consumption. A few net-positive wastewater treatment schemes are presented here.

The municipal wastewater entering the treatment plant contains 10 times the energy required to treat the waste (WERF, 2011; USDOE, 2014). In a AS process, the biodegradable organic carbon contained in a primary sludge is higher than the biological sludge which is highly digested in the aeration tank. Increasing the removal of biodegradable organic carbon in the primary treatment stage could potentially increase biogas production and at the same time reduce the energy consumption by lowering the amount of oxygen required by the heterotrophic microorganism for cell synthesis. Based on this fact, a novel energy efficient WWTP was proposed by Petros Gikas. The proposed configuration consists of a combination of separation processes (i.e., advanced microsieving and filtration) for upfront solids removal, along with downstream low-energy biological filtration process (trickling filter and encapsulated denitrification) for carbon and nitrogen removal. The key to this technology is to remove as much total suspended solids (TSS) as possible and utilize biosolids in energy production (Gikas, 2016).

The proposed technology uses a proprietary rotating fabric belt Micro-Screen with pore size ranges from 100–300 mm, followed by a proprietary Continuous Backwash Upflow Media Filter or cloth media filter. The preliminary/primary treatment achieves about 80–90% reduction in TSS and 60–70% (a 30% to 45% dry solid cake) reduction in BOD_5 (Gikas, 2016). The estimated energy consumption for by micro-sieving with an auger press and primary filtration (cloth or sand media filters) were 0.005 kWh/m^3 and 0.010 kWh/m^3 respectively. As shown in Figure 4.10 energy needed for complete municipal wastewater treatment per unit volume of inlet raw wastewater was estimated to be 0.057 kWh/m^3 (or 0.087 kWh/m^3 if UV disinfection was selected). This energy requirement is about 85% lower when compared to the AST process. The biosolids produce about 0.172 kWh/m^3 of net electric energy through biogas (Gikas, 2016).

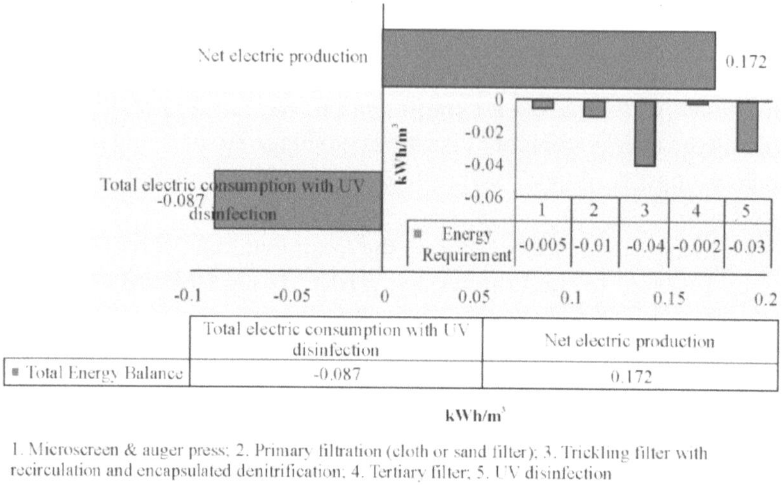

FIGURE 4.10 Energy analysis of novel energy positive WWTP proposed by Petros Gikas.

McCarty proposed a hypothetical anaerobic treatment system for energy recovery and efficient treatment (McCarty et al., 2011). The proposed scheme includes an anaerobic fluidized membrane bioreactor (AFMBR), which combines a membrane process with an anaerobic fluidized bed reactor (AFBR) (McCarty et al., 2011). The AFMBR uses granular activated carbon (GAC) for suspended biofilm attachment. An estimated energy consumption of 0.058 kWh/m^3 is required with the influent wastewater BOD_5 concentration of 500 mg/l and a detention time of 5 hours. The low energy consumption estimated for the AFMBR is similar to the energy consumption rate proposed by Petros Gikas (0.057 kWh/m^3 without UV disinfection). Another similarity between Petros' and McCarty's hypothetical configurations is the high amount of primary sludge going to the AD. For phosphorus (P) removal, chemical precipitation or conversion into struvite (NH_4MgPO_4 $6H_2O$) for recovery as fertilizer has been proposed (Aiyuk et al., 2004; Bashan and Bashan, 2004). Other applications such as source separation of urine has been applied in Europe to remove N (McCarty et al., 2011).

Carbon, nitrogen, phosphorus are the main ingredients in wastewater and are the key nutrients for microalgae growth (Eve et al., 2012). Shoener et al. discussed an integrated heterotrophic (anaerobic) and phototrophic (microalgae) bioprocess as an energy-positive scheme (Shoener et al., 2014). Some of the major limitations of aerobic or anaerobic treatment such as nitrogen and phosphorus removal could be addressed by phototrophic technologies while increasing the energetic potential of wastewater resources by leveraging nutrients for biomass growth and organic carbon storage. Shoener et al. proposed a two stage biological treatment; anaerobic stage for COD removal and microalgae for N and P removal. Various anaerobic technologies such as; anaerobic baffled reactor (ABR), anaerobic membrane bioreactor (AnMBR), AFB, upflow anaerobic sludge blanket (UASB), anaerobic sequencing batch reactor (ASBR), microbial electrolysis cell (MEC), and microbial fuel cell (MFC) have been heavily investigated in recent years. Among these technologies, ABR was preferred with an estimated COD removal of 90.3% and a total energy recovery of 47.5%. For nutrient removal, phototrophic technologies such as high rate algal pond (HRAP), photobioreactor (PBR), stirred tank reactor, waste stabilization pond (WSP), and algal turf scrubber (ATS) were evaluated. HRAP was the preferred choice for N and P removal following the anaerobic treatment. The estimated energy consumption for mixing and harvesting for the HRAP process were 3.2–9.6 KJ/m^3 and 34–170 kJ/m^3 respectively.

A combined anaerobic treatment (ABR) and phototrophic (HRAP) system replacing a biological nutrient removal (BNR) process using the Strass WWTP configuration was proposed as a base model (Shoener et al., 2014). The authors proposed that, if this configuration is employed; the total plant biogas could be increased by 39%. After an extensive review, the following conclusions were made. First, it was concluded that combined anaerobic and phototrophic processes could reduce energy demand and achieve energy recovery and production on the order of 5.0–9.2 kWh/m^3 (using higher values for UASBs and PBRs)-well above the whole-plant energy demand of conventional WWTPs (0.3–0.6 kWh/m^3) (Logan, 2005; Zanoni et al., 1982; Shoener et al., 2014). Secondly, phototrophic processes have the potential to produce 280–400% of the amount of energy as anaerobic processes on a per m^3 basis; and finally, it was stated that using nutrients in phototrophic biomass cultivation may result in 130–510% of excess energy production.

4.6 CONCLUSION

This chapter discussed the energy consumption and recovery trends in wastewater treatment. Various approaches to reduce energy consumption and to enhance energy production were also discussed. Based on the level of technological feasibility, three different wastewater treatment configurations such as *basic*, *moderate*, and *advanced* were proposed which include various approaches for energy reduction, equipment upgrade, and advanced biological processes.

Co-digestion of mixed waste for energy production was suggested. However, there are a few challenges that need be addressed by upgrading existing facilities to incorporate various types of waste, high variability of co-digestion feedstock composition and volume, and digester overloading. Other challenges encountered with co-digestion are the disposal of mixed sludge generated from FOG with inconsistent characteristics. The additional sludge generation creates storage issues in winter season. The low pH of FOG causes corrosion issues. Supernatant from the AD with loaded FOG residue promotes growth of undesirable filamentous microorganism in the AS process which causes effluent problems. Despite these limitations, economic, and energy benefits of FOG addition are still attractive to WWTPs.

Moderate and advanced configurations incorporate anammox and phototrophic and anaerobic treatment technologies along with substantial equipment and process upgrades. Moderate configurations are feasible in the near future with more practical demonstrations of actual pilot-scale units. Some of these changes can be done by retrofitting the new and high-efficiency equipment and by modifying the process configurations. For advanced configurations involving the use of anaerobic and or biofilm technologies, advanced separation processes, co-digestion, and phototrophic treatment systems, more research and practical considerations are warranted for their successful application in the future water resource recovery facility design and operations. Life cycle analysis and techno-economic feasibility studies are also important to determine the overall feasibility of these proposed configurations.

KEYWORDS

- **algal turf scrubber**
- **anaerobic baffled reactor**
- **anaerobic digester**
- **biological nutrient removal**
- **combined heat and power**
- **fat-oil-grease**

REFERENCES

ACUA, (2011). *Atlantic County Utilities Authority, Green Initiatives*. http://www.acua.com/acua//content.aspx?id=474 (accessed on 10 June 2020).

Aiyuk, S., Amoako, J., Raskin, L., Van, H. A., &Verstraete, W., (2004). Removal of carbon and nutrients from domestic wastewater using a low investment, integrated treatment concept. *Water Res., 38*(13), 3031–3042.

Alessandro, S., & Joan, G., (2019). Microalgae and bacteria dynamics in high rate algal ponds based on modeling results: Long-term application of BIO_ALGAE model. *Science of the Total Environment, 650*(2) 1818–1831.

Allen, R. M., &Bennetto, H. P., (1993). Microbial fuel cells: Electricity production from carbohydrates. *Appl. Biochem. Biotechnol., 39/40*, 27–40.

ASERTTI, (2009). *Combined Heat and Power Case Study: Gloversville-Johnstown Joint Wastewater Treatment Facility*.

Bond, T., Brouckaert, C. J., Foxon, K. M., & Buckley, C. A., (2012). A critical review of experimental and predicted methane generation from anaerobic codigestion. *Water Sci. Technol., 65*, 183–189.

Boranyak, S., (2012). *Wastewater Plant Turns Food into Electricity* (p. A2). Civil Engineering. Reston, VA: American Society of Civil Engineers.

Boulder, (2012). *City of Boulder, Colorado, Wastewater Facility Solar Electric System*. http://www.bouldercolorado.gov/index.php?option¼com_content&view¼article&id¼8763&Itemid=3482 (accessed on 10 June 2020).

Browning, (2001). *Wind Power for the Wastewater Treatment Plant*. http://www.browningmontana.com/wind.html (accessed on 10 June 2020).

Camacho, P., Li, T., Martin, S., &Dauthuille, (2009). *Evaluation of New Strategies for the Improvement of Anaerobic Digestion, IWA SIWW09*. Singapore.

Cao, Y. S., (2011). *Mass Flow and Energy Efficiency of Municipal Wastewater Treatment Plants*. London, UK: IWA Publishing.

Cartmell, E., Gostelow, P., Riddell-Black, D., Simms, N., Oakey, J., Morris, J., Jeffrey, P., et al., (2006). Bio-solids a fuel or a waste? An integrated appraisal of five co combustion scenarios with policy analysis. *Environ. Sci. Technol., 40*, 649–658.

Cema, G., Schneider, Y., Beier, M., &Rosenwinkel, K. H., (2013). *Proceedings from WEFTEC2013*. Chicago, USA.

Chae, K. J., & Kang, J., (2013). Estimating the energy independence of a municipal wastewater treatment plant incorporating green energy resources. *Energy Convers. Manage, 75*, 664–672.

Chapman, T., &Krugel, S., (2011). Rapid volume expansion: An investigation into digester overflows and safety. *Proceedings of the Water Environment Federation* (pp. 1016–1040). Sacramento, CA, Water Environment Federation.

Crawford, G. V., (2010). *Best Practices for Sustainable Wastewater Treatment: Initial Case Study Incorporating European Experience and Evaluation Tool Concept* (p. 74). Alexandria, VA: Water Environment Research Foundation; IWA Publishing.

Daigger, G. T., (2011). Implementation of a full-scale anammox-based facility to treat an anaerobic digestion side stream at the Alexandria sanitation authority water source facility. *Water Pract. Technol., 6*, 1, 2.

De-Bashan, L. E., & Bashan, Y., (2004). Recent advances in removing phosphorus from wastewater and its future use as fertilizer (1997–2003). *Water Res., 38*(19), 4222–4246.

Doerr, D., (2011). *ACEEE Case Study—Sheboygan, WI: Energy Efficiency in Wastewater Treatment Plant.* Washington DC http://aceee.org/sector/local-policy/case-studies/sheboygan-wastewater-treatment-plant (accessed on 10 June 2020).
Doerr, D., (2011). *ACEEE Case Study—Sheboygan, WI: Energy Efficiency in Wastewater Treatment Plant.* Washington DC http://aceee.org/sector/local-policy/case-studies/sheboygan-wastewater-treatment-plant (accessed on 10 June 2020).
EBMUD, (2012). *Resource Recovery Program and Biogas Turbine Renewable Energy Project.* PowerPoint Presentation. http://www.energy.ca.gov/2012_energypolicy/documents/2012-02-16_workshop/presentations/05_Hake_East_Bay_Municipal_Utilities_District.pdf (accessed on 10 June 2020).

EBMUD, (2014). *Wastewater Rates, Charges, and Fees.* https://www.ebmud.com/water-and-wastewater/rates-and-charges/wastewater-ratescharges-and-fees#trucked fees (accessed on 10 June 2020).

EBMUD.*East Bay Municipal Utility District's Wastewater Treatment Plant: Wastewater and Energy*https://www.energy.ca.gov/process/pubs/ebmud.pdf (accessed on 10 June 2020).

Edelmann, W., Engeli, H., &Gradenecker, M., (2000). Co-digestion of organic solid waste and sludge from sewage treatment. *Water Sci. Technol., 41*, 213–221.

Elías-Maxil, J. A., Peter, V. D. H. J., Hofman, J., & Rietveld, L., (2014). Energy in the urban water cycle: Actions to reduce the total expenditure of fossil fuels with emphasis on heat reclamation from urban water. *Renew, Sustain, Energy Rev., 30*, 808–820.

Enrica, U., Bruno, S., Jérôme, H., Anaïs, B., & Jean, P. S., (2018). CO_2 addition to increase biomass production and control microalgae species in high rate algal ponds treating wastewater. *Journal of CO_2 Utilization, 28*, 292–298.

EPA & USDE, (1995). *Case Studies in Residential Use and Energy Conservation at Wastewater Treatment Plants.*

EPA, (2007). *Opportunities for and Benefits of Combined Heat and Power at Wastewater Treatment Facilities.*

EPA, (2007). *Opportunities for and Benefits of Combined Heat and Power at Wastewater Treatment Facilities.* EPA-430-R-07-003.

ESMAP, (2008). *Good Practices in City Energy Efficiency: Eco2 Cities.* Integrated resource management in Stockholm. http://www.esmap.org/esmap/sites/esmap.org/files/CS_Stockholm.pdf. accessed on 10 June 2020).

Foley, J. M., Rozendal, R. A., Hertle, C. K., Lant, P. A., &Rabaey, K., (2010). Life cycle assessment of high-rate anaerobic treatment, microbial fuel cells, and microbial electrolysis cells. *Environ. Sci. Technol., 44*, 3629–3637.

Frijns, J., Hofman, J., &Nederlof, M., (2013). The potential of (waste) water as energy carrier. *Energy Convers. Manage, 65*, 357–363.

Funamizu, N., Iida, M., Sakakura, Y., &Takakuwa, T., (2001). Reuse of heat energy in wastewater: Implementation examples in Japan. *Water Sci. Technol., 43*, 277–285.

Galbraith, K., (2011). *How Energy Drains Water Supplies.*New York Times.http://www.nytimes.com/2011/09/19/business/global/19iht-green19.html?_r=1 (accessed on 10 June 2020).

Ganidi, N., Tyrrel, S., &Cartmell, E., (2009). Anaerobic digestion foaming causes: A review. *Bioresour. Technol., 100*, 5546–5554.

Goldstein, R., & Smith, W., (2002). *ICF Consulting Associates, Electric Power Research Institute, Water, and Sustainability: US Electricity Consumption for Water Supply and Treatment-the Next Half Century* (Vol. 4). Electric Power Research Institute.

Goldstein, R., & Smith, W., (2002). *Water and Sustainability US Electricity Consumption for Water Supply and Treatment-the Next Half Century* (Vol. 4). Electric Power Research Institute.

Greer, D., (2011). *Directed Biogas to Power Fuel Cells: A Change in California's Public Utility Commission Rule to Allow Injection of Purified Biomethane into Natural Gas Pipelines Opened the Door for On-Site Power Projects in San Diego* (pp. 47–50). Bio-cycle. Emmaus, PA: The JG Press Inc.

Gude, V. G., (2015a). Energy and water autarky of wastewater treatment and power generation systems. *Renew. Sustain. Energy Rev., 45*, 52–68.

Gude, V. G., (2015b). Energy positive wastewater treatment and sludge management. *Edorium. J. Waste Manag., 1*, 10–15. www.edoriumjournals.com/ej/wm (accessed on 11 June 2020).

Han, G., Yaniv, D. S., & George, F., (2014). Wells. Towards energy neutral wastewater treatment: Methodology and state of the art. *Environ. Sci.: Processes Impacts, 16*, 1223.

He, Q., Wang, H., Xu, C., Zhang, J., Zhang, W., Zou, Z., & Yang, K., (2016). Feasibility and optimization of wastewater treatment by chemically enhanced primary treatment (CEPT): A case study of Huangshi. *Chemical Speciation and Bioavailability, 28*, 209–215.

Heidrich, E. S., Curtis, T. P., &Dolfing, J., (2010). Determination of the internal chemical energy of wastewater. *Environ. Sci. Technol., 45*, 827–832.

Hiroko, Y., Jacob, M., & Charlotte, S., (2014). Plant-integrated measurement of greenhouse gas emissions from a municipal wastewater treatment plant. *Water Research, 61*, 108–118.

Horenstein, B. K., (2013). *East Bay Municipal Utility District Sustainability/Energy Committee Staff Report*. Oakland, CA: Food Waste Update.

Ioannis, S., & David, B. M., (2004). Experimental determination of energy content of unknown organics in municipal wastewater streams. *Journal of Energy Engineering,130*(2), 45–53.

Jin, R. C., Yang, G. F., Yu, J. J., & Zheng, P., (2012). *Chem. Eng. J., 197*, 67–79.

John, C. K., Ulas, T., Spyros, G. P., Mike, E., James, D., Robert, A. G., & Allen, C. T., (2009). Methane recovery from the anaerobic co-digestion of municipal sludge and FOG. *Bioresource Technology, 100*, 3701–3705.

Johnson, T. L., Scanlan, P. A., Yurtsever, D., &Kuchenrither, R. D., (2009). *State of Practice: Biosolids Energy and Resources Recovery, Water Convention, SIWW09*. Singapore.

Joss, A., Cyprien, C., Burger, S., Blunschi, M., Zuleeg, S., & Siegrist, H., (2010). *Sludge liquid Treatment with Combined Nitritation/Anammox*. Neptune Meeting, Ghent.

Joss, A., Derlon, N., Cyprien, C., Burger, S., Szivak, I., Traber, J., Siegrist, H., &Morgenroth, E., (2011). *Environ. Sci. Technol., 45*, 9735–9742.

Kampschreur, M. J., Temmink, H., Kleerebezem, R., Jetten, M. S. M., & Van, L. M. C. M., (2009). Nitrous oxide emission during wastewater treatment. *Water Res., 43*, 4093–4103.

Kang, J., &Chae, K. J., (2013). Estimating the energy independence of a municipal wastewater treatment plant incorporating green energy resources. *Energy Conversion and Management, 75*, 664–672.

Kim, B. H., (2009). *Microbial Fuel Cell* (Vol. 2, pp. 4–8). KISToday.

Kougias, P. G., Boe, K., Kristensen, S. O. T., &Angelidaki, L. A. I., (2014). Anaerobic digestion foaming in full-scale biogas plants: A survey on causes and solutions. A journal of the international association on water pollution research. *Water Sci. Technol.,69*, 889–895.

Kum-Lok, H., Cheon-Hee, B., & Kyung-Duk, Z., (2016). Characteristics of methane and nitrous oxide emissions from the wastewater treatment plant. *Bioresource Technology, 214*, 881–884.

Lackner, S., Gilbert, E. M., Vlaeminck, S. E., Joss, A., Horn, H., & Van, L. M. C., (2014). Full-scale partial nitritation/anammox experiences: An application survey. *Water Res., 55*, 292–303.

Larissa, T. A., Ivet, F., Diederik, P. L. R., Stijn, W. H., Van, H., & Marianna, G., (2019). *The Effect of Primary Treatment of Wastewater in High Rate Algal Pond Systems: Biomass and Bioenergy Recovery*. Bioresource Technology, in press, accepted manuscript.

Liu, H., Ramnarayanan, R., & Logan, B. E., (2004). Production of electricity during wastewater treatment using a single chamber microbial fuel cell. *Environ. Sci. Technol., 38*, 2281–2285.

Logan, B. E., (2005). Simultaneous wastewater treatment and biological electricity generation. *Water Science and Technology, 52*(1/2), 31–37.

Long, J. H., Aziz, T. N., Delos, R. F., &Ducoste, J. J. III. L., (2012). Anaerobic co-digestion of fat, oil, and grease (FOG): A review of gas production and process limitations. *Process Saf. Environ., 90*, 231–245.

Mahmood, T., & Elliott, A., (2006). A review of secondary sludge reduction technologies for the pulp and paper industry. *Water Res., 40*, 2093–2112.

Malin, J., (2007). *Energy Benchmark for Wastewater Treatment Processes-A*.

Mata-Alvarez, J., &Llabres, S. M. P., (2000). Anaerobic digestion of organic solid wastes. An overview of research achievement sand perspectives. *Bioresour. Technol., 74*, 3–16.

Mata-Alvarez, J., Dosta, J., Mace, S., &Astals, S., (2011). Co-digestion of solid wastes: A review of its uses and perspectives including modeling. *Crit. Rev. Biotechnol., 31*, 99–111.

Mata-Alvarez, J., Dosta, J., Romero-Guiza, M. S., Fonoll, X., Peces, M., &Astals, S., (2014). A critical review on anaerobic co-digestion achievements between 2010 and 2013. *Renewable Sustainable Energy Rev., 36*, 412–427.

Mata-Alvarez, J., Mace, S., &Llabres, P., (2000). Anaerobic digestion of organic solid wastes. An overview of research achievements and perspectives. *Bioresource Technology, 74*, 3–16.

Mazanec, F. J., (2013). *Point Loma Wastewater Treatment Plant (PLWTP)* (p. 2013). California Energy Commission staff workshop on challenges to procuring biomethane in California. Sacramento, CA: California Energy Commission.

McCarty, P. L., Bae, J., & Kim, J., (2011). Domestic wastewater treatment as a net energy producer-can this be achieved? *Environmental Science and Technology, 45*(17), 7100–7106.

Menger-Krug, E., Niederste-Hollenberg, J., Thomas, H., & Harald, H., (2012). Integration of microalgae systems at municipal wastewater treatment plants: Implications for energy and emission balances. *Environ. Sci. Technol., 46*, 11505–11514.

Miron, Y., Zeeman, G., Van, L. J. B., &Lettinga, G., (2000). The role of sludge retention time in the hydrolysis and acidification of lipids, carbohydrates, and proteins during digestion of primary sludge in CSTR systems. *Water Res., 34*, 1705–1713.

Monteith, H., Kalogo, Y., &Louzeiro, N., (2007). *Achieving Stringent Effluent Limits Takes a Lot of Energy! WEFTEC 2007*. San Diego, California, USA.

Mulder, A., Van, D. G. A. A., Robertson, L. A., & Kuenen, J. G., (1995). *FEMS Microbiol. Ecol., 16*, 177–185.

Nifong, A., Nelson, A., Johnson, C., &Bott, C. B., (2013). *Performance of a Full-Scale Side-Stream DEMONs Deammonification Installation*. 86th Annual water environment federation technical exhibition and conference.Chicago, IL: Water Environment Federation.

Nowak, O., Enderle, P., &Varbanov, P., (2015). Ways to optimize the energy balance of municipal wastewater systems: Lessons learned from Austrian applications. *J. Clean Prod., 88*, 125–131.

Nowak, O., Keil, S., &Fimml, C., (2011). Examples of energy self-sufficient municipal nutrient removal plants. *Water Science and Technology, 64*, 1–6.

Ortiz, M., Raluy, R. G., & Serra, L., (2007). Life cycle assessment of water treatment technologies: Wastewater and water-reuse in a small town. *Desalination, 204*, 121–131.

Oshita, K., Okumura, T., Takaoka, M., Fujimori, T., Appels, L., &Dewil, R., (2014). Methane and nitrous oxide emissions following anaerobic digestion of sludge in Japanese sewage treatment facilities. *Bioresour. Technol., 171*, 175–181.

Ostapczuk, R. E., (2011). *Achieving Zero Net Energy Utilization at Municipal WWTPs: The Gloversville-Johnstown Joint WWTP Experience.* Water Environment Federation.

Pakenas, L. J., (1995). *Energy Efficiency in Municipal Wastewater Treatment Plants, Technology Assessment.* New York State energy research and Development Authority.

Park, D. H., &Zeikus, J. G., (2000). Electricity generation in microbial fuel cells using neutral red as an electronophore. *Appl. Environ. Microbiol., 66*, 1292–1297.

Parkin, G. F., & Owen, W. F., (1986). Fundamentals of anaerobic digestion of wastewater sludge's. *J. Environ. Eng., 112*, 867–920.

Parry, D., (2013). *Improving Economics of Co-Digestion* (pp. 48–49). BioCycle Emmaus, PA: The JG Press Inc.

Petros, G., (2016). Towards energy positive wastewater treatment plants. *Journal of Environmental Management, XXX*, 1–9.

PG&E, (2003). *New Construction Energy Management Program, Municipal Wastewater Treatment Plant Energy Baseline Study.* http://baseco.com/Publications/Wastewater%20 Treatment%20Baseline%20Study.pdf (accessed on 10 June 2020).

Piate, E., Camacho, P., Ewert, W., Kopp, J., Panter, K., & Perez-elvira, S. I., (2009). *Combined Experience of Thermal Hydrolysis and Anaerobic Digestion Latest Thinking on Hydrolysis of Secondary Sludge Only.* Water Convention, SIWW 09. Singapore.

Proctor, P., (2011). *Achieving Energy Independence at the Gresham Wastewater Treatment Plant* (p. S02). PA. Water World. West Chester.

Randall, A. A., Benefield, L. D., & Hill, W., (1997). Induction of phosphorus removal in an enhanced biological phosphorus removal bacterial population. *Wat. Environ. Res., 31*(11), 2869–2877.

Renan, B. S., Marina, S. M., Regiane, P. R., & Ricardo, F. G., (2017). Comparative analysis of the energy consumption of different wastewater treatment plants. *International Journal of Architecture, Arts and Applications, 3*(6), 79–86.

Roller, S. D., Bennetto, H. P., Delaney, G. M., Mason, J. R., Stirling, J. L., & Thurston, C. F., (1984). Electron-transfer coupling in microbial fuel cells: 1. comparison of redoxmediator reduction rates and respiratory rates of bacteria. *J. Chem. Technol. Biotechnol., 34*, 3–12.

Rosenthal, A., Ramalingam, K., Park, H., Deur, A., Bechmann, K., Chandran, K., &Fillos, J., (2009). *Proceedings from WEFTEC2009.* New Orleans, Louisiana.

Rossle, W. H., & Pretorius, W. A., (2001). A review of characterization requirements for in-line prefermenters-Paper 1: Wastewater characterization. *Water Sa., 27*, 405–412.

Schafer, P., Muller, C., & Willis, J., (2013). *Improving the Performance and Economics of Co-Digestion and Energy Production* (pp. 793–803). 86th Annual water environment federation technical exhibition and conference. Chicago, IL: Water Environment Foundation.

Schwarzenbeck, N., Pfeiffer, W., &Bomball, E., (2008). Can a wastewater treatment plant be a power plant? *A Case Study Water Sci. Technol., 57*, 1555–1561.

Shoener, B. D., Bradley, I. M., Cusick, R. D., & Guest, J. S., (2014). Energy positive domestic wastewater treatment: The roles of anaerobic and phototrophic technologies. *Environ. Sci.: Processes Impacts, 16*, 1204.

Siegrist, H., Salzgeber, D., Eugster, J., & Joss, A., (2008). Anammox brings WWTP closer to energy autarky due to increased biogas production and reduced aeration energy for N-removal. *Water Sci. Technol., 57*, 383–388.

SPG Solar. *Sewerage Commission e Oroville Region*. http://www.spgsolar.com/markets-served/government-and-education/water-and-sanitation/seweragecommission-oroville-region/ (accessed on 10 June 2020).

Steve, F., (2016). *Gresham's Infrastructure: Financing a Sustainable Future*. City of Gresham Public Works, Water Finance Forum.

Stillwell, A. S., Hoppock, D. C., & Webber, M. E., (2010). Energy recovery from wastewater treatment plants in the United States: A case study of the energy-water nexus. *Sustainability 2*, 945–962.

Tang, C. J., Zheng, P., Chai, L. Y., & Min, X. B., (2013). Thermodynamic and kinetic investigation of anaerobic bioprocesses on anammox under high organic conditions. *Chemical Engineering Journal, 230*, 149–157.

Tchobanoglous, G., Burton, F. L., &Stensel, D. H., (2003). *Metcalf and Eddy: Wastewater Engineering, Treatment and Reuse* (4thedn.). McGraw Hill, NY, USA.

Thieszen, S., (2013). Sheboygan's journey to sustainability. In: *Wisconsin State Energy Office Half-Day Seminar: Turning Waste to Cash in Wisconsin*. Madison, WI2013.

USDA, USEPA, USDOE, (2014). *Biogas Opportunities Roadmap: Voluntary Actions to Reduce Methane Emissions and Increase Energy Independence*. Washington DC.

USDOE, (2014). *The Water-Energy Nexus: Challenges and Opportunities*. Washington, DC: US Department of Energy.

USDOE-Oregon, (2012). *Bioenergy Optimization Assessment at Wastewater Treatment Plants*. Portland, OR: Oregon Department of Energy;.

USEPA, (2006). *Biosolids Technology Fact Sheet: Multi-Stage Anaerobic Digestion*. Washington DC: United States Environmental Protection Agency.

Wang, H., Brown, S. L., Magesan, G. N., Slade, A. H., Quintern, M., Clinton, P. W., &Payn, T. W., (2008). Technological options for the management of biosolids. *Environ. Sci. Pollut. Res., 15*, 308–317.

Weiwei, M., &Qiong, Z., (2013). Energy-nutrients-water nexus: Integrated resource recovery in municipal wastewater treatment plants. *Journal of Environmental Management, 127*, 255–267.

Wells, G. F., Shi, Y. J., Joss, A., Burgman, H., Johnson, D. R., &Morgenroth, E., (2013). *Proceedings from 9th International Conference on Biofilm Reactors*. Paris, France.

WERF, (2009). *Energy Conservation in Water and Wastewater Treatment Facilities*. Manual of practice No.: 32. WEF Press, Alexandria, VA, USA.

WERF, (2011). *Energy Production and Efficiency Research: The Road Map to Net-Zero Energy*. Alexandria, VA: Water Environment Research Foundation.

Wett, B., (2007b). Development and implementation of a robust demmonification process. *Wa. Sci. and Technol., 56*(7), 81–88.

Wett, B., Buchauer, K., &Fimml, C., (2007). Energy self-sufficiency as a feasible concept for wastewater treatment systems. *Asian Water*, 0219–2365.

Wett, B., Buchauer, K., &Fimml, C., (2007a). *Energy Self-Sufficient as a Feasible Concept for Wastewater Treatment Systems.* Leading-Edge Conference, Singapore.

Wett, R., (2006). Solved up scaling problems for implementing deammonification of rejection water. *Water Science and Technology, 53*(12), 121–128.

Wilkinson, R., (2000). *Methodology for Analysis of the Energy Intensity of California's Water Systems, and an Assessment of Multiple Potential Benefits Through Integrated Water-Energy Efficiency Measures.* California Institute of Energy Efficiency. No. 4910110.

Williams, D. R., (2012). Renewable energy at EBMUD. In: *2012 CAS A Mid-Year Conference.* Palm Desert, CA.

Willis, J., Stone, L., Durden, K., Beecher, N., Hemenway, C., & Greenwood, R., (2012). *Barriers to Biogas Use for Renewable Energy.* Alexandria, VA: Water Environment Research Foundation.

Wiser, J. R., Schettler, J. W., & Willis, J. L., (2012). *Evaluation of Combined Heat and Power Technologies for Wastewater Facilities.* Atlanta GA; Brown and Caldwell.

Yan, X., Li, L., & Liu, J., (2014/2016). Characteristics of greenhouse gas emission in three full-scale wastewater treatment processes. *J. Environ. Sci., 26*, 256–263, 881–884.

Yanwen, S., Jessica, L. L., Urgun-Demirtas, M., Marianne, M. M., & Seth, W. S., (2015). An overview of biogas production and utilization at full-scale wastewater treatment plants (WWTPs) in the United States: Challenges and opportunities towards energy-neutral WWTPs. *Renewable and Sustainable Energy Reviews, 50*, 346–362.

Zabranska, J., Dohanyos, M., Kutil, J., &Jenicek, P., (2009). Intensification of anaerobic digestion towards the energy self-sufficiency of municipal wastewater treatment. *Water.*

Zanoni, A. E., & Mueller, D. L., (1982). Calorific value of wastewater plant sludges. *J. Environ. Eng. Div. (Am. Soc. Civ. Eng.), 108–109*, 187–195.

Zhang, C., Su, H., Baeyens, J., & Tan, T., (2014). Reviewing the anaerobic digestion of food waste for biogas production. *Renewable Sustainable Energy Rev., 38*, 383–392.

CHAPTER 5

Thermal Energy Recovery in Wastewater Treatment Plants

VIOLA SOMOGYI*, VIKTOR SEBESTYÉN, ENDRE DOMOKOS, and SYED MUHAMMAD HASSAAN ALI

University of Pannonia, Veszprém, Hungary

**Corresponding author: somogyiv@uni-pannon.hu*

ABSTRACT

The wastewater treatment plants (wwtp) of today have two main goals, to meet the quality requirements of the effluent in order to mitigate environmental pollution and to keep operation costs as low as possible. Though the technology is considered to be end-of-pipe, waste heat recovery can play an important role to help plants become sustainable from an energy point of view, along with improving the feasibility of other resource recovery technologies. Detailed energy balance is set up using a benchmark for wastewater treatment plants to determine the sinks and sources of heat. The examined system consists of the biological train removing organic matter and nitrogen completed with an anaerobic digester and auxiliary facilities such as offices but waste heat from machinery was not considered. Three summer and three winter scenarios are considered. For all versions, the effluent temperature of the settler is higher than that of the influent, resulting in excess heat that may be recovered. The following solutions for thermal energy recovery are covered in more detail: biogas produced by digesters and the heat content of treated wastewater. The lowest amount of thermal energy that could be theoretically reclaimed for the effluent is approximately 320 kW, considering the benchmark data, which is higher than the heat from the combined heat and power (CHP) generation plant.

Options to utilize the reclaimed energy directly or with heat exchangers and heat pumps are also discussed. As an example, space heating with excess heat from the effluent with heat pump is calculated showing that a small fraction of the influent flowrate is capable of providing space heating for the auxiliary buildings of a plant leaving sufficient amount of heat that could be fed into an existing district heating system. Other options, such as cooling buildings and stabilizing the water temperature of smaller biological tanks are also examined.

Lastly, the theoretical heat recovery potential of the European wastewater treatment sector is calculated and visualized at NUTS2 level, as this is missing from the current estimations of waste heat potentials. Hotspots, where wastewater heat recovery could be beneficial are identified by GIS based models and the overall potential is calculated, which is 2.65 larger than the renewable municipal waste capacity of North America.

5.1 INTRODUCTION

Conventional wastewater treatment processes are mostly end-of-pipe technologies that only focus on removing pollutants before discharge (Bixio et al., 2006) and resource recovery is in many cases unsolved (Sheik et al., 2014). Reasons for that may be a lack of proper policy (van der Hoek et al., 2016) or high operation costs (Boiocchi et al., 2017). A survey in Italy showed that less than 40% of municipal wastewater treatment plants (WWTPs) implemented some sort of resource recovery option (Papa et al., 2017). Reuse of wastewater also faces challenges (Drechsel et al., 2015) and was considered as an option in the past only where physical water scarcity had to be combated, for example in Israel (Friedler et al., 2006) or Namibia (Lahnsteiner and Lempert, 2007). According to the calculations of Elías-Maxil et al. (2014), 80% of the energy used in the urban water cycle is spent on heating by the end-user. On the other end of the chain, the electric energy demand of activated sludge processes was calculated to be between 0.31 and 0.67 kWh/m^3 (Metcalf-Eddy, 2003) provided by mostly not renewable sources. The two most energy-intensive processes are aeration and additional sludge treatment (Gu et al., 2017). More advanced technologies may be even more energy-intensive (WWAP, 2014). For example, if organic trace materials would be removed by means of granular activated carbon (GAC), additional electric energy besides the plant's needs as high as 1.78 kWh/m^3should be accounted for (Mousel et al., 2017). Also, phosphorous recovery solutions tend to increase energy demand (Amann et al., 2018).

One option for WWTPs to become energy neutral is to reduce energy consumption by implementing energy-efficient technologies while another is to recover energy from renewable and/or untapped energy sources available in WWTPs. For example, by installing state-of-the-art aeration equipment and advanced DO control up to 35% of the total energy can be saved (Longo et al., 2016), though one has to take the risk of greenhouse gas (GHG) formation into account as well (Larsen, 2015). Energy generation within the plant to meet the demand may be achieved by using cogeneration plants fuelled with biogas from the sludge, sludge incineration, installing turbines to utilize effluent hydropower, heat pumps to recover low-temperature heat from the wastewater and microbial fuel cells (MFCs), besides harvesting wind and solar power on the site (Mo and Zhang, 2013). In this chapter, thermal energy recovery will be discussed as it may be a solution to foster energy independence along with improving the feasibility of other resource recovery technologies. First, the different heat sources are introduced so that a relatively detailed heat balance could be set up. For this purpose, the calculation is carried out by using a benchmark for WWTPs. After assessing the amount of heat that could be utilized in the sample utility, solutions to utilize the excess heat of the effluent are examined. Lastly, the potential of the European wastewater treatment sector is calculated and visualized by maps.

5.2 HEAT SOURCES IN A WASTEWATER TREATMENT PLANT

Heat can be regained from the water and the solids line, as well. Wastewater heat falls into the category of low-grade heat, which can be utilized with heat pumps. Several examples of operational systems exist which will be discussed in chapter 5.2.2.

Sludge can be digested to produce biogas, incinerated, pyrolyzed, or gasified after dewatering. In the following subchapter, anaerobic digestion will be discussed as this is a more common approach (Raheem et al., 2018). Also, waste heat from machinery may be utilized via heat exchangers (HE). The temperature of compressed air is around 333 K (Wanner et al., 2005) and the temperature in the compressor house can be around 323 K which may be suitable for direct or indirect heat exchange (Miah et al., 2015).

5.2.1 DIGESTERS

Nowadays, anaerobic digestion (AD) is the most used process to recover energy from wastewater as biogas (Silvestre et al., 2015). AD is the decomposition of organic matter in the absence of oxygen by using anaerobic microorganisms and which results in about 60–70% methane, 30–40% carbon-dioxide, <1% nitrogen and traces of H_2S (0.001–0.2%) in the gas phase (Osorio and Torres,

2009) and organic residues with relatively high nitrogen and phosphorous content in the liquid phase. The conversion rate depends on the design and temperature of the anaerobic digester (AD) (Han et al., 2017).

A primary aim of ADs is to reduce the amount of sludge to be disposed of while producing biogas to partially or fully satisfy the energy demands of removing pollutants (McCarty et al., 2011; Jenicek et al., 2013). Biogas can be efficiently transformed into electricity and heat in cogeneration plants or solid oxide fuel cells (Mosayeb Nezhad et al., 2018) or serve as raw material to produce biofuel or syngas. In the case of gas boiler systems, the biogas can be fed directly to the boilers if the H_2S content is lower than 1000 ppm, while in gas turbines and gas engines higher quality gas is required (De Mes et al., 2003). Using biogas as a source of energy may result in 25–50% less external electric energy demand (Tyagi and Lo, 2013; Fernández-Arévalo et al., 2017). The primary settler plays an important role in achieving the reduction by separating the energy rich suspended fraction of the wastewater (Frijns, 2013). This option is quite common in large WWTPs, while scarce attention is given to exploitation of biogas in medium (12%) and small (not even one) plants based on the research of Papa et al. (2017). The internal demand of electricity at the WWTP makes it interesting to investigate whether it is possible to produce electricity at a lower price than buying it from the energy market.

With regards to operating temperature, two main categories may be distinguished: mesophilic (303–313 K) and thermophilic (323–333 K) regimes (Kim et al., 2002). Thermophilic digestion has certain advantages over mesophilic, such as better biogas production, an increased destruction rate of organic solids, improved solids-liquid separation, and increased destruction of pathogenic organisms (Kardos et al., 2011). Nonetheless, mesophilic processes are preferable due to higher supernatant quality, better process stability, and lower process energy demand for reactor heating (Gebreeyessus and Jenicek, 2016). Working with high-rate mesophilic anaerobic digestion may increase the hydrolysis rate of slowly degradable organic matter (Bakhshi et al., 2018) and also reduce retention time (Bolzonella et al., 2012). Operating ADs at lower temperature (283–298 K) would require lower energy expenditure but would result in low rate of anaerobic digestion and biogas production. A short period of treatment at higher temperature (308 K) would remedy this issue (Zhang et al., 2016).

ADs require energy for maintaining temperature, pumping, and mixing (Panepinto et al., 2016). Fresh digester sludge can be preheated using heat from electricity production (Astals et al., 2012). Puchajda and Oleszkiewicz

(2008) calculated the heat requirement for the digester that works at 308.15 K with sludge retention of 20 days and found that 34% of biogas production would be used for digester heating. Nonetheless, in most of the cases, thermal treatments do not need external energy sources and the energy balance is positive, since the required energy is offered by the electricity production. By applying multicriteria optimization, the total cost rate and environmental impacts may be decreased by 16.9% and 5.3%, respectively (Lee et al., 2017). In some cases though, for example in cold climate, thermophilic process cannot provide sufficient energy to be considered self-heating (Carrère et al., 2010).

To improve biogas production, it is important to improve the rate of sludge degradability. Pre-treatment of sludge by physical, chemical, thermal, mechanical, or biological means can reduce the quantity of macromolecules (Ruffino et al., 2015), and biogas generation can be increased by 10 to 145% (Gianico et al., 2013). While some pre-treatment methods may seem energy intensive, such as ultrasound or lysis-centrifuge the energy output is higher (Tyagi and Lo, 2013). Of course, the biogas yield highly depends on the influent characteristics such as the COD/TN/TP ratio (Fernández-Arévalo et al., 2017). Co-digestion (adding energy-rich organic wastes) also increases biogas production (Cavinato et al., 2013).

5.2.2 WASTEWATER HEAT

The heat content of the wastewater comes from the waste heat from households. Depending on the energy-efficiency measures, up to 40% of total energy used in households is lost with sewage (Hepbasli et al., 2014). This value is close to ten times higher than the combined amount of energy required in the drinking water and wastewater industry (Hofman et al., 2011). The wastewater is released into the sewer system at a temperature between 283 K (Hepbasli et al., 2014) to 313 K (Wong et al., 2010) depending on the water usage. There is substantial heat loss in the pipes depending on the size of the system. This effect is less significant in areas where heat island effect is present (Menberg et al., 2013). In the wastewater treatment facility heat is produced by the biological processes (Grady et al., 2011), while losses occur due to the open water surfaces of the biological train (Fernández-Arévalo et al., 2014). Daily fluctuation in wastewater temperature is small, the coefficient is 0.9–1.05 (Cipolla and Maglionico, 2014a), but the seasonal change is more substantial under moderate climate. Influent temperature in winter time may vary between 276 and 300 K (Tchobanoglous et al., 2013) and is a factor of geographical location, the amount of the wastewater and the length of the sewer system. The lower values may be detected in combined sewer systems during snow-melts (Plósz et al., 2009). The temperature of wastewater plays an important role in the treatment as the microorganisms of activated sludge systems are active within 283–313 K (van Haandel and van der Lubbe, 2012). The rate of nitrification is doubled every 7 K (Henze et al., 2008) but because the saturation concentration of dissolved oxygen is inversely proportional to the temperature, the operation optimum is considered to be at 293–298 K (Tchobanoglous et al., 2013). In continental climate, the microorganisms adapt to seasonally changing circumstances resulting in lower performance, thus reduced efficiency of nutrient removal in the colder periods. Additionally, as there is no regulation for municipal wastewater on the temperature difference between the effluent and the receiving water body, the latter is prone to heat pollution as well.

On the other hand, reclaiming the excess heat may provide a continuously renewed energy source (Neugebauer et al., 2015) which can be utilized with heat-pumps. Wastewater source heat pump (WWSHP) systems can be used for both heating (e.g., area heating, hot water production, etc.), and cooling processes. Utilization of such WWSHP systems reduces carbon emission, heat emission to the environment, air pollution, and electricity consumption (Culha et al., 2015). Wastewater heat recovery can be achieved in three ways:

- In a building (wastewater from bathroom and kitchen);
- From raw wastewater in the urban wastewater distribution line; and
- In a wastewater treatment plant from treated wastewater.

For the first version, HE have to be installed individually which might be difficult in existing buildings and special care has to be taken to avoid biofilm formation (Seybold and Brunk, 2013). Nonetheless, if hot water is produced from warm greywater (GW) with a heat pump, the highest potential for reducing greenhouse gas (GHG) emissions may be achieved at household level (Larsen, 2015), though this would require thoughtfully planned investments from individuals.

In the case of the second version, the water temperature cannot go below a certain minimum to allow the microorganisms to proliferate in the treatment facility. If the temperature decreased by 1°C permanently, the aerobic sludge retention time (SRT) must be increased by 10% to maintain the nitrification capacity (Wanner et al., 2005). Thus, careful design is needed to use this solution to avoid overcooling, but there are several examples for it (e.g., Erhorn et al., 2018). If heat recovery systems were to be installed in the sewer network, the residential heat demand of up to 35% could be satisfied in Brussels (Spriet and Hendrick, 2017). It was also pointed out that the temperature drops up to 4 K on one kilometer (Abdel-Aal et al., 2014), until it reaches the ground temperature, thus in cases of longer sewer systems, heat, that would otherwise dissipate, can be reclaimed near the source achieving a reduction of 40% in heating cost (Culha et al., 2015).

As for the third version, by implementing a heat pump after the treatment processes, the recovered heat may be used for heating communal buildings within the facility, providing hot water, sludge drying, or for district heating/cooling. Effluent water heat recovery may be relatively easier compared to raw wastewater due to lower interference of solid matter (Chae and Kang, 2013), which decreases the amount of biofilm growth on the heat exchanger surface (Wanner et al., 2005). Also, the quantity of wastewater is the largest after the treatment plant and therefore has a substantial amount of thermal energy-in Austria the approximate thermal energy potential was calculated to be 3200 GWh/a (Kretschmer et al., 2016). Swedish studies have shown that decreasing the temperature of wastewater by one degree via recovering heat can bring 720 GWh of energy savings annually (Mikkonen et al., 2014). It is estimated that there are over 500 wastewater heat pump systems worldwide with thermal ratings between 10 kW–20 MW (Hepbasli et al., 2014), for instance in Germany, Switzerland, and the Scandinavian countries (Seybold, 2013). GHG emission may be reduced significantly, for example, the emission of a bivalent wastewater heat pump system is 22% of an oil-fired heating system in CO_2 equivalent (Schmid, 2008). Zhao et al. (2010) showed that the heating coefficient of performance (COP) is about 4.3, and the cooling COP is about 3.5 in actual operating conditions which is comparable to the COP of a ground source heat pump (Wołoszyn and Gołaś, 2017; Bedoić and Filipan, 2018). Regarding Austria, a total of 45 MW of space heating with 9 MW of electricity input may be achieved for district heating by using wastewater as the low-temperature heat source of heat pumps (Nowak et al., 2015).

One of the largest facilities in the world is in Stockholm, Sweden, receiving a wastewater discharge of 4,000–18,000 m^3/h. Seven heat pumps with a total capacity of 225 MW are used to recover heat from treated wastewater, producing 1235 GWh of heat annually. This amount of energy is used to heat and cool 95,000 residential buildings (Mikkonen et al., 2014).

Spatial analysis allows evaluating the feasibility of integrating a wastewater treatment plant (WWTP) into the local energy supply system (Kollmann et al., 2017). Wastewater heat pump systems are usually installed in larger facilities (GEA, 2007). Small municipal WWTPs are prone to decreased water temperature due to relatively long retention time in the sewers and the fluctuation of flow rate as a result of variation in daily and weekly residential water use (Langergraber et al., 2008). Since the volume of wastewater is less but varies more, the system is subject to abrupt changes. Unfortunately, they also often lack sophisticated monitoring systems which may result in inefficient operation or even unwanted environmental pollution. Somogyi et al. (2018a) explored the possibility of applying heat pumps to utilize the excess wastewater heat in the biological train and found that increasing the water temperature by 6 K would improve nitrification and denitrification efficiencies by 60% and 17%, based on the modeling results of a small Hungarian WWTP.

5.3 HEAT BALANCE IN MUNICIPAL WASTEWATER TREATMENT PLANTS (WWTPS)

In this chapter, the heat balance of a typical wastewater treatment plant will be discussed to map the sinks and sources of thermal energy. Though energy usage in WWTPs depends on size, design, operation, the composition of WW and local factors, such as legislation and climate (Silvestre et al., 2015), a pattern can still be drawn. Heat is produced by the biological processes in activated sludge tanks, electric devices and by burning biogas to produce electricity. On the other hand, energy is required to preheat the sludge for digestion and heat losses may occur due to evaporation, convection, and radiation. In this version of energy balance, electricity, and indirect energy (i.e., the energy required to produce equipment) is not accounted for.

The scheme of the wastewater treatment plant used as example is shown in Figure 5.1. Its characteristics are based on the benchmark simulation model no. 2 (BSM2) (Alex et al., 2018) which describes a facility capable of nitrogen and organic matter removal, using digesters to reclaim energy with a size of 137,653 population equivalent (PE). In the figure, auxiliary units are represented beside the devices directly used for treatment. Heat balance is calculated for the biological tanks, the secondary settler, and the digester. The heat exchange in the additional units is neglected as the hydraulic retention time in these is relatively small. The heat demand of the offices and auxiliary buildings are calculated in a simplified manner according to the threshold values of energy performance coefficients stated in Hungarian legislation (7/2006 (V. 24.) Decree of Minister without Portfolio about determination of energy efficiency of buildings). Meteorological data are estimated resembling humid continental climate (Cfb according to the updated Köppen classification (Kottek et al., 2006)).

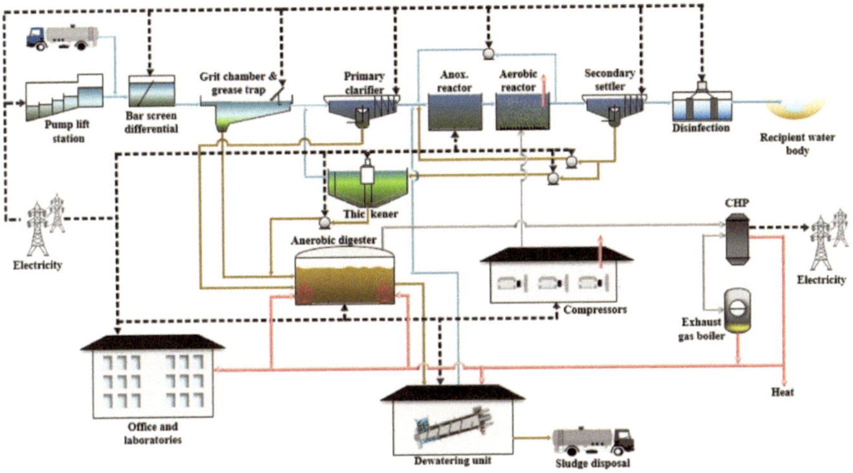

FIGURE 5.1 Schematic representation of the wastewater treatment plant used to assess the heat balance of the facility. Blue and brown lines represent water and sludge, red, and grey are for heat flows and gas streams while dotted lines show electricity demand and supply.

5.3.1 HEAT BALANCE OF THE ACTIVATED SLUDGE PROCESS

Calculations in this subchapter are based on the works of Corbala-Robles et al. (2016) with a few modifications. In their article, they set up the heat balance for a facility treating the liquid fraction of manure after centrifugation. This type of wastewater is extremely rich in nutrients thus the biological processes produce substantial amount of heat and often cooling is required. Since municipal wastewater is more diluted and the danger of overheating is not present, heat recovery or in other words cooling is not considered in the biological train. Also, as foaming is a sign of malfunction in municipal wastewater systems (Fryer and Gray, 2012), the insulating effect of foam was neglected as opposed to the Corbala-Robles et al. (2016) article.

The biological tank is treated as a completely mixed reactor with constant volume. In BSM2, two anoxic and three aerated compartments are distinguished but in this case, they are not calculated separately. Nonetheless, the volume fractions are taken into consideration when calculating the effect of aeration. The change in heat content can be summarized by Eqn. (5.1). The density and specific heat capacity of the influent, basin, and effluent were assumed to be the same and constant through time. Flow rate changes due to evaporation were neglected. The nomenclature along with the chosen values of parameters is shown in Table 5.1.

$$\Delta H = H_{out} - H_{in} = \acute{A}_{w} \cdot V \cdot c_{pw} \left(T_{w} - T_{w,in}\right) = \acute{A}_{w} \cdot V \cdot c_{pw} \cdot \frac{dT_{w}}{dt} \tag{5.1}$$

The net heat exchange is a sum of heat fluxes, such as solar radiation (H_{sr}), heat change due to biological processes (H_b) and aeration (H_{ae}), atmospheric radiation (H_{ar}), evaporation (H_{ev}), surface convection (H_c) and losses through the walls (H_{tw}) (Eqn. (5.2)). If the net heat exchange is positive, heat recovery (H_{hr}) is possible or even required to maintain appropriate temperature for the microorganisms (Figure 5.2).

$$\Delta H = H_{sr} + H_{b} - H_{ae} - H_{ar} - H_{ev} - H_{c} - H_{tw} \tag{5.2}$$

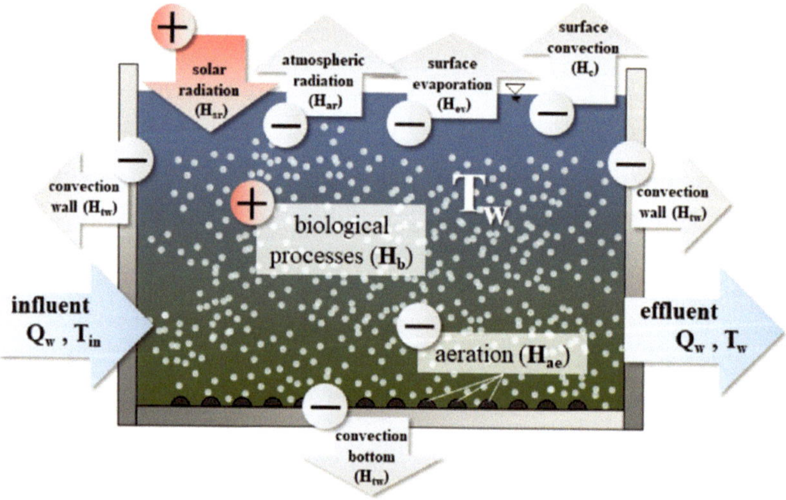

FIGURE 5.2 Heat fluxes in an aerated tank. Solar radiation and biological processes increase the temperature while aeration, atmospheric radiation, surface evaporation, and convection tend to decrease the basin temperature.

Source: Adapted from Corbala-Robles, 2016.

The equations of each heat flux are given by Eqns. ((5.3), (5.5–5.9) and (5.11)). The BSM2 (Alex et al., 2018) was chosen to serve as an example as most of the required data could be obtained from there. For example, the aeration

TABLE 5.1 Parameters of the Equations Used to Calculate the Net Heat Exchange

Symbol	Description	Unit	Summer			Winter			Source/ Comment
			Clear Sky	Cloudy	Night	Clear Sky	Cloudy	Night	
A	Surface area of the basin	m^2	2,500	2,500	2,500	2,500	2,500	2,500	Alex et al., 2018
A_{wall}	Wall area	m^2	984	984	984	984	984	984	Depth 4.8 m
$C^*_{T,ws}$	Saturated water vapor concentration at water temperature	kg/m^3	0.0272	0.0268	0.0255	0.0137	0.0139	0.0135	Calculated based on Eqn. (5.9)
$C^{\infty}_{T,air}$	Ambient air water vapor concentration	kg/m^3	0.0180	0.0134	0.0064	0.0026	0.0021	0.0011	Calculated, Eqn. (5.9)
C_c	Cloud cover	Tenths	0	8	0	0	8	0	Estimate
$c_{p,air}$	Air specific heat	$J/(kg \cdot K)$	1,005	1,005	1,005	1,005	1,005	1,005	
$c_{p,w}$	Water specific heat	$J/(kg \cdot K)$	4184.1	4184.1	4184.1	4184.1	4184.1	4184.1	
$D_{w,a}$	Diffusivity coefficient of water vapor in air	m^2/s	$2.65\ 10^{-5}$	$2.57\ 10^{-5}$	$2.40\ 10^{-5}$	$2.22\ 10^{-5}$	$2.18\ 10^{-5}$	$2.06\ 10^{-5}$	Calculated based on Kreith et al., 1999
f_{COD}	Aerobic COD removal fraction	Fraction	0.93	0.93	0.93	0.93	0.93	0.93	Estimate
f_{denit}	Nitrogen's denitrified fraction	Fraction	0.80	0.80	0.80	0.80	0.80	0.80	Estimate
f_{nit}	Nitrogen's nitrified fraction	Fraction	0.99	0.99	0.99	0.9	0.9	0.9	Estimate

TABLE 5.1 *(Continued)*

Symbol	Description	Unit	Summer			Winter			Source/ Comment
			Clear Sky	Cloudy	Night	Clear Sky	Cloudy	Night	
h_f	Exit air humidity factor	(–)	0.9	0.9	0.9	0.9	0.9	0.9	Corbala-Robles et al., 2016
h_{lat}	Water latent heat of evaporation	J/kg	$2.257\ 10^6$	$2.257\ 10^6$	$2.257\ 10^6$	$2.257\ 10^6$	$2.257\ 10^6$	$2.257\ 10^6$	
h_{sr}	Solar radiation on a flat surface	W/m^2	335.8	183.2	0.0	47.7	26.0	0.0	Thackston, and Parker, 1972
L	Characteristic distance (basin diameter)	m	62.5	62.5	62.5	62.5	62.5	62.5	Assuming 4 parallel lines with 10 m width
M_w	Wastewater molecular weight (taken as water)	kg/mol	0.018	0.018	0.018	0.018	0.018	0.018	
P_{aer}	Aeration energy	W	166,667	166,667	166,667	166,667	166,667	166,667	Alex et al., 2018
p_{air}	Vapor pressure at ambient air temperature	Pa	471.83	354.14	169.83	70.52	56.79	28.67	Calculated
Pr	Prandtl number	(–)	7.0	7.0	7.0	7.0	7.0	7.0	Estimate (for 293.15 K)

Symbol	Description	Unit	Summer			Winter			Source/ Comment
			Clear Sky	Cloudy	Night	Clear Sky	Cloudy	Night	
p_w	Vapor pressure at basin water temperature	Pa	390.94	358.07	334.43	197.03	189.14	181.20	Calculated
Q_{air}	Volumetric airflow rate	m3/s	1.54	1.54	1.54	1.54	1.54	1.54	Calculated based on Batstone et al., 2007
Q_w	Wastewater flow rate	m^3/s	0.24	0.24	0.24	0.24	0.24	0.24	Alex et al., 2018
R	Gas constant	kJ/(kg mol)	8.314	8.314	8.314	8.314	8.314	8.314	
Re	Reynolds number	(–)	96,878.5	96,878.5	96878.52	96,878.5	96,878.5	96,878.5	Calculated
r_h	Ambient air relative humidity	(–)	0.5	0.5	0.5	0.5	0.5	0.5	Estimate
Sc	Schmidt number	(–)	331.82	331.82	331.82	331.82	331.82	331.82	
S_{COD}	Influent COD concentration	kg_{COD}/m^3	0.4	0.4	0.4	0.4	0.4	0.4	Alex et al., 2018
$S_{N\text{-}NH4}$	Influent ammonium concentration	kg_N/m^3	0.04	0.04	0.04	0.04	0.04	0.04	Alex et al., 2018
T_{air}	Ambient air temperature	K	305	300	288	275	272	263	Estimate
T_{ground}	Ground temperature	K	288.15	288.15	288.15	278.15	278.15	278.15	Estimate

TABLE 5.1 *(Continued)*

Symbol	Description	Unit	Summer			Winter			Source/ Comment
			Clear Sky	Cloudy	Night	Clear Sky	Cloudy	Night	
T_w	Basin water temperature	K	299.95	299.71	298.85	289.04	289.16	288.81	Calculated via iteration
$T_{w,in}$	Influent water temperature	K	298.00	298.00	298.00	288.00	288.00	288.00	Estimate
U_{bottom}	Bottom overall heat transfer coefficient	$W/(m^2 \cdot K)$	1.5	1.5	1.5	1.5	1.5	1.5	Estimate
U_{wall}	Wall overall heat transfer coefficient	$W/(m^2 \cdot K)$	4.6	4.6	4.6	4.6	4.6	4.6	Estimate
V	Reactor volume	m^3	12,000	12,000	12,000	12,000	12,000	12,000	Alex et al., 2018
ΔH_{COD}	Reaction enthalpy for COD-removal	J/kg_{COD}	13.9×10^6	13.9×10^6	13.9×10^6	13.9×10^6	13.9×10^6	13.9×10^6	Blackburn and Cheng, 2005
ΔH_{denit}	Reaction enthalpy for denitrification	J/kg_N	41.3×10^6	41.3×10^6	41.3×10^6	41.3×10^6	41.3×10^6	41.3×10^6	Corbala-Robles et al., 2016
ΔH_{nit}	Reaction enthalpy for nitrification	J/kg_N	18.9×10^6	18.9×10^6	18.9×10^6	18.9×10^6	18.9×10^6	18.9×10^6	Corbala-Robles et al., 2016
$\varepsilon_{f,s}$	Surface emissivity	(–)	0.97	0.97	0.97	0.97	0.97	0.97	Taken as water emissivity

Symbol	Description	Unit	Summer			Winter			Source/ Comment
			Clear Sky	Cloudy	Night	Clear Sky	Cloudy	Night	
ε_{sky}	Sky emissivity	(–)	0.78	0.78	0.78	0.78	0.78	0.78	Iziomon et al., 2003
η	Efficiency of blowers	(–)	0.84	0.84	0.84	0.84	0.84	0.84	Estimate
λ_{air}	Air thermal conductivity	W/(m·K)	0.02552	0.02552	0.02552	0.02552	0.02552	0.02552	
ν_w	kinematic viscosity of water	m^2/s	8.9×10^{-7}	8.9×10^{-7}	8.9×10^{-7}	1.14×10^{-6}	1.14×10^{-6}	1.14×10^{-6}	Estimate for influent temperature
ρ'	Basin solar reflection	(–)	0	0	0	0	0	0	Corbala-Robles et al., 2016
ρ_{air}	Air density	kg/m^3	1.223	1.223	1.223	1.223	1.223	1.223	
ρ_w	Water density	kg/m^3	1,000	1,000	1,000	1,000	1,000	1,000	
σ	Stefan Boltzmann constant	$W/(m^2\cdot K^4)$	5.67×10^{-8}	5.67×10^{-8}	5.67×10^{-8}	5.67×10^{-8}	5.67×10^{-8}	5.67×10^{-8}	
Φ	Latitude	°	46	46	46	46	46	46	Estimate

energy (P_{aer}) for a day is given (4000 kWh/d) thus the heat gain from that can be easily calculated. The influent temperature was nonetheless changed according to values typical to a summer or winter day (see Table 5.1) based on measurement experiences in larger (>100,000 PE) WWTPs in Hungary. Data of BSM2 for stabilization period were chosen for each scenario to allow comparison of the effects of ambient air and water temperature along with cloud coverage. Six scenarios are taken into consideration: 2–2 day time and 1–1 night time for summer and winter. The effect of precipitation was not assessed this time.

Solar radiation may be influenced by reflective surfaces that cover the biological tank (Eqn. (5.3)). Also, meteorological circumstances such as cloud coverage, season, and site latitude were taken into account using the formulas provided by Thackston and Parker (1972) as shown in Eqn. (5.4). Calculation was carried out at northern latitude (Φ) 46° as this correlation is accurate between 26–46°. The results are in correspondence for measurements in Hungary as well (at 47°). The day values in Eqn. (5.4) were determined according to the minimum and maximum of the sinus curve: 356 (21st December) for winter and 173 (21st June) for summer.

$$H_{sr} = (1 - \rho') \cdot A \cdot h_{sr} \tag{5.3}$$

$$h_{sr} = \Big[95.1892 - 0.3591 \cdot \Phi - 0.0084537 \cdot \Phi^2 + \left(6.2484 - 1.6645 \cdot \Phi + 0.011648 \cdot \Phi^2\right) \cdot \sin\left(\frac{2 \cdot \pi \cdot day}{366} + 1.44451 + 0.01434 \cdot \Phi - 0.0001745 \cdot \Phi^2\right)\Big] \cdot 3.154 \cdot \left(1 - 0.0071 \cdot C_c^2\right) \tag{5.4}$$

The effect of aeration is twofold; on one hand, the aerators introduce heat to the basin, on the other hand, there is sensible and latent heat loss due to the movement of air and formation of bubbles (Eqn. (5.5))

$$H_{ae} = P_{aer} \cdot (1 - \eta) - Q_{air} \cdot \rho_{air} \cdot c_{p,air} \cdot (T_w - T_{air}) - \frac{M_w \cdot Q_{air} \cdot h_{lat}}{R} \cdot \left(v_w \cdot \frac{r_h + h_f \cdot (1 - r_h)}{T_w} - v_{air} \cdot \frac{r_h}{T_{air}}\right) \tag{5.5}$$

Since the BSM2 only considers removal of organic matter and nitrogen, the heat generation of nitrification, denitrification, and organic matter (Eqn. (5.6)) is taken into account based on the information of Corbala-Robles et al. (2016). Removal efficiencies are given in fractions (f_{nit}, f_{denit}, and f_{COD}) in correspondence with the benchmark.

$$H_b = Q_w \cdot S_{NH_4-N} \cdot f_{nit} \cdot \Delta H_{nit} + Q_w \cdot S_{NH_4-N} \cdot f_{denit} \cdot \Delta H_{denit} + Q_w \cdot S_{COD} \cdot f_{COD} \cdot \Delta H_{COD} \quad (5.6)$$

In case of atmospheric radiation (Eqn. (5.7)), evaporation (Eqn. (5.8)) and convection (Eqn. (5.10)), the air temperature plays an important role. The effect of wind speed in case of evaporation and convection is neglected in this case study as opposed to the findings of Talati and Stenstrom (1990). Reason for that is provided when discussing the results. To calculate the Reynolds number (Eqn. (5.11)) the following assumption was made. As a rule of thumb, the reactors are built so that their lengths are around 2 times their width. Also, in order to facilitate sufficient mixing in the cross section, the width of a tank is limited to around 10 m. Thus, it is assumed that the treatment is done in four parallel lines, each consisting of five compartments according BSM2. This assumption is used when calculating the heat loss through the walls (Eqn. (5.12)). Since biological tanks are designed and built to minimize heat loss, besides a small wall surface/volume ratio, the basins are assumed to be mostly below ground level. Temperature gradient in the ground was neglected and average values of 278.15 and 288.15 K were taken for winter and summer, respectively. For colder climate (e.g., in Norway) these values should be 277.15 K and 287.15 (Wang et al., 2017).

$$H_{ar} = \varepsilon_w \cdot \sigma \cdot A \cdot T_w^4 - \varepsilon_{sky} \cdot \sigma \cdot A \cdot \left(\frac{c_c}{10} \cdot T_{air} + \left(1 - \frac{c_c}{10} \right) \cdot 0.0522 \cdot T_{air}^{1.5} \right)^4 \quad (5.7)$$

$$H_{ev} = \frac{D_w}{L} \cdot 0.037 \cdot \mathrm{Re}^{\frac{4}{5}} \cdot Sc^{\frac{1}{a}} \cdot A \cdot \left(C_{T,w}^{*} - C_{T,air}^{\infty} \right) \cdot = h_{lat} \quad (5.8)$$

Water vapor concentration above wastewater or in the air may be calculated following Tetens (1930) (Eqn. (5.9)).

$$C_T = \rho \cdot 0.622 \frac{r_H \cdot 610.78 \cdot e^{\frac{T-273.15}{(T-273.15)+238.3} 17.2694}}{P_T - r_H \cdot 610.78 \cdot e^{\frac{T-273.15}{(T-273.15)+238.3} 17.2694}} \quad (5.9)$$

$$H_c = \frac{\lambda_{air}}{L} \cdot 0.037 \cdot \mathrm{Re}^{\frac{4}{5}} \cdot \mathrm{Pr}^{\frac{1}{3}} \cdot A \cdot \left(T_w - T_{air} \right) \quad (5.10)$$

$$\mathrm{Re} = \frac{Q_w \cdot L}{u \cdot v_w} \tag{5.11}$$

$$H_{tw} = \left(U_{wall} \cdot A_{wall} + U_{bottom} \cdot A\right) \cdot \left(T_w - T_{ground}\right) \tag{5.12}$$

In Table 5.2, the results of heat fluxes in the activated sludge reactor are given for all scenarios. Solar radiation and biological reactions are net heat producers while the other processes cause mostly heat loss. Surface convection in summer during the day may result in a small heat gain due to the higher ambient air temperature, but it is negligible even if compared to solar radiation. To visualize the extent to how each flux contributes to the overall heat balance, a pie chart was created from the results of the winter scenarios (Figure 5.3). The slices on the right represent heat loss and on the left, the heat gain. It is clear that the largest contributors are biological reactions (+) and atmospheric radiation (–). That also means that determining these fluxes are the most crucial to assess the heat balance of the tank. Surface evaporation on the other hand is only 3–12% of the net heat change, thus even a 100% increase due to strong wind may cause only a maximum of 12% deviation in the result. Solar radiation plays an important role in the summer and this is when unwanted overheating of the system could be experienced.

A similar calculation was carried out for the secondary settler (Table 5.3) but the temperature change in the course of primary settling, dewatering, and disinfection was neglected. In the case of the settler, the equations for aeration and biological processes were not used, as there is no aeration in the sedimentation tank and the biological train is designed so that degradation and oxidation processes should not take place outside the reactors. This results in the solar radiation being the most dominant process in summer days and atmospheric radiation in winter. Heat gain due to convection appears in the two summer day scenarios, though these values are minuscule compared to the others. The net change is positive for these versions; otherwise, the heat content decreases in the settler. Nonetheless, in this given configuration the wastewater in the settler does not cool down below the influent temperature even in the worst case (winter night when solar radiation is zero and the air temperature is 263 K, around 10° below freezing point): the effluent temperature of the settler is 0.33 K higher than the influent arriving to the biological tank.

TABLE 5.2 Results of Heat Balance Calculation for the Biological Tank

Scenario	Solar Radiation	Biological Reactions	Aeration	Atmospheric Radiation	Surface Evaporation	Surface Convection	Heat Exchange Through Walls and Bottom	Net Change
	H_{sr}(+)	H_b (+)	H_{ae}(–)	H_{ar} (–)	H_{ev}(–)	H_c(–)	H_{tw}(–)	ΔH
	W	W	W	W	W	W	W	W
Summer day, clear sky	839,381	1,730,427	18,373	452,180	52,570	–3,275	97,682	1,952,278
Summer day, cloudy	457,966	1,730,427	33,087	280,520	74,147	–191	95,637	1,705,194
Summer night	0	1,730,427	61,379	628,262	98,528	7,038	88,531	846,689
Winter day, clear sky	119,160	1,714,166	39,644	604,680	47,301	7,475	90,164	1,044,062
Winter day, cloudy	65,014	1,714,166	46,755	420,702	49,022	9,135	91,144	1,162,423
Winter night	0	1,714,166	64,063	685,016	49,240	13,740	88,259	813,849

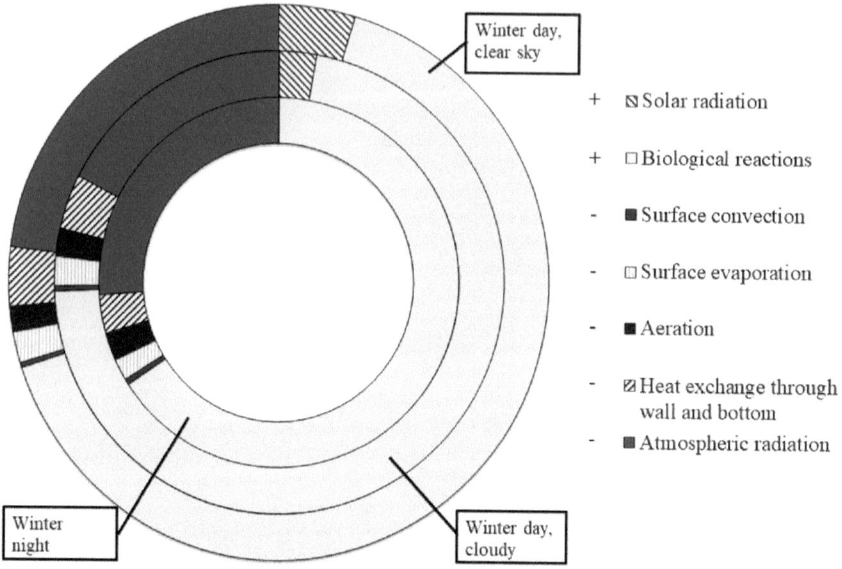

FIGURE 5.3 Ratios of the different heat fluxes in the biological reactors for winter scenarios.

TABLE 5.3 Results of Heat Balance Calculation for the Settler

Scenario	Solar Radiation	Atmospheric Radiation	Surface Evaporation	Surface Convection	Heat Exchange Through Wall and Bottom	Net Change
	$H_{sr}(+)$	$H_{ar}(-)$	$H_{ev}(-)$	$H_{c}(-)$	$H_{tw}(-)$	ΔH
	W	W	W	W	W	W
Summer day, clear sky	503,629	273,066	–21,353	–2,652	57,313	197,254
Summer day, cloudy	274,780	168,771	–901	–133	55,437	51,606
Summer night	0	372,977	25,337	5,684	48,926	–452,924
Winter day, clear sky	71,496	359,949	14,466	6,139	50,316	–359,374
Winter day, cloudy	39,008	250,140	16,303	7,571	51,231	–286,237
Winter night	0	407,161	18,123	11,363	48,615	–485,262

In Figure 5.4, the different temperature values are shown. The largest increase in temperature is witnessed in the first scenario (2.15 K); the increment in the biological tank is between 0.81 and 1.95 K, and in the clarifier –0.49 and 0.20 K. It is interesting that in winter clear sky results in higher loss due to atmospheric radiation as opposed to cloudy weather when the ambient air temperature is even lower by 3 K. In Eqn. (5.7), the second term refers to the observed temperature of the sky which is significantly lower if no cloud coverage is accounted for (around 37 versus 8 K difference).

These results show that heat recovery may be possible as even in case of harsh meteorological circumstances the overall heat balance of the biological tanks and secondary settler is positive. The magnitude of the retrievable energy will be discussed together with the energy recovery options of the digester.

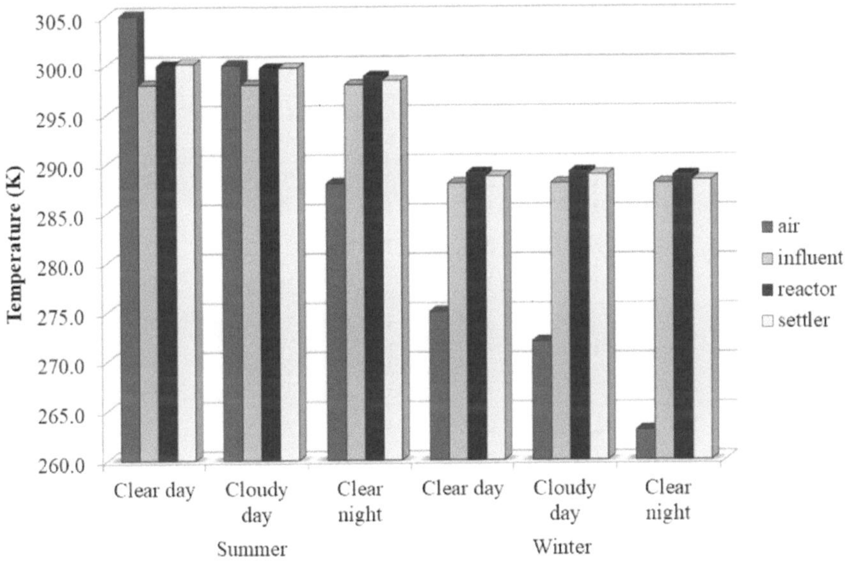

FIGURE 5.4 Temperature values in the example.

5.3.2 HEAT BALANCE OF THE DIGESTER

The heat balance of the digester was drawn based on the information of BSM2. Sludge from the secondary settler goes through a thickening process but the temperature is assumed to be the same as in the settler. The temperature of the primary sludge is identical to that of the influent. These streams are mixed before entering the digester. Heating energy (H_{ad}) is required to raise the sludge temperature to the digester temperature (T_{ad}=308.15 K) and also to cover the heat loss from the digester according to Eqn. (5.12).

$$H_{ad} = \rho_w \cdot c_{p,w} \cdot \left(T_{ad} - T_{ad,in}\right) \cdot Q_{ad} + \Sigma_{i=1}^{3} U_i \cdot A_i \cdot \left(T_{ad} - T_i\right) \qquad (5.13)$$

where U_i is the heat transfer coefficient is for the given A_i surface, i.e., the walls, floor, and roof and T_i is the ambient temperature of air or in case of the floor, the ground. The building material for the digester was considered to be 300 mm concrete with insulation (U=0.8 J/m^2 s K) and floating cover with 25 mm insulating board installed under roofing (U=1.0 J/m^2 s K) based on Turovskiy and Mathai (2006). The temperature values for air and ground are the same as in Table 5.1.

TABLE 5.4 Results of the Digester Heat Balance for the Six Scenarios

	Unit	**Summer**			**Winter**		
		Clear Sky	**Cloudy**	**Clear Night**	**Clear Sky**	**Cloudy**	**Clear Night**
Q_{ad}	m^3/s	2.0656× 10^{-3}	2.0656× 10^{-3}	2.0656× 10^{-3}	2.0656× 10^{-3}	2.0656× 10^{-3}	2.0656× 10^{-3}
$T_{ad,in}$	K	298.37	298.30	298.07	288.12	288.15	288.06
T_{ad}	K	308.15	308.15	308.15	308.15	308.15	308.15
Q_{gas}	m^3/s	3.1346× 10^{-2}	3.1346× 10^{-2}	3.1346× 10^{-2}	3.1346× 10^{-2}	3.1346× 10^{-2}	3.1346× 10^{-2}
c_{CH4}	g/m^3	393.236	393.236	393.236	393.236	393.236	393.236
Heating demand (H_{ad})	W	92,740	97,914	110,960	211,547	214,013	223,088
Produced heat (H_{pr})	W	250,296	250,296	250,296	250,296	250,296	250,296
Surplus heat (H_{pr}–H_{ad})	W	157,555	152,382	139,336	38,749	36,283	27,208
Produced electricity	W	181,249	181,249	181,249	181,249	181,249	181,249

The density and specific heat capacity of the sludge was considered to be the same as water as the solids content is only 7%. Results of the calculation are shown in Table 5.4. Produced heat is calculated according to Eqn. (5.14).

The heating value of biogas ($h_c°$) was taken to be 50,014 J/g (Alex et al., 2018) and assumed that from the overall efficiency (70%) of the cogeneration generator 58% is heat (and 42% is electricity). That is in correspondence with literature data (Carrère et al., 2010). Surplus heat is what remains from the produced heat after satisfying the heating demand of the digester.

$$H_{pr} = Q_{gas} \cdot c_{CH_4} \cdot h_c^{\circ} \cdot 0.7 \cdot 0.58 \tag{5.14}$$

5.3.3 HEAT DEMAND OF BUILDINGS

There is no information in the benchmark on the size and energy demand of auxiliary buildings and offices, thus it had to be estimated. Based on practical experience, a wastewater treatment plant of such a magnitude would need a laboratory, plant administration and control rooms, dining room, locker rooms with showers, toilets, meeting rooms, offices, and a lobby. Garages, plant maintenance and storage areas may or may not require heating but most of the devices need freeze protection. In order to assess the heating and domestic hot water (DHW) demand in these buildings threshold values of energy performance coefficients were used (7/2006 (V. 24.) Decree of Minister without Portfolio about determination of energy efficiency of buildings). For the first set of facilities a two-storey building with 1200 m^2 gross internal area with a ceiling height of 3 m was assigned while for maintenance and storage purposes a 800 m^2 one-storey building (5 m ceiling height) was assumed. Due to the equations used, there is no difference between the heat demand for one big structure and several smaller ones with the same total floor space.

The threshold values (in kWh/m^2 a) are indicators characterizing the efficiency of the building's energy consumption, which takes into account the building's orientation, solar access, the effects of adjacent buildings and other climate factors; the thermal insulation capacity, building construction and other technical properties of the building, the type of energy used, the energy demand for ventilation and the produced energy if present. The energy performance values of a building cannot be higher than the relevant threshold determined by the decree. Depending on the area-volume ratio, two equations can be used to determine the maximum energy demand of the buildings in the example. Eqn. (5.15) is for the office building (H_o), where the ratio is larger than 0.3 (=0.33) and Eqn. (5.16) may be used for the maintenance building (H_m) where that value is only 0.2.

$$H_o = \left(28 \cdot \frac{A_o}{V_o} + 123.6 \right) \cdot \frac{1000}{182.24} \cdot A_o \tag{5.15}$$

$$H_m = 132 \cdot \frac{1000}{182.24} \cdot A_m \tag{5.16}$$

$$H_{DHW} = \frac{9}{182.24} \cdot 1000 \cdot \left(A_o + A_m \right) \tag{5.17}$$

For heating, the days of the heating season (182) are taken into account but for the DHW use were calculated for 365 days a year (Eqn. (5.17)). While this is probably more than the actual working days even in a bigger plant with 24/7 services, the indicator for offices (9 kWh/m^2 a) was used due to a lack of relevant data. Heating energy is calculated to be a total of 60,696 W while DHW energy demand came to be 2055 W.

While this is an overly simplistic method, it allows comparing thermal energy demand with the supply side. In the case of a real system, a more detailed calculation could be carried out in possession of data regarding size, structure, and used materials.

5.3.4 OVERALL HEAT BALANCE FOR THE PLANT

Even with regards to the simplifications taken into account, the heat potential of wastewater is significant. In Table 5.5 the results are summarized for all 6 scenarios. Heat recovery was calculated with two different limit values for summer and winter, 298 K and 288 K, respectively. These correspond to the influent temperatures, and may be chosen according to the water temperature of the receiving water body. Since for all scenarios, the effluent temperature from the settler was higher than the influent's, it is natural that excess heat appears. Heat generation was the highest in summer mainly due to solar radiation; increasing the wastewater temperature by around 2 K. Unfortunately, this is the period when the heat is not needed for other processes either. On the other hand, it can be utilized for cooling purposes but the temperature of the recipient has to be taken into account to avoid overheating.

TABLE 5.5 Heat Supply and Demand in the Benchmark Plant

	Summer			Winter		
	Clear Day	**Cloudy Day**	**Night**	**Clear Day**	**Cloudy Day**	**Night**
Supply						
Wastewater excess heat (W)	2,097,480	1714,258	384,230	668,108	854,969	320,630
CHP (W)	250,296	250,296	250,296	250,296	250,296	250,296
Demand						
Digester heating (W)	92,740	97,914	110,960	211,547	214,013	223,088
Building heating (W)	0	0	0	60,696	60,696	60,696
DHW (W)	2,055	2,055	0	2,055	2,055	0
Sum (W)	2,252,981	1,864,584	523,566	644,106	828,501	287,143

What is striking is that even in the worst case, i.e., when all temperatures are the lowest, the heat reclaimable from the wastewater is higher than that can be retrieved from the CHP. The heat from the generator is enough to maintain the operational temperature in the digester but it has to be completed to meet the heating and hot water demand of the buildings. With a well-designed system of HE and heat pumps it is possible to use the excess heat from the wastewater to cover all thermal needs of the plant but a purpose to the CHP waste heat has to be found. One solution would be to utilize it in heating nearby facilities or industrial buildings.

While there is plenty of thermal energy, the CHP cannot cover the electricity demands. Based on BSM2, the energy requirements of aeration, mixing, and pumping add up to 217 kW. Assuming that the aeration is roughly 60% of the total electricity demand of the whole plant (Gu et al., 2017), 278 kW has to be provided. Unfortunately, the generator can only produce 181 kW, which means that 35% of the electricity has to be provided by external sources.

Additional electric energy is required if the heat content of the wastewater is to be exploited. Assuming that 320 kW of thermal energy would be utilized, installing heat pumps would call for 80 kW of electricity (COP=4).

5.4 DEVICES TO RECOVER THERMAL ENERGY

In the case study, it is already assumed that the energy of the produced biogas is fed to a cogeneration system where the waste heat of the generator or gas turbine producing electricity is utilized in a boiler. Besides the thermal energy of flue gas, the heat content of wastewater can be also put to use.

Depending on the temperature of the waste heat stream, the heat recovery system can be a combination HE, heat pumps, and thermal storage equipment. In the following subsection, HE and heat pumps will be discussed.

5.4.1 HEAT EXCHANGERS (HE)

HEs transfer heat from a hotter source (fluid or gas) to a sink with lower temperature by means of conduction through a wall separating the two fluids. They are often used in WWTPs, for example, to recover heat from the exhaust gas of generators, for preheating sludge to be digested or for building heating purposes. Important design parameters for these systems are: temperature and flow rate of the fluids, pipe geometry, heat exchanger geometry, viscosity, resistance to fouling, heat exchange coefficient and heat transfer surface (Cipolla and Maglionico, 2014b). In this subsection, only those HEs will be discussed that are commonly used for WWTPs.

Three arrangements can be distinguished: concurrent, counter current (Figure 5.5), and the cross-flow type where the two flows are normal to each other. Each configuration has its advantages and disadvantages. The concurrent arrangement has the lowest, while the counter-current version has the highest exchanger efficiency among the single-pass exchangers for the same flow rates, capacity rate (mass × specific heat) ratio, and surface area. On the other hand, concurrent flow can be used for heating viscous fluids as it provides for rapid heating, resulting in reduced pumping power requirements due to a change in viscosity. It is also beneficial where the fluid or heat exchanger material is sensitive to temperature change (Thulukkanam, 2013). In countercurrent flow, the thermal stress in the wall caused by temperature difference is minimal compared to other flow arrangements. However, it is not always practical to use counter flow arrangement due to difficulties in designing inlet and outlet header to separate fluids (Shah and Sekulic, 2003). This can be overcome by using a crossflow arrangement.

HE can be internal or external to the system; in general, external versions are used as they are more flexible and easier to maintain (Appels et al., 2008). Their heat transfer coefficients range from 0.85 to 1.6 kJ/(m^2Ks) (Turovskiy and Mathai, 2006). HE for wastewater heat recovery can also be grouped according to Figure 5.6 (Shen et al., 2018). The primary difference between the types is the internal circulating work fluid, which is usually water or refrigerant. Wastewater (or sludge) may get into contact directly with the heat exchanger of the heat pump or an additional loop is introduced to protect the evaporator from fouling.

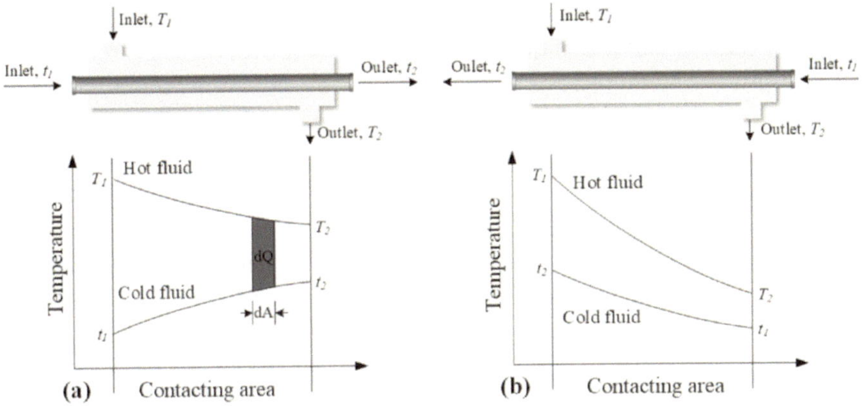

FIGURE 5.5 Simplified diagram of concurrent and counter-current type heat exchangers and corresponding temperature gradients.

Source: Adapted from Shah and Sekulic, 2003.

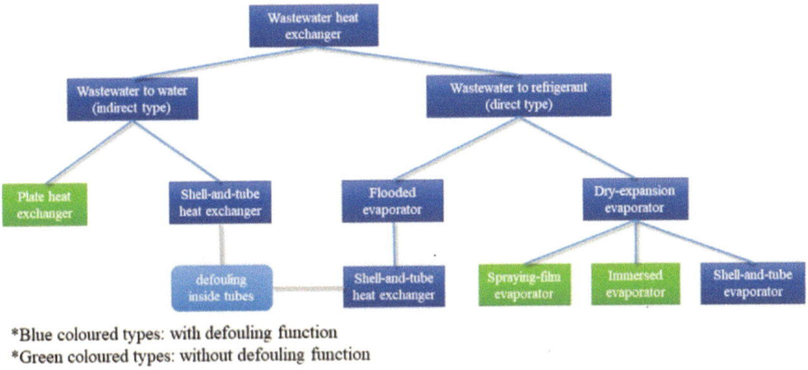

FIGURE 5.6 Classification of wastewater heat exchangers. *Source:* Adapted from Shen, 2018.

Wastewater HEs are subject to fouling, which influences performance and can cause blockages (Cunault et al., 2013a). Fouling takes place when solids deposit on the surface or when microorganisms form a biofilm on solid surfaces (referred to as biofouling); this can decrease the heat transfer ratio by 20–50% (Culha et al., 2015). Trace elements in the water can accelerate biofouling development thus regular maintenance to prevent biofilm from entering the rapid growth stage is necessary (Tian et al., 2012). Fouling can be removed via physical methods, such as flushing/rinsing, taking HE apart for manual cleaning, circulating rubber balls inside heat exchanger tubes or reciprocating brushes inside tubes (Shen et al., 2014), or by chemical methods, such as using acid wash (Cunault et al., 2013b) though all of the solutions are temporary.

According to Culha et al. (2015) spray, plate, pressure pipe, and helical HEs are used in the case of treated wastewater, while for digesters tube-in-tube (water jacket), spiral plate or water-bath types are common (Davis, 2010).

The spray type heat exchanger is a direct type of WWHE without defouling function where the (filtered) wastewater is sprayed onto the evaporator. It can be operated with a very low temperature difference ($\Delta T \approx 1$ K) between the working fluids (Baqir et al., 2016). Vende et al. (2018) concluded that the spraying surface, thus the efficiency primarily depends on the surface temperature as opposed to the airflow. Spray HE was used in a 3.3 MW heat pump without problems of fouling or corrosion in the eighties, though clogging in the water distributor holes occurred often (Lindström, 1985).

Plate HE belongs to the indirect group. Fluids flow in counter-current mode in alternating chambers allowing heat transfer to occur through the largest surface of the metal plates. Plate HE are advantageous for several reasons: high thermal efficiency can be achieved, the units may be expanded, cross-contamination is not possible under normal operation, leaks are easy to see, and maintenance is simple (Lines, 1991). They are compact, up to 93% of the heat can be recovered, and due to the turbulent flow pattern, fouling is less likely to appear (Thulukkanam, 2013). A disadvantage of such a system is the high-pressure drop. The design variables, such as the port diameter, length of compact plates, and the chevron angle have to be carefully chosen to resolve the conflicts of the thermal-hydraulic objectives, i.e., relatively high heat transfer coefficient with a considerably low total pressure drop (Raja et al., 2018). Kumar et al. (2018) found, that the chevron angle significantly affects the fluid flow distribution in the channels and has a huge impact on the thermal performance and power consumption of PHEs. The plate thickness can also improve the energy efficiency (Zhang et al., 2018). In the case of frame-and-plate HE and brazed plate HE it was concluded that the effect of end plates could be significant when the number of plates is small (Jin and Hrnjak, 2017).

There are several varieties of plate HE, spiral plate HE being one of importance with regards to wastewater heat recovery. In this case, two relatively long strips of plates are rolled to form a pair of spiral passages. While the maximal pressure of fluid is limited (Picón-Núñez et al., 2007), they can handle high viscosity and solids content (Thulukkanam, 2013). Since the heat transfer is not pure counter-flow, sizing can be done by using empirical correlations such as in the work of Minton (1970).

The shell-and-tube HE have direct and indirect types as well and can be formed with or without integrated defouling function (Shen et al., 2018). The most common type has a fixed tube sheet with singe segmental baffles and it is highly recommended to applications, where the low pressure drop is a key factor (Zahid et al., 2018). The performance can be improved by helical baffles (the pressure drop depends on the baffle inclination angle) instead, the conventional segmental units and it can reduce the maintenance and operating costs by 20–40% (Shinde and Chavan, 2018). A numerical simulation for shell-and-tube HE with louver baffles was published by Lei et al. (2017). In this case, the heat transfer coefficient per pressure drop was calculated to be better than shell-and-tube HE with segmental baffles. The shell-and-tube HE with continuous helical baffles on shell side were analyzed by Shindle and Chavan (2018). They found that the larger helix angles add to lower heat transfer and lower pressure drop, and smaller helix angles resulted higher heat transfer and pressure drop. By using the constructal theory for improving the efficiency of the STHEs, Mohsen et al. (2017) reached more than 28% growth in the thermal efficiency.

5.4.2 HEAT PUMPS

In order to utilize the low-temperature energy stored in wastewater, heat pumps are required to increase it to a higher temperature level so that it can be supplied for utilization. In order to achieve this, external energy is needed which is often electricity supplied from the grid but it can come from renewable sources, e.g., photovoltaic units, too. The working principle of a heat pump is given here following the description of Meyer (2011). In the evaporator, the heat is transferred from the heat source (wastewater in our case) to the refrigerant that is at lower pressure and temperature. As a result, the working fluid evaporates and becomes a saturated vapor. From here, it goes to the compressor where with the help of external energy the refrigerant undergoes adiabatic compression and both the temperature and pressure is increased. This hot vapor transfers the heat in the condenser to the thermal sink and becomes liquid. The pressure and temperature is lowered using an expansion valve or capillary tube. From this, point on the cycle starts again.

Heat pumps can be categorized based on several aspects. Regarding the operating principle three main types can be distinguished: mechanical, absorption, and thermo-compression. In the first case, electricity is used to mechanically compress the refrigerant to achieve higher condensation temperature as a factor of pressure. Both absorption and thermo-compression heat pumps utilize thermal energy but in the former, the refrigerant is absorbed in a medium while the latter operates similar to mechanical compression heat pumps.

Based on functionality three different categories can be drawn: used for only heating (space and/or water) for heating and cooling, and integrated systems that may include exhaust air heat recovery. The system can be either monovalent, serving as the sole supplier of heat or bivalent when the heating system is complemented by another equipment. In this case, the heat pump is sized for 20–60% of the maximum heat load and in peak periods boilers are utilized as well (Chua et al., 2010). The temperature lift is usually between 15–75 K depending on the temperature of the source and the sink. The higher the difference, the lower is the COP. For example, if the temperature lift is 20 K, the COP would be 9.1 while for 40 K this value is 5.0 (evaporation temperature at constant 283.15 K, Carnot efficiency 60%) (Gasser et al., 2017). To achieve high temperature lift either high pressure or multi-stage compressors may be used.

The type of refrigerant used as working fluid is chosen based on the desired temperature lift and the pressure required for it, but environmental and safety aspects have to be considered, too. This poses several challenges: the source and even the sink temperature may be prone to changes and for certain, refrigerants the pressure is too high at elevated temperatures. These along with concerns of greenhouse effects in case of leakage lead to the research of new refrigerants as well as their mixtures (Chua et al., 2010). Presently the most commonly used refrigerants are R134a ($C_2H_2F_4$, global warming potential (GWP) = 1300), R410a (zeotropic mixture of CH_2F_2 and CHF_2CF_3, GWP = 1700) and ammonia for lower temperature ranges while butane and water are used at higher temperatures. Carbon-dioxide is gaining field in both the lower temperature applications and when high temperature lift is required (Austin and Sumathy, 2011) but due the transcritical nature of the refrigerant, the heat pump has to be operated at high pressure.

5.5 UTILIZATION OF RECLAIMED HEAT

In the previous sections, the processes where thermal energy is generated in a wastewater treatment plant have been discussed along with the equipment for reclamation. In this section, the areas where the reclaimed heat may be utilized are reviewed. Heat energy recovered from wastewater or flue gases of the generator may be used within the treatment plant for digester heating (Tchobanoglous et al., 2013), sludge drying (Tańczuk et al., 2016), hot water supply or space heating (Chae and Ren, 2016; Chae and Kang, 2013).

5.5.1 SPACE HEATING WITH EXCESS HEAT FROM THE EFFLUENT

For the purpose of domestic heating in the wastewater treatment plant, useful heat can be extracted from the effluent using a heat pump and HE. To assess the magnitude of the demand and the supply as well as the feasibility of the solution a simplified version of heat pump sizing was prepared. Figure 5.7 shows a representation of the proposed system. The following assumptions were made: the heat demand calculated in Section 1.3.3 is considered without the DHW (Q_d= 60,696 W) and the distance between the heat source where the heat pump was installed and the sink (the buildings) is 100 m. The temperature values were considered to be: for the input from the wastewater side (T_i) 288.33 K, the value computed for the winter night scenario, cooled down to (T_o) 287.15 K, and the output to the building (T_h) 338.15 K and the temperature drop is 10 K (T_c=328.15 K). The ambient air temperature (T_{air}) is 263.15 K.

Commercially available mechanical heat pumps with R-134a as refrigerant were considered in this study. The sizing of the heat pump was not carried out at this stage, thus the heat exchanger surfaces and the temperatures of the working fluid in the different stages were not calculated, but the efficiencies (η) of the evaporator and the condenser were considered to be 95% and 90% for the compressor. Other properties and tabulated values used throughout the calculation are summarized in Table 5.6.

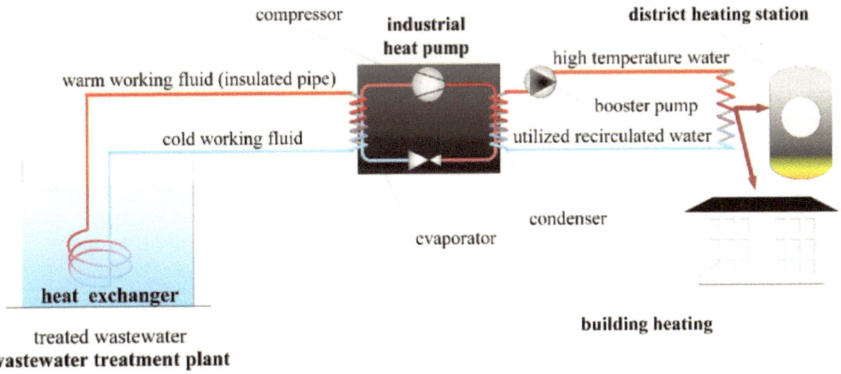

FIGURE 5.7 Schematic representation of wastewater heat reclamation for space heating purposes.

Source: Reprinted with permission from Somogyi et al. (2018b). © 2018 Elsevier.

TABLE 5.6 Constants Used for the Sizing of the Heat Pump

Property	Value	References
Thermal conductivity of PVC, λ_{PVC} (W/m.K)	0.150	Titow, 2012
Thermal conductivity of insulation, λ_s (W/m. K)	0.034	Kristjansson and Bøhm, 2006
Inner pipe radius r_i (m)	0.04125	Kristjansson and Bøhm, 2006
Outer pipe radius r_o (m)	0.04445	Kristjansson and Bøhm, 2006
Outer radius of insulation r_s (m)	0.08125	Kristjansson and Bøhm, 2006
Water heat transfer coefficient, z (W/m^2.K)	2000	Perry and Green, 2008
Thermal conductivity at bulk temperature (T_b), λ_b (W/m. K)	0.0139	Perry and Green, 2008

Based on these assumptions using insulated PVC pipes (Titow, 2012) the heat loss in the pipes can be determined by Eqn. (5.18).

$$Q = \frac{2 \cdot \pi \cdot \left(T_{pipe} - T_{air}\right) \cdot L}{\frac{\text{In}\left(\frac{r_o}{r_i}\right)}{\lambda_{PVC}} + \frac{\text{In}\left(\frac{r_s}{r_o}\right)}{\lambda_s}} \tag{5.18}$$

where T_{pipe} is the temperature inside the pipe (T_h to the building and T_c from the building). Thus, the cumulative heat loss for the delivery and return pipes (Q_1 and Q_2) is 4,824 W (Eqns. (5.18a) and (5.18b)).

$$Q_1 = \frac{2 \cdot \pi \cdot (338.15 - 263.15) \cdot 100}{\frac{\text{In}\left(\frac{0.04445}{0.04125}\right)}{0.15} + \frac{\text{In}\left(\frac{0.08125}{0.04445}\right)}{0.034}} = 2584\ W \tag{5.18a}$$

$$Q_2 = \frac{2 \cdot \pi \cdot (338.15 - 263.15) \cdot 100}{\frac{\text{In}\left(\frac{0.04445}{0.04125}\right)}{0.15} + \frac{\text{In}\left(\frac{0.08125}{0.04445}\right)}{0.034}} = 2240\ W \tag{5.18b}$$

Accounting for the heat losses and a condenser efficiency of 95%, the final heat load at the condenser of the heat pump (Q_{con}) is obtained to be approximately 69 kW (Eqn. (5.19)).

$$Q_{con} = \frac{Q_1 + Q_2 + Q_d}{\eta_{cond}} = \frac{4824 + 60696}{0.95} = 68968\,W \tag{5.19}$$

The COP of heat pumps in the case of 50 K temperature lift can be assumed to be around 4. The maximum theoretical COP would be at (). Assuming an adiabatic compressor efficiency of 90% and using the heat load calculated before, the compressor power required to be added to the refrigerant is calculated by Eqn. (5.20).

$$P_{comp} = \frac{Q_{con}}{\eta_{comp} \cdot COP} = \frac{68968}{0.95 \cdot 4} = 19158\,W \tag{5.20}$$

Consequently, the theoretical heat that should be available at the evaporator, which has an efficiency of 95%, is given by Eqn. (5.21).

$$Q_{eva} = \frac{Q_{con} - P_{comp}}{\eta_{eva}} = \frac{(68968 - 19158)}{0.95} = 52432\,W \tag{5.21}$$

A supplementary coil heat exchanger was decided to be added in order to facilitate the extraction of heat from the effluent of the wastewater treatment plant and to avoid fouling within the heat pump. This type of heat exchanger was chosen to simplify the sizing procedure and because fouling can be easily managed in such a structure.

The enthalpy change of wastewater entering at 288.33 K and leaving at 283.15 K is 21,020 J/kg (Eqn. (5.22)). Therefore, with a heat transfer rate of 52,432 W and enthalpy change as calculated, the required wastewater flow-rate is 2.5 kg/s (Eqn. (5.23)).

$$\Delta E = 63.04 - 42.02 = 21.02\,kJ/kg = 21020\,J/kg \tag{5.22}$$

$$m = \frac{Q}{\Delta E} = \frac{52432}{21020} = 2.494\,kg/s \tag{5.23}$$

The log mean temperature difference ΔT_{LM} (Eqn. (5.24)) for the temperature profile is calculated from the temperature difference on the wastewater side ($T_i - T_o$) and on the working fluid side which was assumed to have a temperature lift of 1 K (from 282.15 to 283.15).

$$\Delta T_{LM} = \frac{\Delta T_1 - \Delta T_2}{\ln\left(\frac{\Delta T_1}{\Delta T_2}\right)} = \frac{(288.33 - 287.15) - (283.15 - 282.15)}{\ln\frac{(288.13 - 287.15)}{283.15 - 282.15}} = 1.09\ K \quad (5.24)$$

The bulk temperature is the arithmetic mean temperature of the hot side of the exchanger, thus it can be computed by Eqn. (5.25). This allows defining the thermal conductivity (λ_b) at this temperature (see Table 5.6) but the Nusselt number is also needed.

$$T_b = \frac{T_i + T_{20}}{2} = \frac{283.15 + 288.33}{2} = 285.74\ K \quad (5.25)$$

Due to availability, light-half hard copper tubes with an inner (d_i,) outer (d_o) diameters of 0.04, 0.044 m, respectively will be used. Since the copper tubes are cylindrical, the overall (d_h) hydraulic diameter is equal to the inner diameter. With these conditions, the Nusselt number was calculated to be 5755.4 (Eqn. (5.26)). Thus the overall heat transfer coefficient (U) can be estimated from Perry and Green (2008) which gave the value of U = 2200 W/(m^{2}K).

$$Nu = \frac{z \cdot d_h}{\lambda_b} = \frac{2000 \cdot 0.04}{0.0139} = 5755.4 \quad (5.26)$$

With the following characteristics, the length of the exchanger (L) can be calculated (Eqn. (5.27)).

$$L = \frac{Q}{U \cdot d_i \cdot \Delta T_{LM} \cdot \pi} = \frac{52432}{2200 \cdot 0.04 \cdot 1.09 \cdot \pi} = 174.0\ m \quad (5.27)$$

With an assumed coil diameter (d_c) of 1 m, the number of turns (N) is calculated to be 56 (Eqn. (5.28)); the overall height of the heat exchanger (H) and the overall heat transfer area (A) are given by Eqns. (5.29) and (5.30)

$$N = \frac{L}{\pi \cdot d_c} = \frac{174.0}{\pi \cdot 1} = 55.4 \approx 56 \quad (5.28)$$

$$H = d_o \cdot N = 0.044.56 = 2.44\ m \quad (5.29)$$

$$A = \frac{Q}{U \cdot \Delta T_{LM}} = \frac{52432}{2200 \cdot 1.09} = 21.86\, m^2 \tag{5.30}$$

Finally, the required mass flow-rate of wastewater in the exchanger can be determined (Eqn. (5.31)). The average available flow of water for heat extraction is 0.24 m^3/s (240 kg/s), which means that as small as 1% of the wastewater flow may be able to fully supply the heating demands of the offices and auxiliary buildings, though the required surface and thus the dimensions of the heat exchanger may seem disheartening.

$$m = \frac{Q}{C_{v(water)} \cdot \Delta T} = \frac{52432}{4184.1 \cdot (288.33 - 283.15)} = 2.42 \frac{kg}{s} \tag{5.31}$$

As it was already indicated, the generated heat can be utilized in district heating as well, if the excess heat is substantially more than the demand in the plant. In case of larger WWTPs, if higher heat density areas are nearby and district heating systems are already in place-even with higher electricity and lower natural gas prices-the return time of such an investment may be within realistic expectations (Somogyi et al., 2018b). Besides that, using wastewater excess heat can help reduce GHG emissions (Pavičević et al., 2017) of heating systems.

5.5.2 KEEPING THE TEMPERATURE CONSTANT IN THE AERATED TANK

The same principle may be used in order to stabilize the temperature of smaller aerated tanks where the temperature fluctuation may cause inefficient removal of nutrients in the winter. The proposal was first explained by Somogyi et al. (2018a). In order to achieve the temperature increase in the aerated tank, a heat pump system would be installed between the biological tank and where the effluent can be accessed (Figure 5.8). Ideally, the distance between the source (effluent) and the sink (reactor) is small. The system is assumed to be open from the effluent side since solids are removed in the settler but closed from the wastewater side. To avoid fouling on the HE, several solutions are available starting from choosing the geometry accordingly to implementing self-cleaning system. Nonetheless, it should not cause insufficient mixing or aeration while achieving effective heat transfer in the reactor. While it is not the scope of this chapter to find an optimal solution, the authors would suggest to submerge the HE to where the influent and the return sludge are mixed or an external solution may be applied if the mixing in the tank could be compromised by the geometry of the heat exchanger.

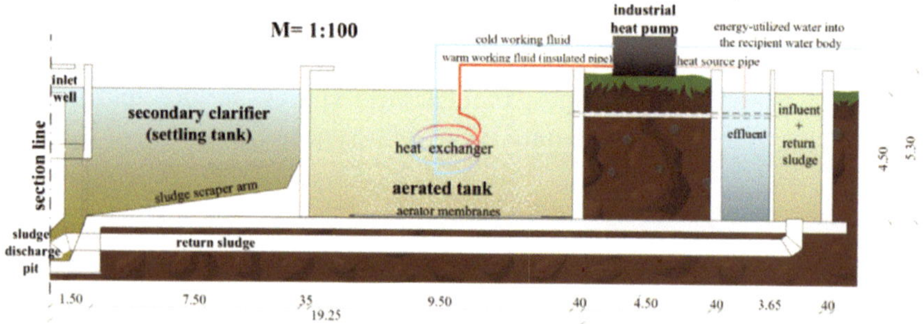

FIGURE 5.8 Schematic design of introducing heat pump and heat exchangers to utilize waste heat in the aerated tank.

Source: Reprinted from Somogyi et al. (2018a). Open access. https://creativecommons.org/licenses/by/3.0/deed.en_US

5.5.3 COOLING WITH EXCESS HEAT

Absorption chillers can use excess heat to produce chilled water. Since the hot water temperature requirement for absorption chiller is relatively high, above 353 K, the waste heat from the CHP serves a better purpose. Wastewater could be used for direct cooling, though it would result in increased temperature of the effluent and may cause elevated risk to public health, so it is crucial to take the water quality requirements or guidelines regarding discharge temperature into consideration and to consult with the water protection authority.

5.6 WASTEWATER HEAT POTENTIAL OF EUROPE

In the previous section, it was shown that the municipal WWTPs have significant energy content even if it falls into the low-grade heat category. To show the magnitude of the reclaimable energy, the wastewater heat potential was calculated for Europe.

First, information on the amount of wastewater, i.e., the sizes of wastewater agglomerations (given in population equivalent) had to be assessed. The dataset reported by Member States in the frame of the Urban Wastewater

Treatment Directive (UWWTD) (EEA, 2017) served as the basis of the analysis but had to be completed with information from the Dissemination Platform of UWWTD (DGENV, 2017) on Finland and Spain and from the EFTA report (EAI, 2013) on Iceland.

The second step was to calculate the theoretical wastewater heat potential for each wastewater treatment plant in the database according to Eqn. (5.32). Since the wastewater flow-rates were not stored in numerous cases the average hydraulic load was assumed to, be 0.12 m^3/d capita uniformly (specific water consumption can vary between 0.06 to 0.40 m^3/d capita). In order to provide continuous operation the minimum flow is taken into consideration (Eqn. (5.33)). A factor of 0.45 defined by Grazer Energie Agentur (2007) was used.

$$Q = \dot{V}_{\min} \cdot \rho_w \cdot c_{p,w} \cdot \Delta T \quad (5.32)$$

where,

$$\dot{V}_{\min} = \dot{V}_{avg} \cdot 0.45 \quad (5.33)$$

The temperature drop (ΔT) was assumed to be 2 K (GEA, 2007), density (ρ_w) and specific heat capacity of water ($c_{p,w}$) was taken to be 1000.0 kg/m^3 and 4184.1 J/kg·K (NIST) in all cases, respectively.

The data of the urban wastewater agglomerations could be linked to a town or city where the WWTP was situated. Locations, population equivalent and theoretical heat potential data were processed in an attribute table using QGIS 3.0 software (QGIS, 2018) and the point data were overlaid with planning and statistical regions (level 2 of Nomenclature of Territorial Units for Statistics-NUTS2). This way the amount of wastewater in compliance with the provisions of legislation in a NUTS2 region could be determined. NUTS2 was chosen because in case of using the first level (large regions) the characteristics of region with regards to wastewater treatment was virtually erased while the NUTS3 level would have meant an incomprehensible amount of polygons to work with.

The following map (Figure 5.9) shows the distribution of wastewater heat potential per capita in each region. In this case, all of the WWTPs were taken into consideration and compared against the overall population of the regions. Austria, Iceland, Denmark, most part of Germany, and about half of Italy and Switzerland have the specific heat potential higher than 9.0 W/capita, while the rest of the regions are below this value. The average specific heat capacity was 7.6 W/capita with a standard deviation of 3.0 W/capita, the median being 7.2 W/capita.

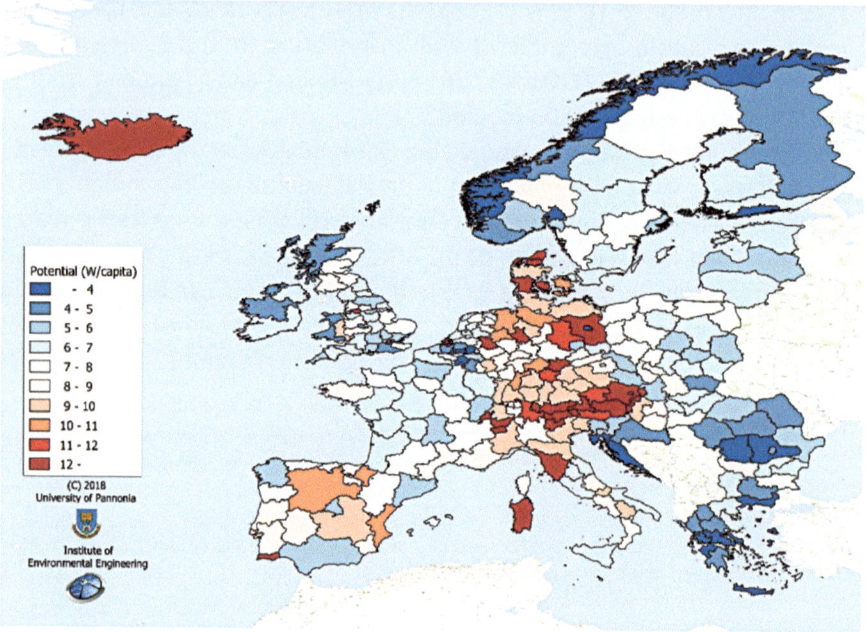

FIGURE 5.9 Distribution of specific potential in the NUTS2 regions.

The heat potential in a region was found to be dependent on two factors: the ratio of generated and treated municipal wastewater and the amount of wastewater entering the treatment plant in comparison to the population number. The results would vary even more if the actual hydraulic load could have been taken into account, instead of a constant specific water consumption value. Nonetheless, since in many cases the wastewater treatment plant of an agglomeration accepts sewage from the industry and service sector as well, the ratio of treated wastewater in population equivalent and the number of residents can be as high as 4.85. This NUTS2 region is Greater Manchester which is one of the biggest economic centers in the UK. The following regions have outstanding (>3.0) PE-number of resident's ratios: Sardinia and South Tyrol in Italy, Vorarlberg in Austria and the Lake Geneva region in Switzerland. All four of these have booming economies and their service industries play a dominant role in that. Greater Manchester has also the largest wastewater heat potential per capita (25.4 W/capita) but the region as a whole has only 70 MW potential. On the other hand, the maximum potential was calculated in the Parisian region (88 MW) where the specific value is 7.6 W/capita.

The threshold of economic feasibility based on the calculations of Gutzwiller et al. (2008) is 1 MW. That can be provided from the wastewater treatment plant of, 30000 PE assuming 6 K reductions in the effluent temperature. As it shown in Figure 5.10, 80% of the WWTPs are below 30,000 PE but responsible for treating only 19% of the wastewater. While Gutzwiller et al. (2008) took 6 K drop into consideration it has to be borne in mind that all of the following evaluations were based on a conservative estimate of 2 K decrease in the effluent temperature.

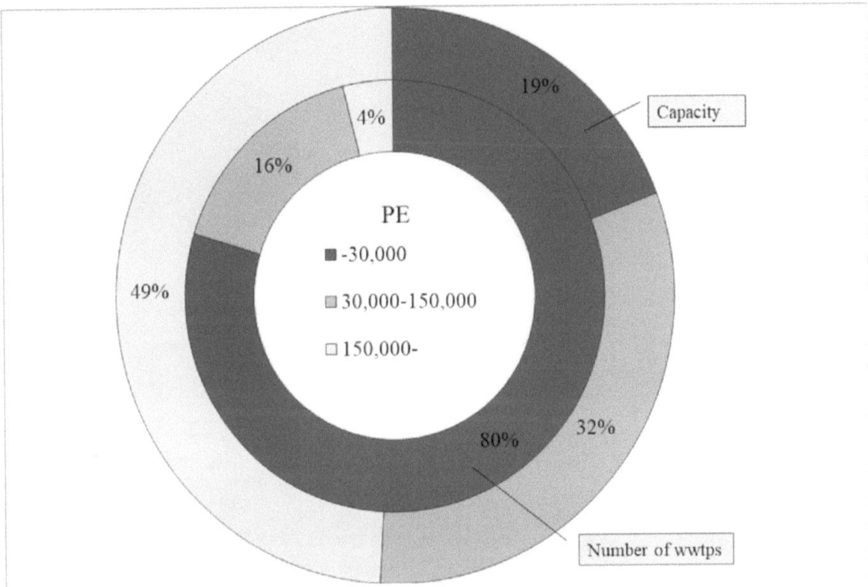

FIGURE 5.10 Percentage of wastewater treated in treatment plants of different size: outside is the number of plant; inside is the cumulated capacity (in PE) in each category.

Source: Data *source:* EEA (2017), DGENV (2017), and EAI (2013).

From Figure 5.11, regions with the highest economically feasible potential (considering WWTPs above 30,000 PE) can be identified. The overwhelming majority (62%) of the 279 regions have capacities of less than 10 MW while only a handful of regions are above 50 MW. Northern Italy could easily take advantage of its wastewater heat resources in district heating due to the population density. On the other hand, some of the southern regions have high values while it is less likely though not unimaginable that heating would be needed. In these cases, the energy could still be used to produce DHW. The authors would also like to point out that a region having a low value of total and/or specific wastewater heat potential does not mean that feasible utilization of the energy could not be achieved. The calculation was based on conservative assumption and did not take into consideration the ambient temperature of air and receiving water body. It is not a coincidence that northern countries started to install systems to reclaim waste heat and use it in district heating. As it was already mentioned, if the circumstances allow it the temperature drop and thus the retainable heat potential may be increased.

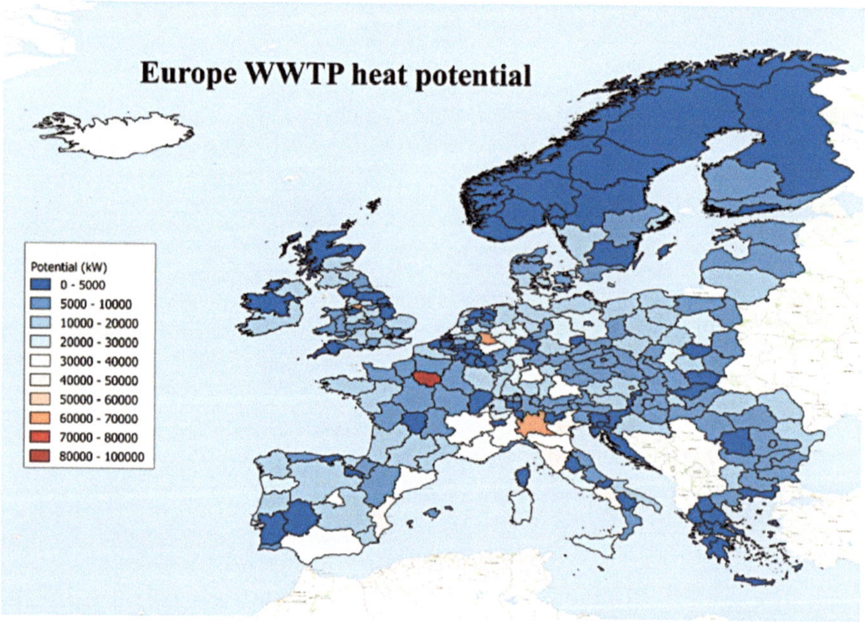

FIGURE 5.11 Distribution of economically feasible potential in the NUTS2 regions.

The overall potential (in case of 2 K temperature drop) for the EU28, Norway, Iceland, and Switzerland was found to be 3,926 MW if all WWTPs are accounted for and 3,166 MW if only the WWTPs above 30,000 PE are considered. While this amount may seem small compared to the total capacity of renewable energy (512,000 MW in Europe, 2017), it is three times more than the total renewable capacity of Hungary and is comparable to Ireland's; meanwhile the European renewable municipal waste capacity is only 1.4 times higher (for North America it is only 1,192 MW) (IRENA, 2018). Options of wastewater heat reclamation should be examined considering the close vicinity of the wastewater plant, the actual heat demand of the region and climatic conditions along with the return time of the investment and abundance of other resources.

5.7 CONCLUSION

The WWTPs of today have two main goals, to meet the quality requirements of the effluent in order to mitigate environmental pollution and to keep operation costs as low as possible. Though the technology is considered to be end-of-pipe, waste heat recovery can play an important role to help plants become sustainable from an energy point of view. The data presented in this book chapter prove that the excess thermal energy of municipal wastewater is abundant and may be utilized for various purposes. The energy balance, which was be set up for a municipal WWTP serving as benchmark, included the biological train, the secondary settler, the AD and the buildings to determine the sinks and sources of heat. Options to utilize the reclaimed energy (directly or with heat pumps) were discussed such as heating/cooling and hot water supply as well as increasing efficiency and robustness of small WWTPs. While there is surplus thermal energy stored in the wastewater, recovering would need additional electric energy. In the example that would be more than the biogas, production could cover.

Additionally, the theoretical wastewater heat potential of Europe has also been assessed to show the magnitude of recoverable heat from municipal wastewater plants on a regional level. This was to give an insight into which areas are the most promising from this aspect. Clearly, besides recovering the chemical energy from the wastewater sludge the waste heat poses opportunities in several regions which should not be overlooked in the long run. The research was supported by Széchenyi (2020) under the GINOP-2.3.2-15-2016-00016 project.

KEYWORDS

- **anaerobic digestion**
- **benchmark simulation model no. 2**
- **coefficient of performance**
- **domestic hot water**
- **population equivalent**
- **Urban Wastewater Treatment Directive**

REFERENCES

Abdel-Aal, M., et al., (2014). Modeling the viability of heat recovery from combined sewers. *Water Sci. and Technol., 70*(2), 297–306.

Alex, J., et al., (2018). *Benchmark Simulation Model No. 2 (BSM2)*. [Online]. http://iwa-mia.org/benchmarking/#BSM2 (accessed on 10 June 2020).

Amann, A., et al., (2018). Environmental impacts of phosphorus recovery from municipal wastewater. *Resources Conservation and Recycl., 130*, 127–139.

Appels, L., et al., (2008). Principles and potential of the anaerobic digestion of waste-activated sludge. *Progress in Energy and Combustion Sci., 34*, 755–781.

Astals, S., et al., (2012). Balancing hygienization and anaerobic digestion of raw sewage sludge. *Water Research, 46*, 6218–6227.

Austin, B. T., & Sumathy, K., (2011). Transcritical carbon dioxide heat pump systems: A review. *Renew. Sustainable Energy Rev., 15*, 4013–4029.

Bakhshi, Z., Jauffur, S., & Frigon, D., (2018). Assessing energy benefits of operating anaerobic digesters at low temperature with solids pre-ozonation. *Renew. Energ., 115*, 1303–1311.

Baqir, A. S., et al., (2016). Heat transfer measurement in a three-phase spray column direct contact heat exchanger for utilization in energy recovery from low-grade sources. *Energy Conv. and Manag., 126*, 342–351.

Batstone, D., Pons, M. N., & Vanrolleghem, P. A., (2007). *Temperature Effects in BSM1_LT and BSM2,* [Online] https://pdfs.semanticscholar.org/f77a/3859db443322d8fa7f186a2235 87a760820d.pdf (accessed on 10 June 2020).

Bedoić, R., & Filipan, V., (2018). Heating performance analysis of a geothermal heat pump working with different zeotropic and azeotropic mixtures. *J. Sustain. Dev. Energy Water Environ. Syst., 6*(2), 240–253.

Bixio, D., et al., (2006). Wastewater reuse in Europe. *Desalination, 187*(3), 89–101.

Boiocchi, R., et al., (2017). Techno-economic analysis of resource recovery technologies for wastewater treatment plants. *Computer Aided Chemical Eng., 40*, 1945–1950.

Bolzonella, D., et al., (2012). High rate mesophilic, thermophilic, and temperature phased anaerobic digestion of waste activated sludge: A pilot scale study. *Waste Manag., 32*(6), 1196–1201.

Carrère, H. C., et al., (2010). Pretreatment methods to improve sludge anaerobic degradability: A review. *J. Hazard. Materials, 183*(1), 1–15.

Cavinato, C., et al., (2013). Mesophilic and thermophilic anaerobic co-digestion of waste activated sludge and source sorted biowaste in pilot-and full-scale reactors. *Renew. Energ., 55*, 260–265.

Chae, K. J., & Kang, J., (2013). Estimating the energy independence of a municipal wastewater treatment plant incorporating green energy resources. *Energ. Conv. and Manag., 75*, 664–672.

Chae, K. J., & Ren, X., (2016). Flexible and stable heat energy recovery from municipal wastewater treatment plants using a fixed-inverter hybrid heat pump system. *Applied Energ., 179*, 565–574.

Chua, K. J., Chou, S. K., & Yang, W. M., (2010). Advances in heat pump systems: A review. *Applied Energ., 87*(12), 3611–3624.

Cipolla, S. S., & Maglionico, M., (2014a). Heat recovery from urban wastewater: Analysis of the variability of flow rate and temperature in the sewer of bologna, Italy. *Energ. Procedia., 45*, 288–297.

Cipolla, S. S., & Maglionico, M., (2014b). Heat recovery from urban wastewater: Analysis of the variability of flow rate and temperature. *Energ. Building.*, *69*, 122–130.

Corbala-Robles, L., et al., (2016). Effect of foam on temperature prediction and heat recovery potential from biological wastewater treatment. *Water Res.*, *95*, 340–347.

Culha, O., et al., (2015). Heat exchanger applications in wastewater source heat pumps for buildings: A key review. *Energ. and Buildings*, *104*, 215–232.

Cunault, C., Burton, C., & Pourcher, A., (2013). The impact of fouling on the process performance of the thermal treatment of pig slurry using tubular heat exchangers. *J. Env. Manag.*, *117*, 253–262.

Cunault, C., et al., (2013). Characteristics and composition of fouling caused by pig slurry in a tubular heat exchanger-recommended cleaning systems. *J. Env. Manag.*, *117*, 17–31.

Davis, M. L., (2010). *Water and Wastewater Engineering: Design Principles and Practice*, McGraw-Hill.

De Mes, T. Z. D., et al., (2003). Methane production by anaerobic digestion of wastewater and solid wastes. *Bio-Methane and Bio-Hydrogen*, 58–102.

Directorate General Environment (DGENV), (2017). *Urban Waste Water Treatment Directive: Dissemination Platform.* [Online] http://uwwtd.oieau.fr (accessed on 10 June 2020).

Drechsel, P., Mahjoub, O., & Bernard, K., (2015). Social and cultural dimensions in wastewater use. In: *Wastewater: Economic Asset in an Urbanizing World* (pp. 75–92). Springer: Dordrecht.

Elías-Maxil, J. A., et al., (2014). Energy in the urban water cycle: Actions to reduce the total expenditure of fossil fuels with emphasis on heat reclamation from urban water. *Renew. Sustainable Energy Rev.*, *30*, 808–820.

Environment Agency of Iceland (EAI), (2013). *Report to the EFTA Surveillance Authority Regarding the Implementation of Directive 91/271/EU on the Treatment of Wastewater from agglomerations.* [Online] https://www.ust.is/library/Skrar/utgefid-efni/Annad/stoduskyrsla_skolpmal.pdf (accessed on 10 June 2020).

Erhorn, H., et al., (2018). "Neckar Park Stuttgart": District heat from wastewater. *Energ. Procedia.*, *149*, 465–472.

European Environmental Agency (EEA), (2017). *Water Base-UWWTD: Urban Waste Water Treatment Directive.* [Online] https://www.eea.europa.eu/data-and-maps/data/waterbase-uwwtd-urban-waste-water-treatment-directive-5 (accessed on 10 June 2020).

Fernández-Arévalo, T., et al., (2014). New systematic methodology for incorporating dynamic heat transfer modeling in multi-phase biochemical reactors. *Water Res.*, *60*, 141–155.

Fernández-Arévalo, T., et al., (2017). Quantitative assessment of energy and resource recovery in wastewater treatment plants based on plant-wide simulations. *Water Research*, *118*, 272–288.

Friedler, E., et al., (2006). Study of urban population attitudes towards various wastewater reuse options: Israel as a case study. *J. Environ. Manag.*, *81*, 360–370.

Frijns, J., Hofman, J., & Nederlof, M., (2013). *The Potential of Wastewater as Energy Carrier: Energy Convers. Manag.*, *65*, 357–363.

Fryer, M., & Gray, N. F., (2012). Foaming scum index (FSI): A new tool for the assessment and characterization of biological mediated activated sludge foams. *J. Envi. Manag.*, *110*, 8–19.

Gasser, L., et al., (2017). High efficiency heat pumps for low temperature lift applications. In: *12th IEA Heat Pump Conference*. Rotterdam, Netherlands: Stichting HPC.

Gebreeyessus, G. D., & Jenicek, P., (2016). Thermophilic versus mesophilic anaerobic digestion of sewage sludge: A comparative review. *Bioeng.*, *3*(2).

Gianico, A., et al., (2013). Reduced temperature hydrolysis at 134°C before thermophilic anaerobic digestion of waste activated sludge at increasing organic load. *Bioresource Technol.*, *143*, 96–103.

Grady, Jr. C. L., et al., (2011). *Biological Wastewater Treatment*. CRC Press.

Grazer Energie Agentur (GEA), (2007). *Technical and Economic Market Potential of Thermal Energy Use from Wastewater in Austria*. Graz, Austria.

Gu, Y., et al., (2017). The feasibility and challenges of energy self-sufficient wastewater treatment plants. *Applied Energ.*, *204*, 1463–1475.

Gutzwiller, S., Reto, R., & Eicher, H., (2008). *Wastewater Heat Utilization Potential, Economy and Advancement*. Swiss Federal Office of Energy, Berne.

Han, D., et al., (2017). Enhanced methane production and wastewater sludge stabilization of a continuous full scale thermal pretreatment and thermophilic anaerobic digestion. *Bioresource Technol.*, *245*, 1162–1167.

Henze, M., et al., (2008). *Biological Wastewater Treatment, Principles, Modeling, and Design*. IWA Publishing, London.

Hepbasli, A., et al., (2014). A key review of wastewater source heat pump (WWSHP) systems. *Energy Conversion and Manag.*, *88*, 700–722.

Hofman, J., et al., (2011). Water and energy as inseparable twins for sustainable solutions. *Water Sci. and Technol.*, *63*(1), 88–92.

IRENA, (2018). *Renewable Capacity Statistics*. International Renewable Energy Agency (IRENA), Abu Dhabi.

Iziomon, M. G., Mayer, H. E., & Matzarakis, A. N., (2003). Downward atmospheric longwave irradiance under clear and cloudy skies: Measurement and parameterization. *J. Atmospheric and Solar-Terrestrial Phys.*, *65*(10), 1107–1116.

Jenicek, P., et al., (2013). Energy self-sufficient sewage wastewater treatment plants: Is optimized anaerobic sludge digestion the key? *Water Sci. and Technol.*, *68*(8), 1739–1744.

Jin, S., & Hrnjak, P., (2017). Effect of end plates on heat transfer of plate heat exchanger. *International Journal of Heat and Mass Transfer*, *108*, 740–748.

Kardos, L., et al., (2011). Comparing of thermophilic and mesophilic anaerobic fermented sewage sludge based on chemical and biochemical tests. *Appl. Ecol. Environ. Res.*, *9*, 293–302.

Kim, M., Ahn, Y. H., & Speece, R., (2002). Comparative process stability and efficiency of anaerobic digestion; mesophilic vs. thermophilic. *Water Res.*, *36*, 4369–4438.

Kollmann, R., et al., (2017). Renewable energy from wastewater-practical aspects of integrating a wastewater treatment plant into local energy supply concepts. *J. Clean. Prod.*, *155*, 119–129.

Kottek, M., et al., (2006). World map of the Köppen-Geiger climate classification updated. *Meteorol. Z.*, *15*, 259–263.

Kreith, F., & Boehm, R. F., (1999). *Heat and Mass Transfer*, *Mechanical Engineering Handbook*. Boca Raton: CRC Press LLC.

Kretschmer, F., Simperler, L., & Ertl, T., (2016). Analyzing wastewater temperature development in a sewer system as a basis for the evaluation of wastewater heat recovery potentials. *Energy and Buildings*, *128*, 639–648.

Kristjansson, H., & Bøhm, B., (2006). Advanced and traditional pipe systems optimum design of distribution and service pipes. *Euro Heat and Power International*, *3*(2), 11.

Kumar, B., Soni, A., & Singh, S. N., (2018). Effect of geometrical parameters on the performance of chevron type plate heat exchanger. *Experimental Thermal and Fluid Sci.*, *91*, 126–133.

Lahnsteiner, J., & Lempert, G., (2007). Water management in Windhoek, Namibia. *Water Sci. Technol.*, *55*(1/2), 441–448.

Langergraber, G., et al., (2008). Generation of diurnal variation for influent data for dynamic simulation. *Water Sci. Technol.*, *57*(9), 1483–1486.

Larsen, T., (2015). CO_2-neutral wastewater treatment plants or robust, climate-friendly wastewater management: A systems perspective. *Water Research*, *87*, 513–521.

Lee, S., et al., (2017). Thermo-environ-economic modeling and optimization of an integrated wastewater treatment plant with a combined heat and power generation system. *Energy Conversion and Manag.*, *142*, 385–401.

Lei, Y., et al., (2017). Design and performance analysis of the novel shell-and-tube heat exchangers with louver baffles. *Applied Thermal Eng.*, *125*, 870–879.

Lindström, H., (1985). Experiences with a 3.3 MW heat pump using sewage water as heat source. *Journal of Heat Recovery Systems*, *5*, 33–38.

Lines, J. R., (1991). Heat exchangers in municipal wastewater treatment plants. *Water Engineering and Manag.*, *138*(9), 28–29.

Longo, S., et al., (2016). Monitoring and diagnosis of energy consumption in wastewater treatment plants. A state of the art and proposals for improvement. *Applied Energy*, *179*, 1251–1268.

McCarty, P. L., Bae, J., & Kim, J., (2011). Domestic wastewater treatment as a net energy producer-can this be achieved? *Environmental Science and Technol.*, *45*(17), 7100–7106.

Menberg, K., et al., (2013). Sub-surface urban heat islands in German cities. *Sci. Total Env.*, *442*, 123–133.

Meyer, J. P., (2011). Heat pumps. In: *Thermopedia A-to-Z Guide to Thermodynamics, Heat and Mass Transfer, and Fluids Engineering, Heat Pumps* [Online] http://thermopedia.com/content/837/ (accessed on 10 June 2020).

Miah, J., et al., (2015). Maximizing the recovery of low grade heat: An integrated heat integration framework incorporating heat pump intervention for simple and complex factories. *Applied Energy*, *160*, 172–184.

Mikkonen, L., et al., (2014). *Heat Recovery from Wastewater: Assessing the Potential in Northern Areas* (pp. 161–164). Water Research at Oulu Univ.

Minton, P. E., (1970). Designing spiral heat exchangers. *Chemical Engineering*, *4*, 103–112.

Mo, W., & Zhang, Q., (2013). Energy-nutrients-water nexus: Integrated resource recovery in municipal wastewater treatment plants. *J. Environmental Manag.*, *127*, 255–267.

Mohsen, M., Hassan, H., & Hadi, F., (2017). Multi-objective optimization of shell-and-tube heat exchanger by constructal theory. *Applied Thermal Eng.*, *125*, 9–19.

Mosayeb Nezhad, M., et al., (2018). Techno-economic assessment of biogas-fed CHP hybrid systems in a real wastewater treatment plant. *Appl. Therm. Eng.*, *129*, 1263–1280.

Mousel, D., Palmowski, L., & Pinnekamp, J., (2017). Energy demand for elimination of organic micro pollutants in municipal wastewater treatment plants. *Sci. Total Envi.*, *575*, 1139–1149.

National Institute of Standards and Technology (NIST). *Isothermal Properties of Water*. [Online] http://webbook.nist.gov/ (accessed on 10 June 2020).

Neugebauer, G., et al., (2015). Mapping thermal energy resource potentials from wastewater treatment plants. *Sustainability*, *7*, 12988–13010.

Nowak, O., Enderle, P., & Varbanov, P., (2015). Ways to optimize the energy balance of municipal wastewater systems: Lessons learned from Austrian applications. *J. Clean. Prod.*, *88*, 125–131.

Osorio, F., & Torres, J., (2009). Biogas purification from anaerobic digestion in a wastewater treatment plant for biofuel production. *Renew. Energ.*, *34*(10), 2164–2171.

Panepinto, D., et al., (2016). Evaluation of the energy efficiency of a large wastewater treatment plant in Italy. *Applied Energ.*, 404–411.

Papa, M., et al., (2017). How far are we from closing the loop of sewage resource recovery? A real picture of municipal wastewater treatment plants in Italy. *J. Environmental Manag.*, *198*, 9–15.

Pavičević, M., et al., (2017). Hourly optimization and sizing of district heating systems considering building refurbishment: Case study for the city of Zagreb. *Energ.*, *137*, 1264–1276.

Perry, R., & Green, D., (2008). *Perry's Chemical Engineers' Handbook* (8th edn.). McGraw-Hill Education,

Picón-Núñez, M., et al., (2007). Shortcut design approach for spiral heat exchangers. *Food and Bioproducts Processing.*, *85*, 322–327.

Plósz, B. G., Liltved, H., & Ratnaweera, H., (2009). Climate change impacts on activated sludge wastewater treatment: A case study from Norway. *Water Sci. Technol.*, *60*, 533–541.

Puchajda, B., & Oleszkiewicz, J., (2008). Impact of sludge thickening on energy recovery from anaerobic digestion. *Water Sci. Technol.*, *57*(3), 395–401.

QGIS Development Team, (2018). *QGIS Geographic Information System.* Open Source Geospatial Foundation Project. [Online] https://www.qgis.org/en/site/ (accessed on 10 June 2020).

Raheem, A., et al., (2018). Opportunities and challenges in sustainable treatment and resource reuse of sewage sludge: A review. *Chem. Eng. J.*, *337*, 616–641.

Raja, B. D., Jhala, R. L., & Patel, V., (2018). Thermal-hydraulic optimization of plate heat exchanger: A multi-objective approach. *Int. J. Thermal Sci.*, *124*, 522–535.

Rezaie, B., & Rosen, M. A., (2012). District heating and cooling: Review of technology and potential enhancements. *Appl. Energ.*, *93*, 2–10.

Ruffino, B., et al., (2015). Improvement of anaerobic digestion of sewage sludge in a wastewater treatment plant by means of mechanical and thermal pre-treatments: Performance, energy and economical assessment. *Bioresource Technol.*, *175*, 298–308.

Schmid, F., (2008). *Sewage Water: Interesting Heat Source for Heat Pumps and Chillers.* Swiss Energy Agency for Infrastructure Plants, Zürich, Switzerland.

Seybold, C., & Brunk, M. F., (2013). In-house waste water heat recovery. *REHVA J.*, 18–21.

Shah, R. K., & Sekulic, D. P., (2003). *Fundamentals of Heat Exchanger Design.* John Wiley and Sons.

Sheik, A. R., Muller, E. E. L., & Wilmes, P., (2014). A hundred years of activated sludge: Time for a rethink. *Front. Microbiol*, *5*(47), 1–7.

Shen, C., et al., (2014). An experimental and numerical study of a de-fouling evaporator used in a wastewater source heat pump. *Applied Thermal Eng.*, *70*(1), 501–509.

Shen, C., et al., (2018). A review on the current research and application of wastewater source heat pumps in China. *Thermal Sci. and Eng. Progress*, *6*, 140–156.

Shenghan, J., & Pega, H., (2017). Effect of end plates on heat transfer of plate heat exchanger. *Int. J. Heat and Mass Transfer*, *108*(A), 740–748.

Shinde, S., & Chavan, U., (2018). Numerical and experimental analysis on shell side thermo-hydraulic performance of shell and tube heat exchanger with continuous helical FRP baffles. *Thermal Sci. and Eng. Progress*, *5*, 158–171.

Somogyi, V., et al., (2018a). Enhanced pollution removal with heat reclamation in a small Hungarian wastewater treatment plant. *J. Sustain. Dev. Energy Water Environ. Syst.*, *6*(3), 494–504.

Somogyi, V., Sebestyén, V., & Domokos, E., (2018b). Assessment of wastewater heat potential for district heating in Hungary. *Energy*, *163*, 712–721.

Spriet, J., & Hendrick, P., (2017). Wastewater as a heat source for individual residence heating: A techno-economic feasibility study in the Brussels capital region. *J. Sustain. Dev. Energy Water Environ. Syst.*, *5*(3), 289–308.

Talati, S. N., & Stenstrom, M. K., (1990). Aeration-basin heat loss. *J. Env. Eng.*, *116*(1), 70–86.

Tańczuk, M., Kostowski, W., & Karaś, M., (2016). Applying waste heat recovery system in a sewage sludge dryer: A technical and economic optimization. *Energy Conversion and Management*, *125*, 121–132.

Tchobanoglous, G., et al., (2013). Metcalf and Eddy, Inc. *Wastewater Engineering: Treatment and Resource Recovery* (5thedn.). McGraw-Hill Education.

Tetens, O., (1930). About some meteorological terms. *Z. Geophys.*, *6*, 297–309.

Thackston, E. L., & Parker, F. L., (1972). Geographical influence on cooling ponds. *J. Water Pollution Control Federation*, *44*(7), 1334–1351.

Thulukkanam, K., (2013). *Heat Exchanger Design Handbook* (2ndedn.). CRC Press.

Tian, L., et al., (2012). Effect of calcium ions on the evolution of biofouling by *Bacillus subtilis* in plate heat exchangers simulating the heat pump system used with treated sewage in the 2008 Olympic Village. *Colloids and Surfaces B: Biointerfaces*, *94*, 309–316.

Titow, M. V., (2012). *PVC Technology*. Springer Science and Business Media.

Turovskiy, I. S., & Mathai, P. K., (2006). *Wastewater Sludge Processing*. Wiley.

Tyagi, V. K., & Lo, S. L., (2013). Sludge: A waste or renewable source for energy and resources recovery. *Renewable and Sus. Energ. Reviews*, *25*, 708–728.

Van, D. H. J., De Fooij, H., & Struker, A., (2016). Wastewater as a resource: Strategies to recover resources from Amsterdam's wastewater. *Resources, Conservation and Recycl.*, *113*, 53–64.

Van, H. A. C., & Van, D. L. J. G., (2012).*Handbook of Biological Wastewater Treatment: Design and Optimization of Activated Sludge Systems*. IWA.

Vende, P. E., et al., (2018). Efficiency of water spraying on a heat exchanger: Local characterization with the impacted surface. *Applied Thermal Eng.*, *128*, 684–695.

Wang, X., Kvaal, K., & Ratnaweera, H., (2017). Characterization of influent wastewater with periodic variation and snow melting effect in cold climate area. *Computers and Chem. Eng.*, *106*, 202–211.

Wanner, O., et al., (2005). Effect of heat recovery from raw wastewater on nitrification and nitrogen removal in activated sludge plants. *Water Research*, *39*(19), 4725–4734.

Werner, S., (2017). International review of district heating and cooling. *Energ.*, *137*, 617–631.

Wołoszyn, J., & Gołaś, A., (2017). Coefficient of performance stabilization in ground source heat pump systems. *J. Sustain. Dev. Energy Water Environ. Syst.*, *5*(4), 645–656.

Wong, L. T., Mui, K. W., & Guan, Y., (2010). Shower water heat recovery in high-rise residential buildings of Hong Kong. *Appl. Energ.*, *87*(2), 703–709.

WWAP (United Nations World Water Assessment Programme), (2014). *The United Nations World Water Development Report 2014: Water and Energy*. Paris.

Zahid, H. A., et al., (2018). Performance characteristics of a novel shell and tube heat exchanger with shell side interstitial twisted tapes for viscous fluids application. *Applied Thermal Eng.*, *134*, 248–255.

Zhang, L., et al., (2016). Hydrolysis rate constants at 10–25°C can be more than doubled by a short anaerobic pre-hydrolysis at 35°C. *Water Res., 104*, 283–291.

Zhang, Y., et al., (2018). A quantitative energy efficiency evaluation and grading of plate heat exchangers. *Energ., 142*, 228–233.

Zhao, X. L., et al., (2010). Study of the performance of an urban original source heat pump system. *Energy Convers. Manag., 51*, 765–770.

CHAPTER 6

Advances in Nitrogen Removal and Recovery Techniques in Wastewater

DOLORES HIDALGO, JESÚS M. MARTÍN-MARROQUÍN, and FRANCISCO CORONA

CARTIF Technology Center, Parque Tecnológico de Boecillo – 205, 47151, Boecillo, Valladolid, Spain, Phone: +34 983 546504, E-mail: dolhid@cartif.es (D. Hidalgo)

6.1 INTRODUCTION

In the current transition to a Circular Economy, it has become a critical challenge to maximally close nutrient cycles and migrates to a more effective and sustainable resource management, both from an economical and an ecological perspective (Puyol et al., 2017).

The situation today is clear: the agricultural demand for mineral fertilizers is continuously increasing and this dependency of agriculture on fossil reserve-based mineral fertilizers must be regarded as a very serious threat to future human food security. According to FAO (2016), world demand for total fertilizer nutrients ($N + P_2O_5 + K_2O$) is estimated to grow at 1.6% per annum, reaching 199 million tons (expected) by the end of 2019. Also, estimates of phosphorus reserves are highly uncertain, but based on population growth and future demand for nutrients, it is expected that depletion will occur in a short time.

Despite these circumstances, large amounts of nutrients are dispersed in the environment every day, in a controlled or uncontrolled way, through the disposal of waste streams. In addition, the intensification of animal production and the resulting manure excesses, combined with a limited availability of arable land for the disposal of waste (manure, sludge, etc.), and the excessive use of chemical mineral fertilizers, has led to surplus fertilization and nutrient accumulation in many soils worldwide. These facts have frequently caused environmental pollution.

As a consequence, it is clear that a new global effort is needed to draw a new scenario where improved nutrient use efficiency (NUE) and, at the same time, reduced nutrient losses provide the bases for a greener economy to produce more food and energy while reducing environmental impact.

There are many options to improve nutrients management. Agriculture is a key sector where acting. Improving NUE in crop production and in animal production is a good strategy, as it aims at increasing global food productions, optimizing the use of external resources, and minimizing environmental pollution due to these activities.

In reference to animal production, this was traditionally centered around homes and was conducted at small scales. However, as the demand for animal products, such as eggs, milk, and meat has grown worldwide, livestock farming has gradually become more separate, and animal production has become more intensive, particularly in developed countries. This has led to geographic concentration of animal production systems to link the feed, production, processing, distribution, and marketing components more closely, particularly for the production of poultry and swine. Consequently, animal numbers on farms have grown, farms have consolidated, and manure production has increased, often exceeding the capacity of nearby cropland to efficiently recycle manure nutrients. This over-application of manure has exacerbated problems in vulnerable areas with nitrate leaching to groundwater, ammonia, and nitrous oxide emissions to air and the saturation of soils with phosphorus to the point that phosphorus losses in surface flow and leaching are serious concerns.

In the group of key actions for waste and recycling, there are also several opportunities to optimize the management of nutrients. Recycling nitrogen and phosphorus from waste streams, such as municipal sewage systems, manure, or industrial effluents. The technology to do that exists, but it is not equally implemented around the world.

Several technologies are available today, for example, for wastewater treatment and nutrients recovery, however, from all these technologies, only a limited number is fully used at industrial scale. Indeed, the amount of nutrients recovered in Europe from wastewaters represents only a minimum percentage of the potential of these streams. The reluctance of wastewater treatment plants (WWTPs) managers for some techniques relies on several parameters that vary depending on the European region considered.

One of the greatest challenges is, thus, to implement existing technologies, especially considering the infrastructure that may be needed, or redesigning and upgrading existing treatment systems. This is often a matter for governments due to the large costs associated with these actions.

The present chapter shows an overview of the available and emerging processing techniques for nutrient recovery in wastewater illustrated with some success stories of utilization at industrial scale in order to clarify the gap existing between available technologies and society demands.

6.2 TECHNOLOGIES FOR NUTRIENTS RECOVERY

Addressing global nutrient needs, as well as recovering energy and water from wastewater streams are powerful drivers for change in the wastewater industry. This has led to two major position papers for novel domestic wastewater processes that are low energy or energy generating, designed to produce wastewater fit for reuse (given specific purposes), and designed to allow recovery of nutrients (Batstone et al., 2015).

In recent years, a number of groundbreaking developments in the field of wastewater treatment and nutrients recovery have paved the way for utilities to better manage and improve operations. Biological technologies, advanced reuse and recycling techniques and progressive green-based practices have led to various economic, environmental, and societal benefits that can help reduce costs, conserve energy, sustain the environment, and improve customer service.

Figure 6.1 presents the main steps required for nutrients recovery and reuse. Nutrient removal can be in the form of concentration in the biomass such as in activated sludge or algae, or physicochemical concentration into precipitate or adsorption on media.

Wastewater typically contains a large amount of nutrients that can pose a harmful threat to infrastructure and the environment, causing problems such as eutrophication in water bodies and a buildup of struvite in mechanical systems. By utilizing nutrient recovery, wastewater plants can mitigate these challenges while improving water quality and meeting stringent discharge limits. The recovery process also offers municipalities an opportunity to generate revenue while providing agricultural businesses with refined, usable nitrogen and phosphorus—an increasingly scarce natural resource. Moreover, it enables wastewater entities to serve as more than just treatment facilities but ultimately as resource recovery agents, transforming the perception of traditional wastewater treatment (Haddaway, 2015).

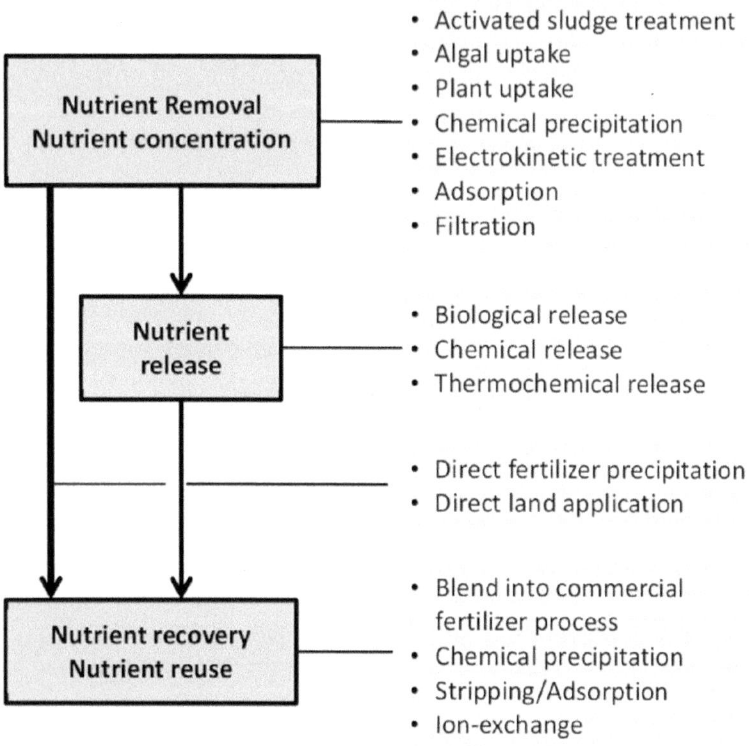

FIGURE 6.1 The three main steps in nutrient recovery and reuse. *Source:* Reprinted from Oleszkiewicz et al., 2015. Open access.

Domestic wastewater contains approximately 0.8–1.5 kg of phosphorus per capita per year, of which 25–60% is originated by synthetic detergents. The average concentration of phosphorus within raw wastewater is approximately 10 mg/l, of which 70% is in the soluble inorganic form.

Effluent standards commonly require a reduction of more than 90% of the total P initially present in the raw wastewater. Since the majority of the phosphorus cannot be embodied by the biological sludge it much be removed by other processes.

Faeces, urine, and food-processing discharges are the primary sources of nitrogen in domestic wastewater, with a per capita contribution in the range of 3.5–5.5 kgN/y. Around 40% is in the form of ammonia and 60% bound in organic matter. After the biological treatment almost all (around 92%), the remaining nitrogen in the effluent is converted into the inorganic form.

There are many nutrients recovery technologies currently available or under development in Europe and North America. These technologies vary considerably in their process, cost, application, and nutrient recovery capabilities. The greatest reason for this variability is because of the contrasting nutrient management needs in different areas of Europe and North America. Despite this variability, in general mechanical, chemical, and biological recovery techniques can be used as standalone technologies, or in combination with other technologies, to recover nutrients from wastewater.

6.2.1 BASIFICATION

Basification is a common step previous to several techniques of nitrogen and phosphorous recovery. The main technique to recover phosphorus from wastewater in a form which can be processed by the industry is to precipitate it with hardness ions (e.g., calcium). Nitrogen can be recovered from wastewater as dissolved ammonium. In both cases, the pH of the wastewater must be increased to very high levels (around 11). Under these conditions, phosphorus precipitates with hardness ions (calcium, magnesium) resulting in a solid compound which can be separated from the liquid wastewater. If the concentration of hardness ions exceeds the concentration of phosphorus, the normal presence of hardness within the wastewater can be used for phosphorus removal without addition of any chemicals (Cohen, 2010).

At very high pH, ammonium nitrogen is also transformed to gaseous ammonia, which can be stripped off the wastewater and concentrated in an acid liquid phase.

The method for increasing the pH of wastewater is mainly based on chemical addition of lime or sodium hydroxide. Lime is the cheapest option but its use is problematic due to the calcium content of lime. Usually, very high doses of lime are required to reach pH above 10, which causes severe scaling problems and leads to the accumulation of excess calcium within the treated wastewater. On the other hand, the use of sodium dioxide for increasing the pH of wastewater is expensive and increases the sodium content of the treated wastewater.

Trends in basification are focused now on increasing the pH at high levels without the addition of chemicals and at low costs, making this operation feasible and economical.

6.2.2 ION EXCHANGE AND ADSORPTION-BASED METHODS

Ion exchange is the process of transfer of an ion on a solid surface for a similarly charged ion in solution. This exchange occurs in equivalent amounts of charge and is reversible. Ion exchange and adsorption-based processes are highly relevant because of their unique properties such as high selectivity for $NH4^+$, high removal, fast uptake kinetics, and regeneration, less space requirements and simplicity of application and operation (Sengupta et al., 2015). The most popular ion exchanger/adsorbent for nitrogen recovery is zeolite.

Once adsorbents/ion-exchangers are exhausted, nitrogen recovery and reuse opportunities are exploited. The loaded zeolites can be applied directly onto agricultural fields as slow-release fertilizers (Taddeo et al., 2017). Most popular is the regeneration technique using NaCl solution where NH_4^+ is desorbed and exchanged with Na^+ in solution. This provides a concentrated stream of NH_4^+. Other regeneration techniques such as acid regeneration (Lee et al., 2018), heating regeneration (particularly useful for exothermic adsorption process) (Sengupta et al., 2015), and biological regeneration (Von Eckstaedt et al., 2016) can also be used, depending on the recovery process employed subsequently.

PO_4^{-3} can be removed from wastewater using ion exchange just as discussed with ammonia however; synthetic resins are typically used for PO_4-P removal. This process has been researched for PO_4-P removal due to its ability to work in variable wastewater composition, flow-rate, and temperature; however, the challenges associated with PO_4-P ion exchange include the following: poor selectivity towards PO_4-P over other ions, inefficient regeneration, and loss of loading capacity due to fouling (Willians, 2013).

Adsorption and ion exchange can be used either, in the form of a fixed-bed column system through which wastewater is passed, or it can be dispersed throughout wastewater to be settled out in a downstream clarifier. Dispersal is best used when the sorbent material cannot be reused, if recovery of the solute is not necessary, or if increasing sludge volume is not a problem; powdered activated carbon is often used in this manner. When appropriate sorbents are used, fixed-bed column systems have several major advantages over other methods of phosphorus removal and recovery. Downstream from the column, the wastewater effluent contains non-detectable or very low phosphorus concentrations until exhaustion of the sorbent is reached (Sengupta et al., 2015).

With many different resins to choose from, efficiency of regeneration could be optimized by testing various media with multiple regenerants. Regenerant could be beneficially used to recover PO_4-P: thereby providing a valuable product to offset process costs. Previous work has been performed using PO_4-P ion exchange with struvite precipitation of the regenerant (Sengupta and Pandit, 2011). Zeolite regenerant has been proposed as a N or K source in struvite precipitation of this process (Petruzzelli et al., 2003).

6.2.3 BIOELECTROCHEMICAL SYSTEMS (BES)

In bioelectrochemical systems (BES), chemical energy contained in the organic matter is directly converted to electrical energy by certain microorganisms (Figure 6.2). Bio-oxidation of organic matter generates electrons and, simultaneously, produces other valuable compounds. At the anode, anaerobic bacteria oxidize the organic matter. Microorganisms transfer these generated electrons to anode, which is connected over an external circuit to cathode—the site of reduction reaction. Based on the cathodic reaction, BESs can be classified into two types: galvanic and electrolytic cells (Sengupta et al., 2015).

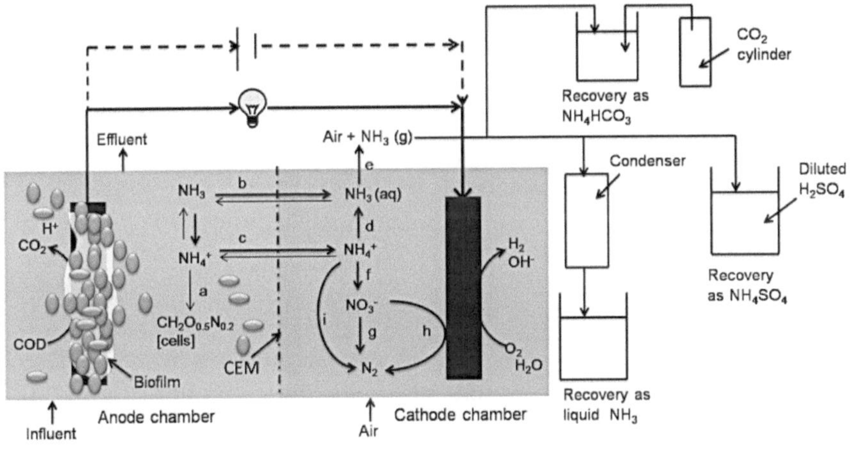

FIGURE 6.2 Schematic diagram showing possible nitrogen transformation processes in bioelectrochemical systems. *Source:* Reprinted with permission from Nancharaiah et al., 2016). © Elsevier.

BES scheme is potentially a sustainable way of treating wastewater as it produces electricity and recovers ammonia and utilizes low-grade substrates such as wastewater itself as an electron source, thus reducing the use of carbon source.

In BES, the ammonium ion gets transported from the anode chamber to the cathode chamber chiefly by two distinct processes such as diffusion and migration. Diffusion of NH_3 is induced by the concentration gradient across the cation exchange membrane, while the migration of $NH4^+$ ion is driven by the electric field. The alkaline pH generated in the cathode chamber converts the $NH4^+$ ion that has entered from the anode chamber to NH_3(aq). The NH_3(aq) present in the catholyte is then partitioned into the gaseous phase (NH_3(g)) for recovery by stripping. Thus, BES offers an opportunity for direct recovery of ammonium nitrogen in the form of NH_3.

Ammonia that is vented from the cathode chamber can be harvested in the form of ammonium sulfate (NH_3SO_4) by passing it through dilute sulfuric acid. Alternatively, recovered NH_3 can be condensed and collected in the form of liquid ammonia as well.

Ammonium sulfate has the potential for reuse as a fertilizer in agriculture, a raw material in manufacturing nitrogen polymers in industry and food production (Iskander et al., 2016).

BES schemes are new and need optimization of various parameters. Ammonia recovery greatly depends upon current density, catholyte pH, concentration of ammonium ions in the waste stream, type of membrane, and equilibrium among ions. These parameters are inter-related, e.g., current density determines catholyte pH which in turn is affected by type of membrane and ions in the solution (Sleutels et al., 2013). In 2013, Wetsus has set up a BES pilot plant at the water board of Fryslân, Leeuwarden, The Netherlands.

6.2.4 *AMMONIA STRIPPING*

Air stripping of ammonia has been in use for nitrogen removal. It is a pH-dependent scheme where at high pH, ammonium nitrogen from solution converts to ammonia gas. For high stripping efficiency, the process is carried out in a packed tower, as it provides large mass transfer area.

Ammonia can be stripped by air, steam, or vacuum through the liquid fraction in a packed tower (Figure 6.3). Ammonia stripping can be obtained directly from the wastewater by heating at 80°C. Nevertheless, a pH lifting (with lime or sodium hydroxide) up to 10.5 and moderate temperature, allow 85–95% of ammonia to be stripped (Hidalgo et al., 2016). The first full-scale ammonia stripping tower was operated at South Lake Tahoe (USA) in the late, 1970s.

Stripped gas rich in ammonia is then recovered by washing air flux with strong acid solution (H_2SO_4), which produces ammonium sulfate (N = 3–8% w/w). Besides H_2SO_4 as sorbent, also nitric acid (HNO_3) can be applied to obtain ammonium nitrate, as it is done at Oslo WWTP (Sweden) or at the Flemish company Detricon. Also capture of NH_3 in water yields in a commercial product-the ammonia water NH_3(aq) with concentrations up to 25% (Fangueiro et al., 2017).

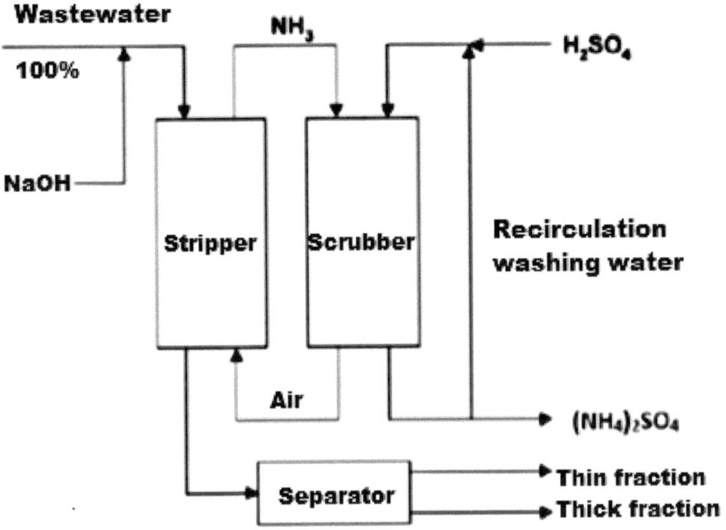

FIGURE 6.3 Ammonia stripping basic scheme.

Another solution is represented by "cold ammonia stripping" to be operated on mineral concentrate. In this case, N stripping can be operated at ambient temperature by adjusting pH with CaO (or similar). Performance reported for full scale plant is of 80–90% of ammonia stripped. Such technology has technology readiness levels (TRL) of 9 and total cost for S/L separation and successive ammonia stripping (hot stripping) were reported for full scale plant between 3 to 6 € m^{-3} (Fangueiro et al., 2017).

6.2.5 *NUTRIENTS PRECIPITATION*

According to Cornel and Schaum (2009), an average of approximately 11% of the incoming phosphorus to sewage treatment and 20% of the incoming nitrogen load are removed with the primary sludge during primary settlement. In biological wastewater treatment (activated sludge), approximately 20–30% of the incoming phosphorus load are incorporated into the biomass and removed with the surplus sludge, even without specific phosphorus removal processes (Parsons and Smith, 2008). In conventional wastewater treatment, plants the remaining nutrients are mainly eliminated by chemical precipitation with metal salts or by enhanced biological nutrients removal.

In this sense, an extended option to recover nitrogen and phosphorus in a controlled way is to create the physical-chemical conditions in the bulk solution (pH, ion strength, etc.), to favor the formation of different salts with low solubility products, and its subsequent precipitation to recover it. Among others, magnesium ammonium phosphate, usually called struvite ($MgNH_4PO4.6H_2O$), is the most common salt enabling the recovery of phosphorous and nitrogen from wastewaters. Other salts that enable to recover phosphorus are calcium phosphate and K-struvite ($KMgPO_4.6H_2O$).

6.2.6 STRUVITE PRODUCTION

Struvite production is a well-known process for many wastewater treatment operations especially for nutrient rich wastewater streams from agricultural, farming, and dairy industries (Gude, 2017). Nitrogen and phosphorus can be recovered together through struvite formation. This can be achieved at multiple locations in a wastewater treatment plant such as the waste activated sludge stream, and anaerobic digester (AD) effluents. Sludge may be dewatered and incinerated to recover phosphorous.

Struvite recovery has been demonstrated at pilot and in full scale with more than 40 installations worldwide mainly with municipal sludge digester effluent and industrial wastewater using biological P removal. Limiting factor for implementation is a minimum concentration required of 100 mg/l dissolved P (ortho-phosphate).

6.2.7 CALCIUM PHOSPHATE AND K-STRUVITE RECOVERY

Calcium phosphate precipitation is very complex and involves various parameters. It depends on calcium and phosphate ions concentration, ionic strength, temperature, ion types, and pH but also on time (Desmidt et al., 2015). When calcium hydroxide ($Ca(OH)_2$) is added to the liquid fraction and the pH increases above 10, phosphorus precipitates as hydroxyapatite ($Ca_5(PO_4)_3OH$) or brushite ($CaHPO_4 \cdot 2H_2O$). Depending on dosage, three different Ca-phosphates can be obtained: the highly water-soluble monocalcium phosphate (MCP), the citric acid soluble di-calcium phosphate (DCP) and the barely soluble tri-calcium phosphate (TCP). For fertilizer application, MCP, and DCP are favored.

Alternatively, K-struvite ($KMgPO_4 \cdot 6H_2O$) can be recovered from wastewater or manures. It has a similar structure as struvite; the only difference is the replacement of NH_4^+ into K^+ ion. From their solubility constants, it can be stated that if both ammonia and potassium are present in excess, struvite will precipitate instead of K-struvite. Therefore, K-struvite will be only precipitated if the excess of potassium is much higher than ammonium (e.g., denitrified wastewaters).

6.2.8 MEMBRANES

Membrane-based separation processes have been utilized in many wastewater operations with microfiltration (MF), ultrafiltration (UF), nanofiltration (NF), and reverse osmosis (RO) being the predominant technologies.

Membrane filtration and UF techniques consist in physical separation by forcing stream input (i.e., wastewater), through membrane by means of pressure. Membranes used to process wastewater can be classified as follows, according to pore size: MF-(pores > 0.1 μm, 0.1–3 bar), UF-(pores > nm, 2–10 bar) and RO-membranes (no pores, 10–100 bar). Membranes used can be either organo-polymeric or ceramic. The first ones are less expensive but they are difficult to be cleaned and they do not support high pressure. Ceramic membranes, used above all for UF, are easier to be cleaned (they have resistance to chemicals) and they allow higher performance because of the high pressure used. Nevertheless, the higher the separation performance, the higher the energy consumption, what might be the main limitation for the implementation of such technique.

UF plus RO has been reported to be able to produce mineral concentrate, i.e., 0.5–1% w/w (95% ammonia) to be used directly as NK-fertilizer. The permeate of RO, that still contains small ions, can be discharged, maybe after a 'polishing' step, or used as process water (Fangueiro et al., 2017).

Mineral concentrate can be successively treated by N stripping technology to remove ammonia producing clean water to be directly discharged in shallow water and ammonia sulfate (7–8% N).

Forward osmosis (FO) has been recently recognized as a highly suitable technological building block to facilitate nutrient and energy recovery from wastewater. Numerous recent studies have demonstrated the capability of FO-based processes to improve the recovery of energy and nutrients from various wastewaters (Ansari et al., 2016; Xie at al., 2014; Xue et al., 2014; Zhang et al., 2014). Despite these promising demonstrations of simultaneous wastewater treatment and resource recovery by FO-based processes, a number of key technical challenges require further development. Further research is needed to optimize the integration of FO with anaerobic processes for biogas production, to overcome issues of salinity accumulation and membrane fouling. Also, according to Ansari et al. (2017), it is necessary to focus efforts to develop nutrient recovery using FO to address the key issues of product purity and membrane fouling/scaling during long-term operation.

Gas-permeable membranes are components of new processes to capture and recover the ammonia in wastewater. The process includes the passage of gaseous ammonia contained in the liquid through a microporous hydrophobic membrane and capture and concentration with circulating diluted acid on the other side of the membrane. The membranes can be assembled in modules or manifolds. For liquid manure applications, the membrane manifolds are submerged in the liquid and the ammonia is removed from the liquid manure in barn pits or storage tanks and lagoons before it goes into the air (Vanotti and Szogi, 2011).

6.2.9 NUTRIENT RECOVERY BY BIOLOGICAL ORGANISMS

Biological organisms generally used for nutrient recovery include bacteria, microalgae, duckweed, wetland, and crops. The recovering efficiency of these biological organisms is primly dependent on the potential biomass growth, as nutrient generally recovered through biomass production (Musfique et al., 2015). Table 6.1 presents the potential uses of nutrients recovered by various biological organisms along with their biomass content.

The use of a duckweed pond is an energy-efficient process as the ammonia is converted into plant protein directly in this system. Another effective approach is constructed wetland (CW) system that is generally implemented with emergent macrophytes which are adapted to grow up through the water column with their root zone and stems submerged. In aquatic macrophyte based treatment systems, the sewage nutrients are recovered and changed into simply harvested protein-rich by-products.

An emerging method of nutrient extraction from wastewater is the production of proteinaceous biomass by cultivating microalgae. This increases the value and manageability of the nutrients.

TABLE 6.1 Biological Organisms for Nutrient Recovery

Biological Organisms	Annual Biomass Yield ton/ha	Uses	Nutrient Recovered (%)	Energy and Environment Benefits
Microalgae	69–91	Livestock feed Biofuel production Fertilizer	Over 60% N and P	• Growing rapidly in brackish water. • Competition with other crops of arable and freshwater is avoided. • Cost and energy-efficient compared to other conventional water treatment technologies. • Reducing costs for harvesting, transporting, and spreading the biomass as a fertilizer by over 60% with respect to algae 'standard' phosphorus content.
Macrophyte (Duckweed)	35–106	Fish biomass Biogas production Alcohol-based fuel production Plant food	Over 50% N and P	• High productivity, high protein content, low fiber content, large nutrient uptake, easy handling, harvesting, and processing and extensive growing period. • Energy efficient process as the ammonia is converted into plant protein directly. • Satisfaction of the irrigation and aquaculture reuse criteria and ensuring annual yield of about 55 t/ha dry matter under sufficient conditions.
Macrophyte (Constructed wetland)	35–106	Cattle feed Food for aquatic organisms Human food supplements Cosmetics	30%–70% P	• Reducing a large amount of pollutants from wastewater prior to flowing into the water body, groundwater, or natural wetland. • High uptake in macrophytes with phosphorus contents of up to 2.9%.

Source: Adapted from: Musfique et al. (2015).

Recycling the nutrients from polluted effluents and assimilating them into algal biomass can result in high quality fertilizers without incurring the environmental and monetary costs of using chemical fertilizers while simultaneously remediating the waste effluent from this process. Effluents from the agro-food industry are an especially attractive feedstock to grow microalgae for biofertilizers production, as it is less contaminated than untreated effluents and rich in nitrogen and phosphorus. Microalgae could be used to recover nutrients from the liquid fraction of these streams and as microalgae incorporate these nutrients into their biomass, a fertilizer is created that is less prone to nutrient losses towards the environment. By reducing the volume of the liquid digestate, the nutrients become more manageable and some reclaimed water (RW) may be produced. Several authors have shown that dried algal biomass produced could be a good substitute for commercial fertilizers (Uggetti et al., 2014; Veronesi et al., 2015). According these authors, dry algae do not contain free ammonia or nitrate that can leach into the environment or volatize at the time of application. Furthermore, concentration of heavy metals in microalgae grown on agro-food effluents is low enough not to reduce its value as soil or feed amendment.

A number of environmental factors affect the rate of nutrient uptake for various species of microalgae, including initial nutrient concentration, light intensity, extracellular pH, temperature, and inoculation density.

The overall composition of the nutrient source affects the nutrient uptake. Nitrogen or phosphorus is often the limiting nutrient in the substrate medium, and thus an optimal N/P ratio exists where the maximum nutrient removal is found. However, the optimal N/P ratio differs by species. In general, the absolute amount of nitrogen or phosphorus was seen to impact the metabolism of algal cells in a similar manner (Cai et al., 2013).

The major challenge associated with culturing algae in nutrient-rich wastewater comes from the design of the cultivation system. Further research is needed to identify algae species and optimize operating parameters for controlled biomass production.

6.3 NUTRIENTS RECOVERY EXAMPLES AND CASE STUDIES

6.3.1 AMMONIA STRIPPING

Ammonia stripping has been successfully used for large-scale treatment of municipal WWTPs.

The VEAS WWTP (650,000 P.E.) for Oslo, Norway (Figure 6.4) has operated a full-scale closed-loop air stripper ammonia recovery unit for two decades in cooperation with Yara, a mineral fertilizers and industrial/environmental solutions company. A substantial fraction (12–15%) of the total N load entering the VEAS plant is recovered-after anaerobic digestion (AD) of sludge, lime conditioning, and filter pressing stages-via ammonia stripping and subsequent capturing (scrubbing) of the ammonia gas (NH_3) with a concentrated nitric acid solution (HNO_3-62% w/w). This stripping and scrubbing treatment yields an industrially reusable ammonium nitrate side stream. Some 350–500 Mt of N/year is recovered from the wastewater at VEAS WWTP, yielding a concentrated (54% w/w) solution of some 4,000 Mt/year.

FIGURE 6.4 Closed-loop ammonia stripping at VEAS WWTP (Oslo).

6.3.2 *NITROGEN AND A PHOSPHORUS JOINT RECOVERY*

Based on existing literature, it is possible to distinguish 22 essentially different approaches focused on the recovery of phosphorus from wastewater or wastewater sludge (Sartorius et al., 2012). These processes, although with the recovery of phosphorus as the main objective, tend to also drag other nutrients, such as nitrogen, producing a joint recovery.

Phosphate recovery techniques developed for wastewater treatment can be applied at various points in the treatment process. Phosphate can be extracted from the liquid phase, from the sludge phase, and from mono-incinerated sludge ash. Table 6.2 gives an overview of the full-scale techniques that can be applied with this aim.

TABLE 6.2 Overview of Full-Scale Phosphate Recovery Processes for Wastewater

Process	Input Flow (m^3/d)	Influent P Concentration (mg/L)	Final Product	Production (ton/d)	Removal Efficiency (%)
ANPHOS®	100/4800	580/58	Struvite	0.45/2	80–90
PHOSPAQ™	2400–3600	60–65	Struvite	0.8–1.2	80
NuReSys®	1920–2880	60–150	Struvite Biostru	1.43–1.58	85
Phosnix	650	100–110	Struvite	0.50–0.55	90
Ostara Pearl	500	100–900	Struvite Crystal Green®	0.50–4	85
Crystalactor®	100–150	60–80	Calcium phosphate	0.55–0.82	70–80
Airprex	1680–2000	150–250	Struvite	1–2.5	80–90
Seaborne	110	600	Struvite	0.58	90

Source: Adapted from Desmidt et al. (2015).

The ANPHOS process (Langeveld and Wolde, 2013) is developed by Colsen and is operated in batch in two separate reactors. In the first reactor, the wastewater is aerated, which results in a pH increase. In the second tank, magnesium oxide is added to the wastewater to recover nitrogen and phosphate as struvite. After the reaction, the struvite is precipitated, dewatered, and dried. The ANPHOS technology has been first implemented on a full scale at the wastewater treatment plant of a potato processing company at the Kruiningen (the Netherlands). The installation is able to produce 2 tons of struvite per day.

The PHOSPAQ™ process (Remy et al., 2013), developed by Paques, takes place in one aerated continuous stirred tank reactor (CSTR). As a result of aeration, the pH increases. Additionally, magnesium oxide is added to the reactor to remove phosphate as struvite. A patented separator system at the top of the reactor is applied to retain the struvite into the system (Driessen et al., 2009). Since 2006, the PHOSPAQ™ process is successfully applied at full scale by Waterstromen in Olburgen (the Netherlands).

NuReSys® stands for Nutrient Recycle System and is developed by the Belgian company Akwadok and is operated also in two reactors, but in continuous mode. This system uses $MgCl_2$ as a magnesium source and includes the addition of a 29% NaOH solution to the crystallization reactor instead of using MgO. A first plant, with a capacity of 1580 kg struvite/day, was taken into operation mid-2006 in Northern Germany by a dairy processing company.

The Phosnix process was developed in Japan by Unitika Ltd. It is a side stream process that enables effective nitrogen and phosphate removal and recovery from the digester wastewater of the sludge treatment process in the sewage treatment plant as granulated struvite (Nawa, 2009). Since 2001, two full-scale struvite recovery plants are operational in Japan: one at the Fukuoka City West Wastewater Treatment Centre and the other at Shimane Perfecture Lake Shinji East Clean Centre.

Through its proprietary Pearl® process (Sharp et al., 2013), Ostara diverts nutrient-rich wastewater streams into a fluidized bed reactor and combines it with compounds such as magnesium chloride, and, on occasion, sodium hydroxide, where small struvite "seeds" are formed (Figure 6.5). This reactor controls the chemistry formation of the seeds and ultimately recovers them in the form of highly pure crystalline pellets, which can grow in diameters of 1.0 to 3.5 mm. This finished material is a high-grade, slow-release end-product that Ostara harvests, dries, packages, and markets as a commercial fertilizer called Crystal Green®. Currently, four full-scale plants have been 16 implemented in the USA. The first industrial scale reactor opened in Edmonton, Canada in 17 May 2007.

The Metropolitan Water Reclamation District (MWRD) of Greater Chicago's Stickney Water Reclamation Plant, located in the city of Cicero (Illinois), is the largest secondary wastewater treatment plant in the world. This facility is reclaiming since 2015 large amounts of nitrogen and phosphorus using an innovative nutrient recovery system from Vancouver-based Ostara Nutrient Recovery Technologies Inc.

One of the existing engineered processes for P recovery is the Crystalactor@ This technology is based on the precipitation of nitrogen and phosphorus upon sand grains in a fluidized bed reactor, due to the high pH achieved (10–10.5) by the addition of lime. Carbon dioxide is removed in a cascade degassifier after acidification with sulfuric acid prior to the addition of lime, in order to reduce the formation of $CaCO_3$.

This technology was implemented by the first time as a full-scale system at Westerbork, the Netherlands, in 1988. In this case, the equipment was located downstream of a biological sewage treatment plant. Another full-scale facility of the Crystalactor® was constructed years later (1994) at Geestmerambacht (the Netherlands), side stream of an enhanced biological nutrient removal (BNR) plant.

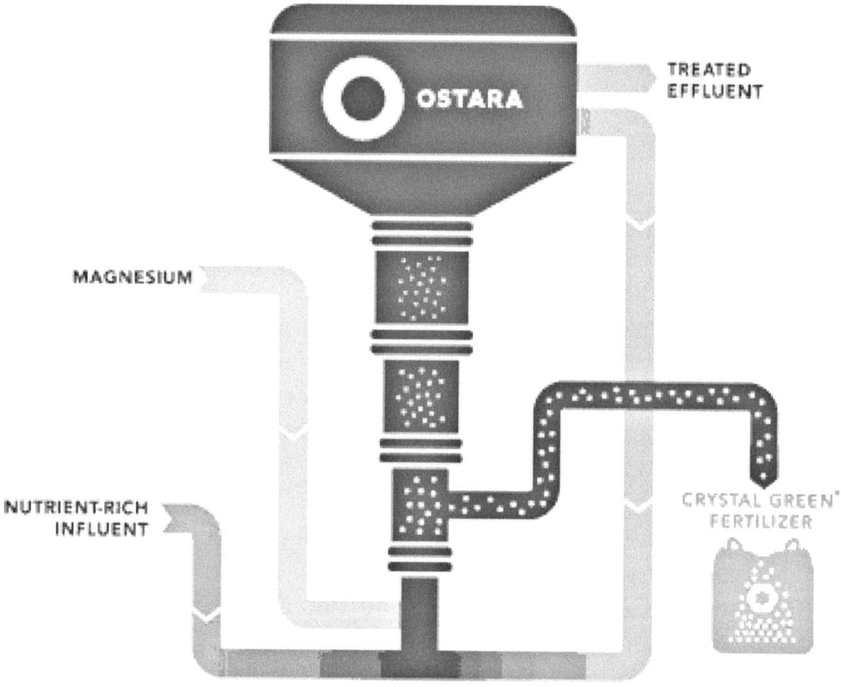

FIGURE 6.5 Schematic overview of the pearl technology.

The AirPrex® technology was developed and patented by the Berliner Wasserbetriebe. The digested sludge is led through a cylindrical 9 reactor, with an inner cylindrical zone mixed by air upflow and a settling zone between this 10 inner cylinder and the outer cylinder. At the moment, three full scale 15 plants are operational, one in Mönchengladbach (Germany), one in Wamansdorf (Germany) 16 and one in Emmen (the Netherlands).

In the Seaborne process, nutrients are separated from the sewage sludge and processed to a fertilizer containing no heavy metals or organic pollutants (Withers et al., 2015). In the first process step, an acidification of the sludge occurs in order to dissolve the solids and to release heavy metals and nutrients. The remaining solids are separated from the flow by using a centrifuge and a filter system. In the next treatment step, the sulfur-rich gas is used to precipitate the heavy metals from the effluent liquor. In the following process step, the nutrients are recycled. Nitrogen and phosphate are precipitated as struvite by the addition of sodium hydroxide and magnesium oxide.

6.3.3 *MEMBRANES*

Ultrafiltration (UF) + reverse osmosis (RO) have been reported to be able producing mineral concentrate, i.e., 0.5–1% w/w (95% ammonia) to be used directly as NK-fertilizer (Figure 6.6). The permeate of RO, that still contains small ions, can be discharged, maybe after a 'polishing' step, or used as process water (Fangueiro et al., 2017).

FIGURE 6.6 Membranes plant used for nutrients recovery.

Since 2009, the agricultural, economic, and environmental effects of the production and use of mineral concentrates are studied in the Netherlands. The mineral concentrates from this pilot can be applied as mineral fertilizer in Dutch agriculture. In 2011, eight installations of the pilot had a processing capacity of 207,500 tons of manure. Meanwhile, different installations in the north of the Netherlands closed because of economic and technical reasons, but, in the south of the Netherlands, the pilot has extended to a capacity of 263.000 tons of manure at the 10 installations in use in 2014 (Luesink et al., 2016).

McLanahan (Pennsylvania, USA) supplies an UF and air stripping system for nutrients recovery. The system is installed in the Car-Min-Vu Farms' 850 milking cows in Webberville, Michigan to treat the liquid waste fraction. Feedstock is fed to a 0.03 μm UF membrane where all suspended solids are captured to produce a ~7% DM nutrient-rich cake containing 50% of the N, 95% of the P and 30% of the K in the feedstock. Once all suspended solids are captured, air stripping and absorption is used to volatilize ammonia before it is absorbed with a solution of sulfuric acid to create an ammonium sulfate solution. Ammonia recovery rates can vary from 40–80% depending upon conditions, while average ammonium sulfate concentration is 35%. The ammonium sulfate solution can be further dehydrated to create a dry product. If cleaner water is required, RO can be used to remove potassium, remaining P, most metals, and all pathogens (Hallbar, 2015).

6.3.4 MICROALGAE CULTIVATION

The fact that microalgae provide potential for autotrophic or heterotrophic nutrients removal speaks to the positive impact they can provide municipal WWTPs (Oleszkiewicz et al., 2017).

The EU funded All-Gas project in Chiclana, Spain, is taking the first major steps towards full scale biofuel production and wastewater treatment (Freyberg, 2012). With a process train consisting of upflow anaerobic sludge blanket (UASB) digesters, racetrack-type high rate algal ponds (HRAP) and algal separation by lamella clarifiers, proponents are claiming economic benefits in addition to high levels of treatment. The facility is able to generate biogas from the start by utilizing anaerobic pretreatment instead of destroying organic matter. For this reason, aeration is not required and approximately 0.5 kWh for every m^3 of wastewater is saved. Furthermore, there will be a net output of energy from algae conversion to either oils or gas, resulting in, approximately, 0.4 kWh positive outputs per m^3 of wastewater. By coupling wastewater treatment with biofuel production, proponents believe that cost-neutral treatment can be achieved (Arbib et al., 2015).

Boelee et al. (2012) estimated different scenarios of nutrient removal from municipal wastewater by microalgal biofilms: (1) as a post-treatment; (2) as a second stage of wastewater treatment, after the first stage where COD was removed by activated sludge; and (3) in a symbiotic microalgal/heterotrophic system. The results indicated that it was not possible to simultaneously remove all nitrogen and phosphorus from the wastewater due to N:P ratio: either N or P was limiting, depending on scenario studied.

Microalgal biomass analysis revealed an increasing nitrogen and phosphorus content with increasing loading rates until the maximum uptake capacities. The internal nitrogen to phosphorus ratio decreased from 23:1 to 11:1 when increasing the loading rate. It was estimated that a full-scale microalgal biofilms post-treatment system for 100,000 inhabitants would be around 10 ha, producing 2 tons of biomass per day (Boelee et al., 2011).

6.4 NUTRIENTS REUSE IN AGRICULTURE

In its current state of practice, agriculture is still characterized by a high overall contribution to GHG emission, a sub-optimal retrieval of valuable organic carbon and an insufficiently efficient re-use of major plant nutrients (NPK) within agro-system boundaries.

Between 2 and 5 Mt of N and 0.6 Mt of P are currently not being recovered for agricultural use from major waste streams (manure, sewage waste, and food chain waste). These quantities represent 18–46% of the 11 Mt of mineral nitrogen currently applied to EU crops, and 43% of the 1.4 Mt of mineral-based phosphorus applied to crops (Buckwell and Nadeu, 2016).

Nitrogen has been highlighted as one of the three "planetary boundaries" that have been exceeded beyond supportable levels alongside climate change (CC) and biodiversity loss (O'Neill et al., 2018). For example, one of the most potent greenhouse gases (GHG) is nitrous oxide (N_2O) whose concentrations have increased in the past 50 years from 270 to 330 ppb (IPPC, 2018). This is significant taking into account that N_2O has a global warming potential (GWP) which is 300x that of CO_2 and for which agriculture is responsible for 60% of its emission. In addition, ineffective management of manure and other fertilizing products results in eutrophication of water bodies with nitrates as well as elevated volatile ammonia emissions threatening biodiversity adjacent to agricultural activities. Phosphorus was placed on the "Critical Raw Material" list by the European Commission, considering the European continent is dependent on import (currently EU28 is 92% dependent) to secure its own agricultural production (and hence food security). Carbon has been not enough investigated and even overlooked in past nutrient oriented research, yet N-, P-, and C-cycles are intertwined and need to be examined within a single methodological framework. Carbon is of primary importance as (effective) carbon in organic matter returned to soils, which plays a vital role in abatement of soil degradation and maintaining soil fertility. In addition, closing the carbon cycle is highly relevant from the perspective of renewable energy potential from agro residues (e.g., biogas) and reduction of GHG emissions from agricultural practices (e.g., CH_4 losses from animal husbandry) (Buckwell and Nadeu, 2016).

The efficiency of nutrient use through the whole food chain is very low. For every five tons of nitrogen entering the agricultural system, only one ton is converted to finished products for human consumption, that is a 20% NUE. For phosphorus, the corresponding figure is 30%. While crop production shows a relatively high NUE due to advances in crop genetics and management and fertilizer application techniques (53% for N and 70% for P), livestock makes a particularly inefficient use of nutrients (18% NUEN and 29% NUEP). These low efficiencies result in large leakage of nutrients into the environment with negative impacts on soils, water, and air, and are associated with unacceptable health and environmental costs.

It is then, clearly necessary to address the current gaps in the N, P, and C cycles of different agricultural systems and the related environmental problems by implementing optimized management systems whilst having a positive trade-off with productivity, quality, and environmental impact.

Recent research (NUTRIMAN project, NUTRI2CYCLE project) elucidated that processing agro-waste and manures with on-farm applicable techniques (e.g., integrated anaerobic digestion, manure separation, etc.), enhances value of processed flows above required costs. Moreover, depending on the envisaged agro-process, renewable energy can be obtained (for on-farm self-consumption), fugitive GHG emissions from conventional farm practices can be mitigated, organic carbon can be channeled to more optimal re-integration in soil stock and NUE and ratios of (macro) nutrients contained within processed residues/manures can be better aligned with plant production requirements.

Agro-processing has emerged as an agricultural activity on its own over the last 15 years. Activities such as manure processing, composting of agro-residues and/or the generation of renewable energy via anaerobic digestion has increased dramatically both, in financial and employment terms, and has provided added value to the conventional agro-activities.

Focusing on wastewater, the proper use of treated wastewater in agriculture benefits (in general terms) human health, the environment and the economy. This use represents an alternative practice for water and nutrients recovery that is being adopted in different regions, mainly those confronted with water shortages and growing urban populations with increasing water needs (Jaramillo et al., 2017).

The nutrients naturally present in wastewater allow economic and environmental savings on fertilizer expenses (Hatami et al., 2018; Wielemarker et al., 2018), thus ensuring a closed and environmentally favorable nutrient cycle that avoids the indirect return of macro-(especially nitrogen and phosphorous) and microelements (Ca, Mg, B, Mg, Fe, Mn or Zn) (Maiolo et al., 2018; Simha et al., 2018) to water bodies. Indeed, wastewater reuse has been recently proven to improve crop yield in a safe way and result in the reduced use of fertilizers in agriculture (Libuti et al., 2018; Shilpi et al., 2018). Also, as another positive consequence of this practice, eutrophication conditions in water bodies would be reduced, as would the expenses for agrochemicals used by farmers (Salgot et al., 2018; Sgroi et al., 2018).

The planning and management of agricultural reuse projects need to consider institutional and legal, socio-economic, financial, environmental, technical, and psychological aspects (Massoud et al., 2018). Israel, Jordan, and Tunis are among the leading countries in wastewater reuse but the reasons are more related with water-saving than with nutrients recycling.

6.5 CONCLUDING REMARKS

There is significant potential to increase nitrogen (and other nutrients) use efficiency at the farm level, because concepts and tools needed to achieve it are already available.

Increasing public awareness of the threat that nutrient discharges pose to surface waters is creating pressure on municipalities to introduce or improve the removal of N and P from wastewater. It is evident that not only the future nutrient limits will be increasingly more stringent but also greater performance reliability will be required as statistical-based criteria become more popular. Greater emphasis will be placed on nitrogen and phosphorus recovery due to its worldwide dwindling supplies.

Spreading biosolids generated during treatment on farmlands is the simplest and most common way to reuse nutrients. However, there is a rising public scrutiny as to the quality of solids generated in the process of wastewater treatment that are land applied. This is fed in particular by insufficient scientific information about accumulation and impact of emerging substances of concern on the environment and crops.

This chapter has focused on technologies which enable nutrients recovery. The drivers are clear, and they are to translate technologies which would normally remove contaminants into a liquid or waste concentrate stream (or reactively dissipate them) into products that feed into the circular economy.

This is not a massive shift from current practices, but instead of focusing the process on removal, it focuses on recovery. That is, multiple candidate technologies that would otherwise remove a contaminant are instead screened to those that allow the byproducts to be reused.

Further studies are required to examine the potentiality and sustainability of hybrid systems gathering several technologies, as each technology has its advantages and limits for the amount of nutrient it can recover. There is also no previous work that has examined which wastewater stream is most effective for nutrient recovery by using innovative hybrid systems. These knowledge gaps are needed to be addressed to ensure maximum nutrient recovery with sustainable methods. Furthermore, new technologies need to be cost-effective and user-friendly, so that these become attractive to farmers.

With current technology and incentives for implementing the circular economy for nutrients, the most promising three categories of substrate to work on are: animal manures, sewage waste and food chain waste, and wastewater especially from the slaughterhouse industry. The potential volumes of recoverable nutrients from these three waste streams are estimated to be in total, approximately, 12 Mt N and 2.5 Mt P annually.

It is necessary to establish more C-, N-, and P-efficient agroecosystems by introducing effective combinations of existing technologies and high-potential innovations in nutrient recycling technologies, thus improving the sustainability of farm systems, reducing negative impacts on water, air, climate, and soil quality.

Better nutrient stewardship engaging all actors across the value chain has the potential to increase the C, N, and P recycling rate significantly. Providing a good roll-out and wide implementation of the project results, this could lead to increased NUE from 60% (raw manures) to 80–100%, as well as reduced agricultural N and P leakage into ground and shallow water by respectively, 30, and 50% associated to enhanced nutrients use efficiency. In addition, restoration of soil C-stock is envisaged, targeting a 10% increase over current values in references to the 4 per mil initiative. By optimized management, 20–40% reduction of GHG emission could be obtained and over 60% NH_3-emission avoided (for example, dairy production accounts for a large proportion of the GHG footprint in European agriculture), based on farming fugitive emissions from manure storage models 30–60% improvement on GHG emissions could be obtained at farm-scale). Until now mitigation, solutions, and novel concepts have been developed for single nutrients and/or on a very specific level without a thorough reference benchmark and transparent trade-off taken into account in the whole value chain. In addition, local conditions in terms of nutrient status, climate, and soil greatly influence the operational practice and can thus hamper the use and geographic spreading of promising technologies and initiatives.

The substitution of fossil-based mineral fertilizers by low organic matter content P and N bio-based fertilizers is not yet self-evident, at least not from a technical, legal, and market point of view. N and P in bio-based products with slow or controlled nutrient release performance generally have a lower availability to crops versus rapid release N and P in mineral fertilizers. Therefore, bio-based products should be evaluated to make sure that sufficient N and P are available for crop growth and to avoid leaching. Both, agronomic yield assurances and sufficient environmental safeguards need to be considered. Specific NUE, as well as nutrient losses, needs to be quantified for the bio-based fertilizers.

ACKNOWLEDGMENTS

The authors gratefully acknowledge support of this work by the Agencia de Innovación, Financiación e Internacionalización Empresarial de Castilla y León. Project: Economía circular en el sector agroalimentario (Circular Economy in the Agri-Food Sector).

KEYWORDS

- **biofertilizers**
- **circular economy**
- **nitrogen**
- **nutrients recovery**
- **phosphorus scarcity**
- **wastewater valorization**

REFERENCES

Ansari, A. J., Hai, F. I., Price, W. E., & Nghiem, L. D., (2016). Phosphorus recovery from digested sludge centrate using seawater-driven forward osmosis. *Sep. Purif. Technol., 163*, 1–7.

Ansari, A. J., Hai, F. I., Price, W. E., Drewes, J. E., & Nghiem, L. D., (2017). Forward osmosis as a platform for resource recovery from municipal wastewater-A critical assessment of the literature. *J. Membr. Sci., 529*, 195–206.

Arbib, Z., Godo, S. I., Fernández, M., Lara, E., Llamas, B., & Rogalla, F., (2015). Wastewater as biofuel: Nutrient and energy recovery with algae treatment. *Proceed. Water and Energy Water Environment Fed. Confer*. Washington DC.

Batstone, D. J., Hülsen, T., Mehta, C. M., & Keller, J., (2015). Platforms for energy and nutrient recovery from domestic wastewater: A review. *Chemosphere, 140*, 2–11.

Boelee, N. C., Temmink, H., Janssen, M., Buisman, C. J. N., & Wijffels, R. H., (2011). Nitrogen and phosphorus removal from municipal wastewater effluent using microalgal biofilms. *Wat. Res., 45*, 5925–5933.

Boelee, N. C., Temmink, H., Janssen, M., Buisman, C. J. N., & Wijffels, R. H., (2012). Scenario analysis of nutrient removal from municipal wastewater by microalgal biofilms. *Water, 4*, 460–473.

Buckwell, A., & Nadeu, E., (2016). *Nutrient Recovery and Reuse (NRR) in European agriculture: A Review of the Issues, Opportunities, and Actions*. http://www.risefoundation.eu/images/files/2016/2016_RISE_NRR_Full_EN.pdf (accessed on 10 June 2020).

Cai, C., Park, S. Y., & Li, Y., (2013). Nutrient recovery from wastewater streams by microalgae: Status and prospects. *Renew. Sust. Energ. Rev., 19*, 360–369.

Cohen, Y., (2000). *Nutrient Recovery from Wastewater.* http://www.vaxteko.nu/html/sll/slu/ex_arb_vaxtnaringslara/EVN119/EVN119.HTM (accessed on 10 June 2020).

Cornel, P., & Schaum, C., (2009). Phosphorus recovery from wastewater: Needs, techniques and costs. *Wat. Sci. Technol., 59*(6), 1069–1076.

Desmidt, E., Ghyserlbrecht, K., Zhang, Y., Pinoy, L., Van, D. B. B., Verstraete, W., Rabaey, K., & Meesschaert, B., (2015). Global phosphorus scarcity and full-scale P-recovery techniques: A review. *Crit. Rev. Env. Sci. Tec., 45*, 336–384.

Fangueiro, D., Snauwaert, E., Provolo, G., Hidalgo, D., et al., (2017). *Available Technologies for Nutrients Recovery from Animal Manure and Digestates*. EIP-AGRI Focus Group-Nutrient recycling. https://ec.europa.eu/eip/agriculture/sites/agri-eip/files/fg19_minipaper_1_state_of_the_art_en.pdf (accessed on 10 June 2020).

FAO. *World Fertilizer Trends and Outlook to 2020*. http://www.fao.org/3/a-i6895e (accessed on 10 June 2020).

Gude, V. G., (2017). Resource recovery from wastewater. *Austin Environ. Sci., 2*(3), 1024.

Haddaway, A., (2015). *Nutrient Recovery Technology Transforms World's Largest Wastewater Treatment Plant* (Vol. 31, No. 2). Water World, http://www.waterworld.com/articles/print/volume-31/issue-2/features/nutrient-recovery-technology-transforms-world-s-largest-wastewater-treatment-plant.html (accessed on 10 June 2020).

Hallbar Consulting Inc., (2015). *Evaluation of Nutrient Recovery Technologies for Dairy Manure and Digestate*. Growing Forward 2 report. https://www2.gov.bc.ca/assets/gov/farming-natural-resources-and-industry/agriculture-and-seafood/agricultural-land-and-environment/waste-management/manure-management/evaluation_of_nrt_dairy_manure_digestate.pdf (accessed on 10 June 2020).

Hatami, T., Nadali, A., Roshanaei, G., & Shokoohi, R., (2018). Feasibility of reuse of effluent from the extended aeration process of wastewater treatment plant in the Bojnoord city for agricultural and irrigation uses. *Pajouhan Scientific J.*, *16*(3), 20–28.

Hidalgo, D., Corona, F., Martín-Marroquín, J. M., Del, Á. J., & Aguado, A., (2016). Resource recovery from anaerobic digestate: Struvite crystallization versus ammonia stripping. *Desalin. Water Treat.*, *57*(6), 2626–2632.

IPPC: Global Warming of 1.5°C, (2018). https://www.ipcc.ch/site/assets/uploads/sites/2/2018/07/SR15_SPM_version_stand_alone_LR.pdf (accessed on 10 June 2020).

Iskander, S. M., Brazil, B., Novak, J. T., & He, Z., (2016). Resource recovery from landfill leach ate using bioelectrochemical systems: Opportunities, challenges, and perspectives. *Bioresour. Technol., 201*, 347–354.

Jaramillo, M., & Restrepo, I., (2017). Wastewater reuse in agriculture: A review about its limitations and benefits. *Sustainability*, *9*(10), 1734–1753.

Langeveld, C. P., & Ten, W. K. W., (2013). *Phosphate Recycling in Mineral Fertilizer Production.* International Fertilizer Society. http://www.prueck-bw.de/cms/content/media/ten%20Wolde.pdf (accessed on 10 June 2020).

Lee, W., Yoon, S., Choe, J. K., Lee, M., & Choi, Y., (2018). Anionic surfactant modification of activated carbon for enhancing adsorption of ammonium ion from aqueous solution. *Science Total Environ., 639*, 1432–1439.

Luesink, H., Postma, R., Smits, M. J., Van, S. L., & De Koeijer, T., (2016). *Effect afzet mestverwerkings producten bij wettelijke status kunstmest of EG-meststof.* LEI Wageningen UR, Wageningen, mei.

Maiolo, M., & Pantusa, D., (2018). A proposal for multiple reuses of urban wastewater. *J. Water Reuse Desal., 8*(4), 468–478.

Massoud, M. A., Kazarian, A., Alameddine, I., & Al-Hindi, M., (2018). Factors influencing the reuse of reclaimed water as a management option to augment water supplies. *Environ. Monit. Assess, 190*(9), 531.

Nancharaiah, Y. V., Mohan, S. V., & Lens, P. N. L., (2016). Recent advances in nutrient removal and recovery in biological and bioelectrochemical systems. *Biores. Technol., 215*, 173–185.

Nawa, Y., (2009). *P-Recovery in Japan-the PHOSNIX Process.* BALTIC 21-Phosphorus Recycling and Good Agricultural Management Practice. Berlin.

NUTRI2CYCLE Project, (2018). www.nutri2cycle.eu (accessed on 10 June 2020).

NUTRIMAN Project, (2018). https://nutriman.net/project (accessed on 10 June 2020).

O'Neill, D. W., Fanning, A. L., Lamb, W. F., & Steinberger, J. K., (2018). A good life for all within planetary boundaries. *Nature Sustainability*, *1*(2), 88.

Oleszkiewicz, J., Kruk, D. J., Devlin, T., Lashkarizadeh, M., & Yuan, Q., (2015). *Options for Improved Nutrient Removal and Recovery from Municipal Wastewater in the Canadian Context.* Waterloo, ON.

Parsons, S., & Smith, J. A., (2008). Phosphorus removal and recovery from municipal wastewaters. *Elements*, *4*(2), 109–112.

Petruzzelli, D., De Florio, L., Dell'Erba, A., Liberti, L., Notarnicola, M., & Sengupta, A. K., (2003). A new phosphate-selective sorbent for the rem nut process. Laboratory investigation and field experience at a medium size wastewater treatment plant. *Wat. Sci. Tech., 48*(1), 179–184.

Puyol, D., Batstone, D. J., Hülsen, T., Astals, S., Peces, M., & Krömer, J. O., (2017). Resource recovery from wastewater by biological technologies: Opportunities, challenges, and prospects. *Front. Microbiol., 7*, 2106.

Remy, M., Kruit, J., Hendrickx, T., Haarhuis, R., & Van, L. M., (2013). Phospaq: Full scale experience with phosphorus recovery via controlled struvite precipitation. *Proceedings of the Water Environment Federation, 4*, 311–317.

Salgot, M., Folch, M., & Unit, S. S., (2018). Wastewater treatment and water reuse. *Curr. Opin. Environ. Sci. Health, 2*, 64–74.

Sartorius, C., Von, H. J., & Tettenborn, F., (2012). Phosphorus recovery from wastewater: Expert survey on present use and future potential. *Wat. Environ. Res., 84*(4), 313–322.

Sengupta, S., & Pandit, A., (2011). Selective removal of Phosphorus from wastewater combined with its recovery as a solid-phase fertilizer. *Wat. Res., 45*, 3318–3330.

Sengupta, S., Nawaz, T., & Beaudry, J., (2015). Nitrogen and phosphorus recovery from wastewater. *Curr. Poll. Rep., 1*(3), 155–166.

Sgroi, M., Vagliasindi, F. G., & Roccaro, P., (2018). Feasibility, sustainability and circular economy concepts in water reuse. *Curr. Opin. Environ. Sci. Health, 2*, 20–25.

Sharp, R., Vadiveloo, E., Fergen, R., Moncholi, M., Pitt, P., Wankmuller, D., & Latimer, R., (2013). A theoretical and practical evaluation of struvite control and recovery. *Water Environ. Res., 85*(8), 675–686.

Shilpi, S., Seshadri, B., Sarkar, B., Bolan, N., Lamb, D., & Naidu, R., (2018). Comparative values of various wastewater streams as a soil nutrient source. *Chemosphere, 192*, 272–281.

Simha, P., Zabaniotou, A., & Ganesapillai, M., (2018). Continuous urea-nitrogen recycling from human urine: A step towards creating human excreta based bio-economy. *J. Clean. Prod., 172*, 4152–4161.

Sleutels, T. H., Ter, H. A., Buisman, C. J., & Hamelers, H. V., (2013). Steady state performance and chemical efficiency of microbial electrolysis cells. *Int. J. Hydrogen Energy, 38*(18), 7201–7208.

Taddeo, R., Prajapati, S., & Lepistö, R., (2017). Optimizing ammonium adsorption on natural zeolite for wastewaters with high loads of ammonium and solids. *J. Porous Mat., 24*(6), 1545–1554.

Uggetti, E., Sialve, B., Latrille, E., & Steyer, J. P., (2014). Anaerobic digestate as substrate for microalgae culture: The role of ammonium concentration on the microalgae productivity. *Bioresour. Technol., 152*, 437–443.

Urso, J. H., & Gilbertson, L. M., (2018). Atom conversion efficiency: A new sustainability metric applied to nitrogen and phosphorus use in agriculture. *ACS Sustain. Chem. Eng., 6*(4), 4453–4463.

Vanotti, M. B., & Szogi, A. A., (2011). *Systems and Methods for Reducing Ammonia Emissions form Liquid Effluents and for Recovering Ammonia*. US Patent Appl. SN 13/164, 363, filed June 20. US Patent and Trademark Office, Washington, DC.

Veronesi, D., Ida, A., D´Imporzano, D., & Adani, F., (2015). Microalgae cultivation: Nutrient recovery from digestate for producing algae biomass. *Chem. Eng. T., 43*.

Von, E. S. V., Charles, W., Ho, G., & Cord-Ruwisch, R., (2016). Novel process of bio-chemical ammonia removal from air streams using water reflux system and zeolite as filter media. *Chemosphere, 144*, 257–263.

Wielemaker, R. C., Weijma, J., & Zeeman, G., (2018). Harvest to harvest: Recovering nutrients with new sanitation systems for reuse in Urban Agriculture. *Resources, Conservation and Recycling, 128*, 426–437.

Williams, A. T., (2013). *Ion Exchange Nutrient Recovery from Municipal Wastewater*. Marquette University.

Withers, P. J., Elser, J. J., Hilton, J., Ohtake, H., Schipper, W. J., & Van, D. K. C., (2015). Greening the global phosphorus cycle: How green chemistry can help achieve planetary P sustainability. *Green Chemistry, 17*(4), 2087–2099.

Xie, M., Nghiem, L., Price, W., & Elimelech, M., (2014). Toward resource recovery from wastewater: Extraction of phosphorus from digested sludge using a hybrid forward osmosis-membrane distillation process. *Environ. Sci. Technol. Lett., 1*, 191–195.

Xie, M., Shon, H. K., Gray, S. R., & Elimelech, M., (2016). Membrane-based processes for wastewater nutrient recovery: Technology, challenges, and future direction. *Wat. Res., 89*, 210–221.

Xue, W., Tobino, T., Nakajima, F., & Yamamoto, K., (2015). Seawater-driven forward osmosis for enriching nitrogen and phosphorous in treated municipal wastewater: Effect of membrane properties and feed solution chemistry. *Water Res., 69*, 20–130.

Yasin, A. M., (2012). *Investigation of Treatment Efficiency, Operation Conditions, and Cost for Nitrogen Removal and Catching by Closed Loop Ammonia Stripping at VEAS.* Norwegian University of Life Sciences.

Ye, Y., Ngo, H. H., Guo, W., Liu, Y., Li, J., Liu, Y., & Jia, H., (2017). Insight into chemical phosphate recovery from municipal wastewater. *Science of the Total Environment, 576*, 159–171.

Zhang, J., She, Q., Chang, V., Tang, C., & Webster, R., (2014). Mining nutrients (N, K, P) from urban source-separated urine by forward osmosis dewatering. *Environ. Sci. Technol., 48*, 3386–3394.

CHAPTER 7

Phosphorus Removal and Recovery in Water Resource Recovery Facilities

CHRISTIAN SCHAUM, CHRISTIAN HUBERT, STEFFEN KRAUSE, and BETTINA STEINIGER

Bundeswehr University Munich, Chair of Sanitary Engineering and Waste Management, Werner-Heisenberg-Weg – 39, 85577 Neubiberg, Germany, E-mail: christian.schaum@unibw.de (C. Schaum)

ABSTRACT

Boon and bane of phosphorus are close together, as it is on the one side essential nutrient for humans, animals, and plants, especially as an indispensable component of fertilizers to provide crop plants with it by applying on agricultural land. On the other side, ending up in water bodies, high amounts of phosphorus cause eutrophication. Water resource recovery facilities (WRRF) with modern processes of biological, chemical, and advanced phosphorus removal are able to reduce the discharged amount of phosphorus and nitrogen, so the nutrient input from wastewater of WRRF can be minimized.

For several years, processes for recovery of phosphorus in WRRF have been developed to bring a phosphorus-rich fertilizer to the market according to the concept of a circular economy. Currently, there are several processes available based on different methods for recovery from effluent, sludge liquors, (digested) sludge/biosolids, or sewage sludge ash, which are still tested on full-scale. Considering the bunch of processes with their specifications, the general success of phosphorus recovery essentially depends on the recovery quota, quality, and price of the product. Prospective challenges are a worldwide establishment for phosphorus removal and recovery to further reduce the nutrient input into water bodies and the still increasing mining of phosphate rocks.

7.1 PHOSPHORUS: RESOURCE OR POLLUTER?

Phosphorus is one of the most enthralling substances of our time. Even the composition of the name "phos" from Greek "light" and "phorus" from Greek "bringer" indicates the peculiarity of the element, cf. Schaum (2018a).

Phosphorus has always been both a curse and a blessing, Schaum (2018a):

- Phosphorus is essential for all life forms and cannot be replaced by anything. It is part of many organic compounds, such as the universal energy currency ATP, the DNA, cell membranes, or bones. Even plants need phosphorus for their growth, for which reason it is used as a fertilizer. Phosphorus is also utilized for various industrial purposes. It is employed, e.g., in surface treatment in the automotive industry, in the production of flame retardants and pesticides as well as in the electrical industry.
- The use of phosphorus in agriculture is targeted to increase plant growth. By contrast, wastewater treatment aims to minimize the phosphorus concentration in the wastewater in order to minimize its discharge into rivers and lakes, where eutrophication caused by high phosphorus concentrations would lead to excessive plant growth. There are also numerous negative examples of the use of phosphorus in the industrial sector, ranging from incendiary bombs to its use as a neurotoxin.

It is precisely this ambivalence that shows the possibilities and limitations of the removal of phosphorus during the wastewater treatment. In Europe, considering the European Water Framework Directive, a significant tightening of the limit values of phosphorus in treated and discharged wastewater into rivers or lakes is being discussed. As a result, the focus is on the implementation of additional process stages (in particular filtration) on water resource recovery facilities (WRRF) and realization of phosphorus recovery in treatment plants or from sewage sludge/ash.

The issue of resource conservation is the focus of attention worldwide-from fossil fuels, across a variety of metals and fertilizers to water itself. For example, the idea of careful and efficient use of natural resources is identified as a key competence of sustainable societies (Schaum, 2016). These concerns in a special way also the (waste) water treatment and accordingly, phosphorus should be used sustainably. This includes recycling of secondary phosphates, efficient extraction, and treatment of raw phosphate as well as its efficient use.

7.2 PHOSPHORUS IN WASTEWATER AND REMOVAL TECHNOLOGIES

7.2.1 *ORIGIN OF PHOSPHORUS IN WASTEWATER*

The major portion of phosphorus in wastewater originates from human excreta and the use of domestic laundry as well as cleaning detergents. Phosphorus is especially contained in protein-rich food like milk and dairy products, cereals, and partly complemented by additives containing ortho-phosphates and polyphosphates (Kalantar-Zadeh et al., 2010; Calvo et al., 2014; USDA, 2017). Daily intake of phosphorus for children is recommended to amount 500 mg P/d, for adolescents 1,250 mg P/d, and for elderly persons 700 mg P/d (Kalantar-Zadeh et al., 2010). Between 20 to 60% of the phosphorus intake releases the body as feces and urine and end up in wastewater (Kalantar-Zadeh et al., 2010). The phosphorus load excreted per person is about 1.9 g P/d. Of these, 67% are contained in urine and 33% in feces (Eastham et al., 1981; Kroiss et al., 2011; Krause, 2018).

A study of UK Water Industry Research characterizes the phosphorus concentrations in domestic wastewater by feces, urine, laundry, and cleaning detergents and corrosion inhibitors added to potable water (UKWIR, 2009), cf. Table 7.1.

TABLE 7.1 Phosphorus Loads in Domestic Wastewater

Source	P Load [g P/d]
Laundry	0.39
Automatic Dishwasher	0.15
Feces	0.50
Urine	0.90
Mains supply	0.12
Total	**2.06**

Source: UKWIR (2009).

Internationally, the daily phosphorus load per capita varies, cf. Table 7.2. Thus, the concentrations of phosphorus in wastewater depending on loads stemming from excrement as well as laundry and cleaning detergents. As domestic wastewater, precipitation water, extraneous water discharged into sewer systems, there are phosphorus concentrations ranged from 5 to 30 mg/L in wastewater (Mara and Horan, 2003; Tchobanoglous et al., 2014). Regional differences caused by nutrition amounts, water consumption, and drainage systems and the disposal of food remnants leads to variations of phosphorus concentration in wastewater for various regions and nations cf. Table 7.2. Reaching the WRRF, approximately one half of the phosphorus in wastewater is in particulate form, while the other half is dissolved. Due to hydrolysis processes during the cleaning stages, the portion of dissolved phosphorus increases (Jardin, 2002).

TABLE 7.2 Daily Phosphorus Load Per Capita and Phosphorus Concentrations in Wastewater in International Context

Region	P in g P/d	Region	P in mg/L
Austria[d]	3.0	Bolivia, Santa Cruz[c]	12
Brazil[e]	2.1	China, Haikou[c]	5.0
Commonwealth of Independent States[f]	1.6	Egypt, Sanhur al M.[c]	19
Denmark[e]	2.7	Ethiopia, Adama[c]	11
Developed Countries[f]	1.6	Georgia, Batumi[c]	5.0
Developing Countries[g]	2.7	Germany, Austria[a]	6.0–16.8
Eastern Asia[f]	1.6	Iran, Teheran[c]	7.0
Egypt[e]	1.4	Macedonia, Prilep[c]	9.0
Germany[e,h]	1.8–3.0	New Zealand, Invercargill[c]	8.0
Israel[f]	2.1	Nicaragua, Managua[c]	7.0
Latin American and the Caribbean[f]	1.5	Norway, Oslo[c]	3.0
Northern Africa[f]	1.7	Poland, Klodzko[c]	4.0
Oceania[f]	1.3	Spain, Sevilla[c]	9.0
South-Eastern Asia[f]	1.1	Tunisia, Nabeul[c]	13
Southern Asia[f]	1.2	Turkey, Adana[c]	9.0
Sub-Saharan Africa[f]	1.1	Turkey, Konya[c]	13
Turkey[e]	1.4	United Arab Emirates, Fujairah[c]	7.0
United States of America[e]	2.7	Wider Caribbean Region[b]	5.0–10.0
Western Asia[f]	1.6		

[a](DWA, 2014a); [b](UNEP, 2015); [c](DWA, 2016); [d](LeBlanc et al., 2009); [e] (Henze, 2008); [f] (Mihelcic et al., 2011); [g](Sperling, 2007); [h](DWA, 2014b).

Source: cf. Krause (2018).

7.2.2 PHOSPHORUS BALANCE FOR A MUNICIPAL WATER RESOURCE AND RECOVERY FACILITY

Based on an average phosphorus load of 1.8 g P/(PE·d) (ATV-DVWK, 2000) in the raw wastewater exemplary with German boundary conditions and a per population equivalent (PE) wastewater flow of 200 L/(PE·d), the influent concentration is around 9 mg/L. An average of approx. 11% of the incoming phosphorus load is removed with the primary sludge during primary settlement (ATV-DVWK, 2000). In biological wastewater treatment, approx. 28% of the incoming phosphorus load are incorporated into the biomass and removed with the surplus sludge, even without specific phosphorus removal processes (Cornel and Schaum, 2009).

Based on the permitted discharge concentrations of 1 or 2 mg/L, respectively, approx. another 50% of the incoming phosphorus load has to be removed specifically, either by biological or chemical phosphorus removal processes or their combination. In summary, this means approx. 90% of the incoming phosphorus load is incorporated into sewage sludge. In Figure 7.1, the phosphorus balance for a typical German municipal WRRF with phosphorus removal is illustrated schematically.

7.3 TECHNOLOGIES FOR PHOSPHORUS REMOVAL IN WATER RESOURCE AND RECOVERY FACILITY

7.3.1 BIOLOGICAL PHOSPHORUS REMOVAL

All biological processes remove phosphorus from wastewater for microbial growth to varying degrees. Within the biological treatment in WRRF, the degradation of organic material causes the incorporation of ammonium and phosphate into cell biomass for growth of biomass. In average, bacteria in WRRF incorporate about 2 mg P/L for 200 mg/L degraded biochemical oxygen demand (BOD) (b.is, 2017).

The total removal of phosphorus in WRRF, regarding removal of primary sludge as well as biological phosphorus removal, is about 20 to 30% referred to concentrations in the influent. Hence, biological phosphorus removal is mainly regulated by removal of solid matter in the secondary sludge. As the impact of biological phosphorus removal is too small for compliance with requirements for discharge in receiving waters, further efforts are required. Therefore, state of the art in WRRF includes enhanced biological phosphorus removal (EBPR) and chemical phosphorus removal. The principal advantages of biological compared to chemical phosphorus removal are reduced chemical costs and less sludge production.

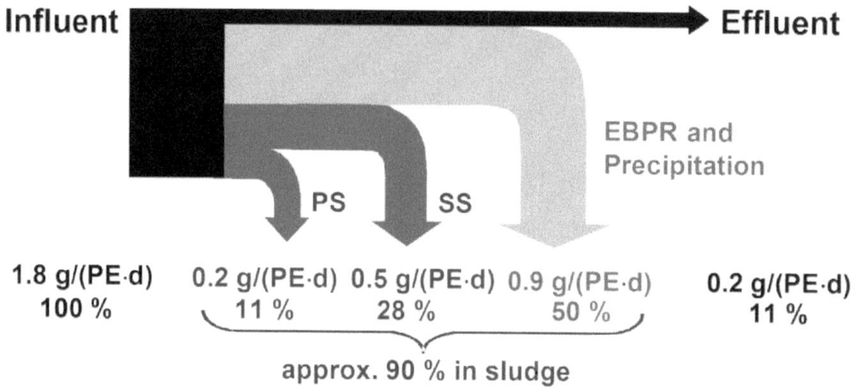

FIGURE 7.1 Phosphorus balance for a typical municipal WRRF in Germany with biological phosphorus removal and/or precipitation (PS: primary sludge. SS: surplus sludge).
Source: Reprinted from Cornel and Schaum (2009) with permission from the copyright holders, IWA Publishing.

7.3.1.1 ENHANCED BIOLOGICAL PHOSPHORUS REMOVAL (EBPR)

For EBPR a specified, heterotrophic population of microorganisms named as phosphorus accumulating organisms (PAO) is encouraged to grow and consume excess amounts of phosphorus as polyphosphates to keep the metabolism in adverse milieu alive. Thus, PAO are provided with a competitive advantage over other bacteria. EBPR is based on the following observations (Figure 7.2); cf. Tchobanoglous et al. (2014):

- Under anaerobic conditions, PAO assimilate fermentation products (e.g., short-chained fatty acids) and store them within the cells. Concomitantly, the consumption of polyphosphates and the release of ortho-phosphate out of the cell take place. Polyphosphates act here as storage of biochemical achievable energy, which is released during the process.
- Under aerobic/anoxic conditions, PAO reincorporate ortho-phosphates and refill the energy storage by forming polyphosphates. The previously stored organic material is used for bacterial growth. Hence, PAO gain a growth advantage over bacteria not following this practice.

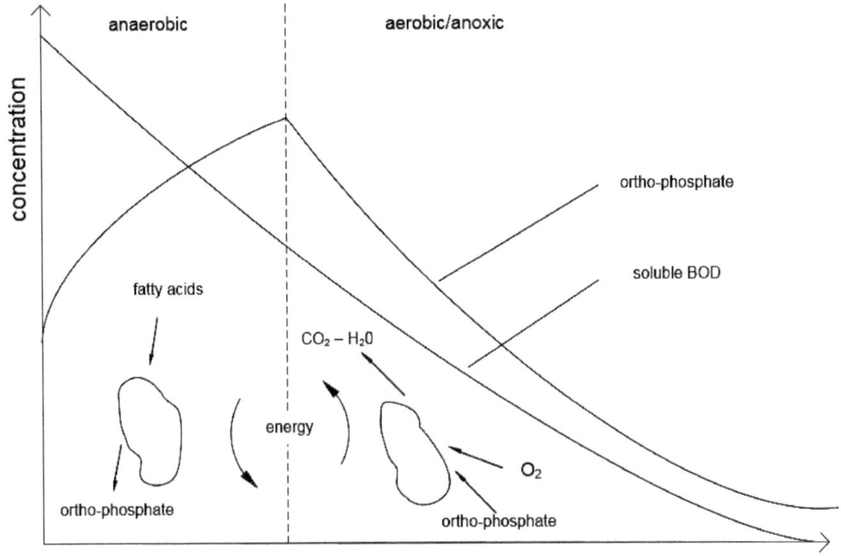

FIGURE 7.2 Simplified biochemical model for EBPR of PAO (BOD: biochemical oxygen demand), cf. (Tchobanoglous et al., 2014; b.is, 2017).

The order of process engineering for EBPR is a combination of anaerobic, anoxic, and aerobic tanks. EBPR is reached by subtracting excess sludge for further sludge treatment. Figure 7.3 exemplarily shows a process combination following the University of Cape Town (UCT) process. Further, diverse combinations of these processes are documented, e.g., in Tchobanoglous et al. (2014) and Bratby (2018).

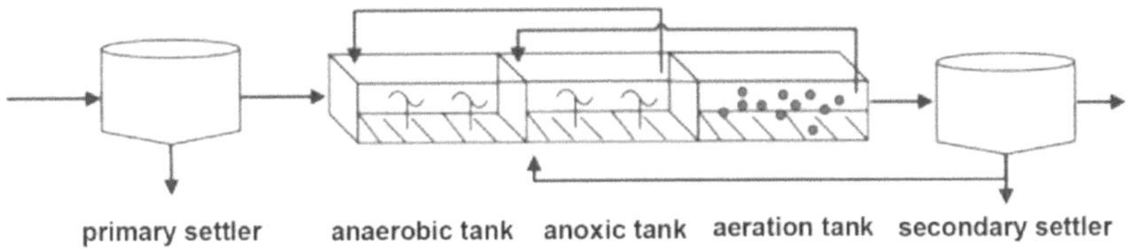

FIGURE 7.3 Simplified flow sheet of enhanced biological phosphorus removal.

The mixed liquor from an EBPR system can contain 0.06 to 0.15 mg P/mg VSS (volatile suspended solids) (Wentzel et al., 2008), in comparison mixed liquor in conventional activated sludge (AST) processes typically contains 0.02 mg P/mg VSS.

7.3.2 CHEMICAL PHOSPHORUS REMOVAL

The principle of chemical phosphorus removal is based on the method of precipitation. Chemical precipitation utilizes the poor solubility of compounds in certain solvents to separate the target substance gravimetrically as a solid compound. The substance to be separated is initially dissolved in the medium. After adding a precipitant beyond the solubility product of the substances a solid is formed which sediments according to the particle size and density and can be drawn off at the bottom of the tank. It should be noted that as the complexity of the mixture increases, further interactions take place between the substances involved. To assist sedimentation of the formed precipitates, flocculants can be added to the solution. Due to their long polymer chains, flocculants lead to a clotting of the particles formed which results in larger particle sizes and improved settlement of the solids.

The efficiency of phosphorus precipitation also depends on the phosphorus species. While ortho-phosphates can be separated by means of chemical precipitation, phosphonates can be regarded as non-precipitable (Krause, 2018).

The precipitant used are multivalent metal and alkaline earth metal ions. For phosphorus removal in WRRF especially Fe (II), Fe (III), Al (III), and Ca (II) play an important role. The reaction equations widely used in the literature for iron (7.1) and aluminum (7.2) are (Tchobanoglous et al., 2014):

$$Fe^{3+} + H_nPO_4^{3-} \leftrightarrow FePO_4 + nH^+ \tag{7.1}$$

$$Al^{3+} + H_nPO_4^{3-} \leftrightarrow AlPO_4 + nH^+ \tag{7.2}$$

The equations above give a stoichiometric ratio for Me:P of 1:1. In reality, the ratio will differ from this theoretical value due to competing reactions. Indeed the reactions and partial reactions are of a much more complexity than described in Eqns. (7.1) and (7.2), cf. Stumm and Morgan (1996). The complex mechanisms consist of precipitation of hydroxides and carbonates, complexations of organic substances, co-precipitation of phosphate, adsorption reactions, diffusion, and coagulation and flocculation processes (Stummand Morgan, 1996; Szabó et al.,

2008). Although the equation above does not reflect the real conditions, it can be used for an initial estimation of precipitant consumption.

Due to the difficult handling, the use of calcium hydroxide is of minor importance in WRRF nowadays (Parsonsa, 2004).

The choice of a precipitant depends primarily on the availability and the expected costs. In particular, between Fe and Al there are only minor differences in the handling and the achievable P-concentrations in the effluent so that the monetary factor comes even more into the foreground.

On the biological, chemical, and physical processes and thus on the overall efficiency of the phosphorus precipitation, various factors besides the precipitant and the dosing quantity, have a significant influence. The following aspects provide an overview of the most important parameters (DWA, 2011):

- Phosphorus species;
- pH and redox value, temperature;
- Concentration of competing substances, e.g., organics, sulfates;
- Mixing system (turbulence), retention time;
- Dosing point.

From pre-precipitation to advanced phosphorus removal the dosing point has a significant influence on the phosphorus precipitation. Basically, three different dosing points can be distinguished:

- Primary treatment (pre-precipitation);
- Secondary treatment (simultaneous precipitation); and
- Tertiary treatment (post-precipitation).

The tertiary treatment leads to the advanced treatment for phosphorus removal.

Figure 7.4 shows the three possible dosing points for chemical phosphorus removal. As 1-point and 2-point precipitation are established, several combinations are possible.

7.3.2.1 PRE-PRECIPITATION

The precipitants are added before the primary clarifier as shown in Figure 7.4. The dosing point is chosen so that formed flocs are not destroyed. The precipitated flocs are separated in the primary settler. Care must be taken to ensure sufficient residual phosphorus content for subsequent biological treatment.

In general, a content of 1–2 mg/l is usually sufficient to supply the bacteria in the subsequent biological stage (DWA, 2011). Adding the precipitants at the primary treatment has the advantage that additionally relevant amounts of organic substances are excreted, so that the organic load for the subsequent biological step is reduced (be carefully if denitrification is needed).

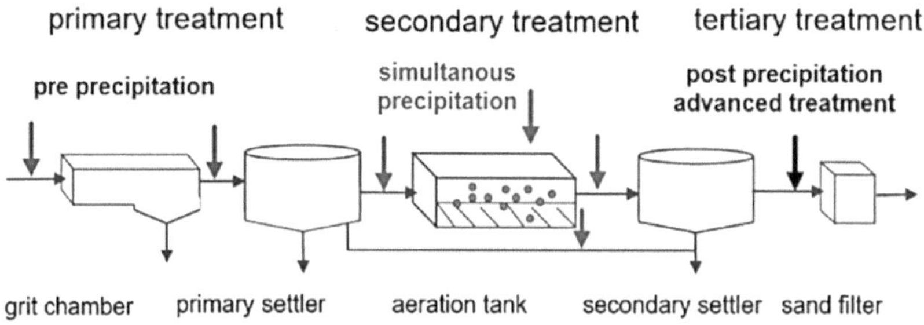

FIGURE 7.4 From pre-precipitation to advanced phosphorus removal.

7.3.2.2 *SIMULTANEOUS PRECIPITATION*

In simultaneous precipitation, the precipitate is added into the biological reactor, before secondary sedimentation tank or in return sludge. The recirculation of the return sludge improves the utilization factor of the precipitant. It is important to maintain the floc structure. It should be noted, that the addition of precipitant increases the inorganic content of the solids in the aeration tank. This must be considered in the design of the biological stage with regard to efficiency, sludge load, and sludge age (DWA, 2011). When selecting the precipitant care must be taken to ensure that, no biological side effects such as pH shifts occur.

7.3.2.3 *POST-PRECIPITATION*

After secondary settler, the precipitants are dosed in a mixing tank which is followed by a flocculation tank. The flocs are separated in a sedimentation basin or by flotation in a lamella separator (DWA, 2011). In addition, filtration may be used to maintain low levels in the effluent. In general, the use of flocculation filtration is recommended. The combination of flotation and filtration exists as well.

7.3.2.4 ADVANCED PHOSPHORUS REMOVAL

Based on the requirements of the European Water Framework Directive (EU, 2000), there is a discussion about processes for advanced phosphorus removal to meet more stringent discharge concentrations of phosphorus for WRRF in the future (Englmannand Müller, 2017; Seel, 2017). In Switzerland, for example, there are discharge values of 0.2 mg/L total phosphorus required (Lake Zurich) and 0.3 mg/L total phosphorus for states bordering Lake Constance in the 24-hour composite sample (Böhlerand Siegrist, 2008). Achieving such low concentrations in effluents of WRRF is significantly influenced by discharge of total solids (TS) from secondary settler (Barjenbruchand Geyer, 2018). For example, with an average content of phosphate in sludge of 30–35 g P/kg TS and concentrations of 10–20 mg/L TS in the effluent, this results in a concentration of particulate phosphate of 0.3–0.7 mg/L in the discharge (Figure 7.5). Furthermore, phosphoric dissolved substances, e.g., phosphonates from industrial dischargers, can contribute to the effluent as these substances are not removable from wastewater neither with biological nor chemical processes, cf. Krause (2018).

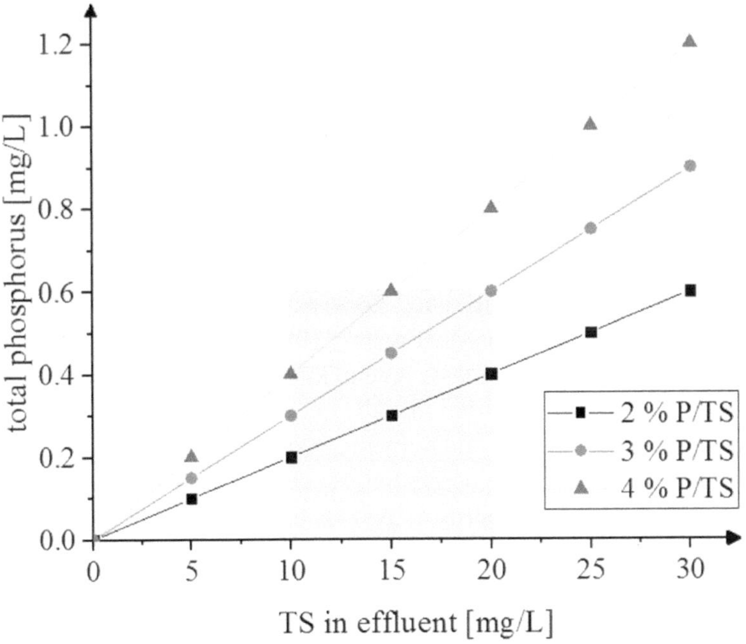

FIGURE 7.5 Relation between TS in effluent and particulate phosphorus.

Advanced phosphorus removal with targeted concentrations below 0.8 mg/L total phosphorus in the effluent require further processes such as combination of precipitation and flocculation, filtration or ultrafiltration (UF). In order to reach concentrations less than 0.1 mg/L membrane filtration can be used. Such low discharge values raise the usage of precipitants (Bratby, 2018). In contrast to advanced phosphorus removal, targeted concentrations of phosphorus of less than 2 mg/L with EBPR is only possible in combination of chemical phosphorus removal (Janssen et al., 2002). The application of chemical phosphorus removal reduces phosphorus concentrations to 1 to 2 mg/L in the effluent (Tchobanoglous et al., 2014). Besides the targeted phosphorus concentrations, advantages, and disadvantages of technologies for EBPR, chemical, and advanced phosphorus removal are shown in Table 7.3.

7.4 PHOSPHORUS: FROM CYCLE TO FINITENESS

Phosphorus is an essential element for all organisms. Beside carbon, hydrogen, oxygen, and nitrogen, phosphorus is one of the vital components of the DNA and the key element of the energy supplier ATP (adenosine triphosphate). As a vital cell component, phosphorus cannot be replaced by any other element. This is why phosphorus is different from other resources, such as fossil fuels, where there are potential alternatives, or from nitrogen fertilizers, which can be technically produced from air nitrogen via the Haber-Bosch process (Cornel and Schaum, 2009).

In nature, phosphorus passes through several interconnected cycles. The inorganic cycle describes the cycle from erosion, transport to the oceans, sedimentation, tectonic uplift, and alteration of phosphate-containing rocks into plant-available phosphates (Emsley, 1980, 2000; Filippeli, 2002). The cycle time of these processes lasts several million years, i.e., in human spaces of times, phosphate transported into the oceans can be considered as "lost" for agricultural use (Cornel and Schaum, 2009).

Besides the inorganic phosphorus, cycle there is two organic cycles attached describing phosphorus as part of the food chain. One cycle takes place on land (soil-plants-humans/animals-organic waste-soil), while the other one occurs in water. The cycle time of these is between a few weeks and up to one year (Emsley, 1980, 2000; Bennett and Carpenter, 2002).

These originally "natural" closed cycles are interrupted when phosphorus compounds in animal and human excrements are not used in fertilization.

TABLE 7.3 Advantages and Disadvantages of Technologies for Phosphorus Removal

Treatment	Application	Advantages	Disadvantages
EBPR	Activated Sludge process	No usage of chemicals Less sludge production Usage of sludge as fertilizer, plant availability of phosphorus Lower salinity in the effluent compared to chemical removal	Limited elimination of P, lower stability and flexibility of total P in effluent Dependence on COD:P Nitrite from nitrification can inhibit the EBPR Increase of sludge volume index Phosphorus release in anaerobic sludge treatment (sludge liquor, struvite scaling)
Chemical	Pre-precipitation	Increased removal of BOD and suspended solids	Removal of BOD, BOD limited for denitrification Needed residual P for biological processes High precipitant consumption Increased primary sludge production
	Simultaneous precipitation	Improved stability of activated sludge, depending of the precipitant, Al or Fe Optimization by multipoint precipitation	Difficult for low alkalinity wastewaters (depending on the precipitant) Increased secondary sludge production
	Post-precipitation	Low effluent phosphorus achievable Enhanced control over P concentration in effluent	High (capital) costs Production of precipitation sludge Impact on disinfection with UV
Advanced	Effluent of secondary clarifier	Low effluent phosphorus achievable Enhanced control over P concentration in effluent removal of suspended solids (filtration process), combination with further treatment processes possible, e.g., activated carbon, ozone	High (capital) costs High energy demand Production of precipitation sludge

Source: (cf. Janssen et al., 2002; cf. Tchobanoglous, 2014; cf. Barjenbruch and Geyer, 2018; cf. Bratby, 2018).

Then, phosphate contained in wastewater is partly transported to the oceans via the discharge systems, partly fixed in sewage sludge, which is deposited in landfill sites or incinerated; in the latter case, the phosphorus contained in the ash is deposited in landfill sites or subterranean storage. The procedure can be similar with organic fertilizers (solid and liquid manure) from intensive stock-rearing. The deficit is balanced by "chemical fertilizers," i.e., the mining of phosphate-rich deposits in the earth's crust. In Figure 7.6, the geological and biological cycles are illustrated, including changes due to human impact (Cornel and Schaum, 2009).

Concerns about pollutants (anthropogenic trace substances, microplastics, etc.), is questioning agricultural utilization of sewage sludge especially in Central Europe. Due to these concerns on the one hand and the desire to preserve the phosphorus cycle on the other, in some countries the obligation to recycle phosphorus is being discussed. In Germany and Switzerland, corresponding legal regulations have already been issued.

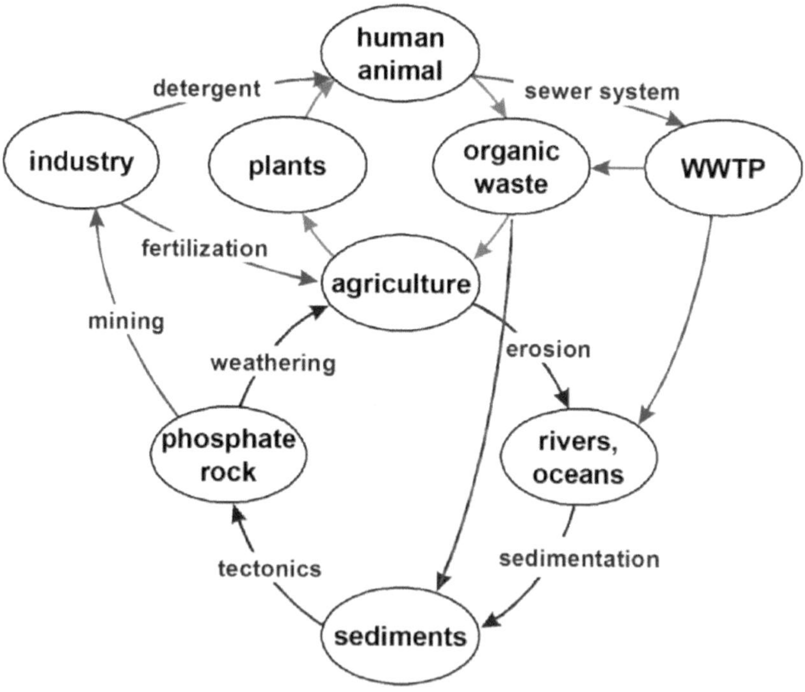

FIGURE 7.6 Geological (inorganic) and organic (land) phosphorus cycles (Pinnekamp, 2002, modified), cf. Bennett and Carpenter (2002), including human impacts (phosphorus cycle in water not included). *Source:* Reprinted from Cornel and Schaum (2009) with permission from the copyright holders, IWA Publishing.

Phosphorus is extracted from rock phosphate deposits. The worldwide occurrence of low-emission and easily accessible rock phosphate is limited. Global phosphate reserves are currently estimated at 16 billion Mg (currently economically degradable), of which 84% are located in five countries: Morocco and the Western Sahara (36%), China (23%), South Africa (9%), Jordan (9%) and the US (7%), cf. (USGS, 2010). These are both sedimentary and magmatic phosphorus deposits and accordingly finite and nonrenewable.

Even if a static lifetime of more than 300 years for phosphorus reserves and more than 1,400 years for resources is assumed, cf. Udert (2014), issues concerning accessibility to phosphorus resources and ecological aspects are becoming more and more important. At European level, various abiotic raw materials were analyzed with regard to supply risks (country risk, market risk, structural risk) and vulnerability (quantity relevance and strategic relevance), cf. EU (2010), Erdmann et al. (2011), EU (2014). With regard to phosphorus a high supply risk was attested due to strong dependency on only few mining countries (e.g., China, Russia, Congo, and Brazil), cf. EC (2014, 2017), EU (2014). In this context, phosphorus was classified as a "Critical Raw Material" by the European Union (EU) in 2014, cf. EU (2014).

Furthermore, global demand for fertilizers steadily increased due to the world's growing population and associated food needs during the past decades (FAO, 2019). In addition to nitrogen and potassium, phosphorus is a major component of fertilizers and essential for plant growth. For humans and animals, phosphorus is an important building block for both deoxyribonucleic acid (DNA) and ATP which is significantly involved in metabolic processes. It is expected, that the demand for fertilizers will continue to increase in the future. In 2017, the Food and Agriculture Organization (FAO) of the United Nations estimates that the need for phosphorus used in fertilizers will increase worldwide during the period between 2015 and 2020 by 1.9% (FAO, 2017).

Due to its importance, phosphorus should be used sustainably regardless of any static lifetime assumptions. This includes recycling of secondary phosphates, efficient extraction, and treatment of raw phosphates as well as efficient use of phosphates. By recovering phosphorus from wastewater, WRRF can make an important contribution to the sustainable use of phosphorus.

7.5 TECHNOLOGIES FOR PHOSPHORUS RECOVERY FROM WASTEWATER AND BIOSOLIDS

More than 50 different processes have been developed worldwide for the recovery of phosphorus in the field of wastewater technology in the last two decades. Though, only a few technologies have been realized at a full scale yet (Schaum, 2018).

Phosphorus recovery can mainly be implemented in three different flow streams at the WRRF (Schaum and Cornel, 2016; Kabbe and Kraus, 2017):

- Sludge liquor from (sludge-) dewatering;
- Biosolids;
- Residues (ash) after thermal sludge treatment (especially mono-incineration).

Furthermore, there are few technological approaches that focus on the recovery of phosphorus in the mainstream of WRRF, especially in combination with technologies for phosphorus removal. As these processes play a subordinate role, they are not further focused.

Table 7.4 compares the three flow streams. Obviously, there is a significant difference between these streams concerning the volume and the phosphorous content. In comparison, the ash has a very low volume and the highest phosphorus concentration at the same time. In contrast to the ash, sludge liquor has a high volume and a low P-concentration. It should be noted, that the theoretical recovery rate in Table 7.4 does not correspond to the technical feasible recovery which depends on the technology used. Figure 7.7 shows various possibilities of phosphorus recovery during sewage sludge treatment.

TABLE 7.4 Comparison of Sludge Liquor, Digested Sludge and Ash

	Sludge Liquor	Digested Sludge	Ash
Volumetric Flow	1–10 L/(PE·d)	0.2–0.8 L/(PE·d)	approx. 0.03 kg/(PE·d)
Total Solid (TS)	< 1%	approx. 3%	100%
Ash Content	–	40–50%	100%
Phosphorus (P_{tot})	20–100 mg/L	2–5% referred to TS	5–10% referred to TS
Recovery Rate*	approx. 10–20%	approx. 90%	approx. 90%

* It is based on the inflow load of the WRRF; theoretical recovery potential; due to the efficiency of the process, the actual potential can be considerably lower.

Source: Schaum and Cornel (2016).

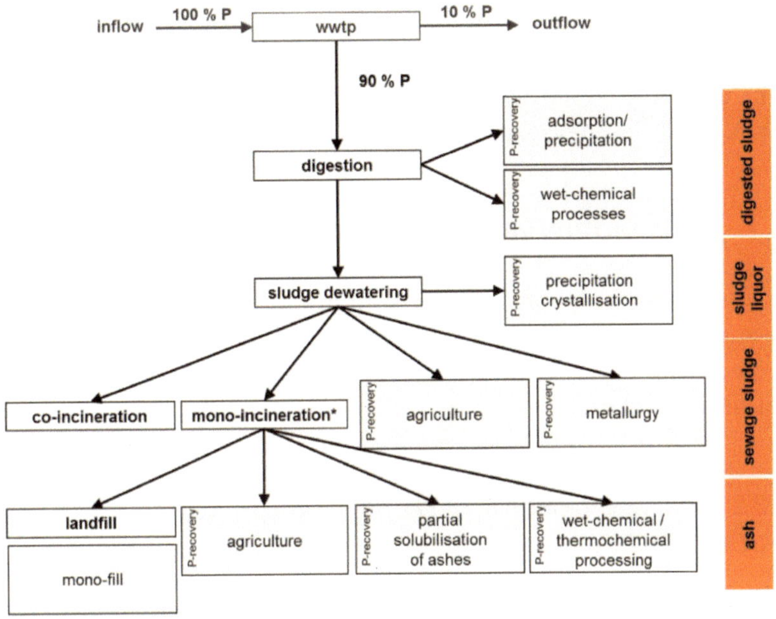

FIGURE 7.7 Possibilities of phosphorus recovery from sewage sludge, modified according to Schaum and Cornel (2013).

7.5.1 PHOSPHORUS RECOVERY FROM SLUDGE LIQUOR

The implementation of phosphorus recovery in the sludge liquor allows separation of already dissolved phosphorus by applying relatively simple technologies. In particular, phosphorus-rich sludge liquors with a P content of>50 mg/L are suitable. One big advantage of phosphorus recovery from sludge liquor is the possibility of combining it with advanced phosphorus removal (Cornel and Schaum, 2009).

The recovery from this stream is particularly efficient in combination with an EBPR in side streams (supernatant liquor of the anaerobic stripper) or from sludge liquor during sludge treatment. The phosphorus-rich water is fed into a precipitation/crystallization tank, where phosphorus is removed as calcium phosphate or magnesium ammonium phosphate (struvite) by adding calcium or magnesium salts and if needed seed crystals (Cornel and Schaum, 2009).

There are various full-scale processes for phosphorus recovery from sludge liquor, especially in the Netherlands and North America (Kabbe and Kraus, 2017; Gysin, 2018; Ochi, 2018). In North America, the Pearl process of the company Ostara was implemented on several WRRF among others (Gysin, 2018; Jeyanayagam, 2018).

7.5.2 PHOSPHORUS RECOVERY FROM BIOSOLIDS

In the case of phosphorus recovery from biosolids, different approaches can be distinguished. On the one hand, process immanent technologies, on the other hand, wet-chemical extraction of phosphorus from sewage sludge with subsequent precipitation as well as thermal-chemical approaches.

7.5.2.1 PROCESS IMMANENT TECHNOLOGIES-PRECIPITATION/ CRYSTALLIZATION

In the AirPrex process, the dissolved phosphorus in the sewage sludge is crystallized as struvite, which prevents scaling effects and improves the dewaterability of sewage sludge. For this purpose, magnesium chloride is dosed into a reactor containing digested sewage sludge. In order to achieve a higher pH value for an improved crystallization of struvite, the digested sludge is aerated. Subsequently, struvite is partly separated from the sludge by sedimentation. The AirPrex process is implemented on full-scale at different WRRF, e.g., at the WRRF Berlin-Waßmannsdorf (Germany) or Amsterdam (The Netherlands) (Ortwein, 2018).

7.5.2.2 WET CHEMICAL PROCESSES

In the wet chemical approach, the biologically and chemically bound phosphorus is dissolved from sewage sludge by an extractant, e.g., acids. Because of higher amount of inorganic bounded phosphorus in digested sludge, this flow stream is more suitable for phosphorus recovery in terms of chemical consumption than secondary sludge or raw sludge (Petzet, 2013). Since 2007, the Gifhorn process (modified Seaborne process) has been tested in full-scale at the WRRF Gifhornin Germany (Günther, 2011; Esemen, 2012). The Stuttgarter-Verfahren (Weidelener, 2010) has been operated at the WRRF Offenburg (Germany) since, 2011/2012 on a semi-industrial scale.

The WRRF Mainz-Mombach (Germany) is currently testing the ExtraPhos process on a semi-industrial scale. In contrast to other processes, ExtraPhos uses CO_2 to extract phosphate (Opitz, 2017; Schnee, 2018). Figure 7.8 shows a simplified scheme of a wet chemical phosphorus recovery process.

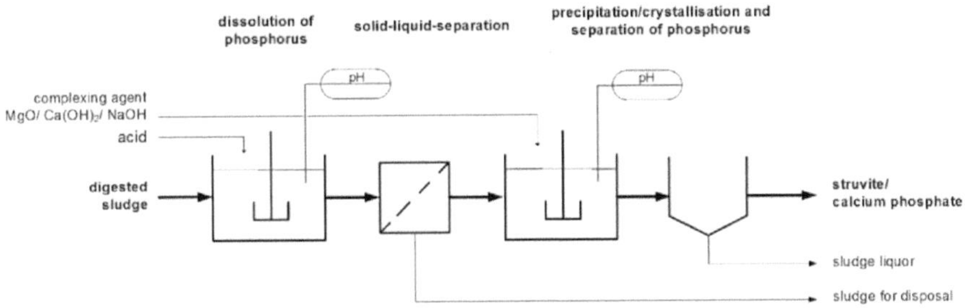

FIGURE 7.8 Simplified flow diagram of a wet chemical process for phosphorus recovery from biosolids.

7.5.2.3 THERMAL-CHEMICAL PROCESSES (METALLURGY)

In the 50s and 60s of the last century, the Thomas process was widely used in the steel industry. The purpose of the process was to reduce the carbon content in the steel. The process was particularly suitable for the melting of phosphorus-rich iron ores. Phosphorus pentoxide which emerges by oxidation of phosphorus was slaked adding limestone into the melter. This technology was adapted to recover phosphorus from sewage sludge. The organic content of the sewage sludge is gasified under reductive conditions. The resulting gas can be used energetically in the latter course.

A semi-industrial pilot plant was operated in Nuremberg (Germany). Furthermore, a pilot plant from Kubota is operated in Japan (Hosho, 2018; Takaoka, 2018).

7.5.3 PHOSPHORUS RECOVERY FROM SEWAGE SLUDGE ASH

The advantage of ash recovery is the low volume that has to be processed. In contrast, there is the need for a (mono-) incineration plant which has to be implemented before the recovery process. Due to the higher phosphorus concentration, sewage sludge ashes (SSA) from mono-incineration are more suitable for phosphorus recovery than sewage sludge treated in co-incineration plants like coal-fired power stations.

7.5.3.1 DIRECT USE/PRETREATMENT OF SSA IN AGRICULTURE

As shown in Figure 7.7, the ash can is used directly on agricultural land as fertilizer. However, in case of direct recovery, the higher heavy metal contents in the ashes should be taken into account. Due to the low phosphorus availability of the ashes, various methods rely on partial acidification of the ashes, e.g., through the addition of phosphoric acid or sulfuric acid, which increases the water solubility and thus the plant availability of the phosphorus (Hiller, 2018).

7.5.3.2 WET CHEMICAL PROCESSES FOR SSA

Heavy metals are contained in the ashes, so that an accumulation of pollutants has to be taken into account in case of agricultural utilization. In order to get rid of the contaminants and to provide a phosphorus fertilizer with a higher quality standard, several processes had been developed during the last decade. In most cases, the ashes are acidified in order to dissolve the phosphorus. The dissolution depends on the process conditions especially the pH value. After a solid-liquid separation, phosphorus, and metals can be separated from the filtrate, e.g., by precipitation, crystallization, membrane technology, liquid-liquid extraction, etc. In some cases, a complexation of heavy metals is executed. Various process technologies have already been developed for this purpose (Schaum, 2007; Petzet, 2013; Pinnekamp et al., 2013).

In Basel (Switzerland), the LeachPhos process was tested in full-scale in 2013 (Bühlerand Schlumberger, 2014; Eicher, 2018). Since, 2012/2013, a semi-industrial pilot plant (TetraPhos process) for the production of phosphoric acid from sewage sludge ash is in operation at the WRRF Hamburg-Köhlbrandhöft (Germany) (Lebek, 2016, 2018). During the PARFORCE process, the ash is being acidified with hydrochloric or nitric acid. After separation of the solids and enrichment, a marketable phosphoric acid is received (Freiberg, 2017; Fröhlich, 2018); a semi-industrial pilot plant is running in Freiberg (Germany) (Figure 7.9).

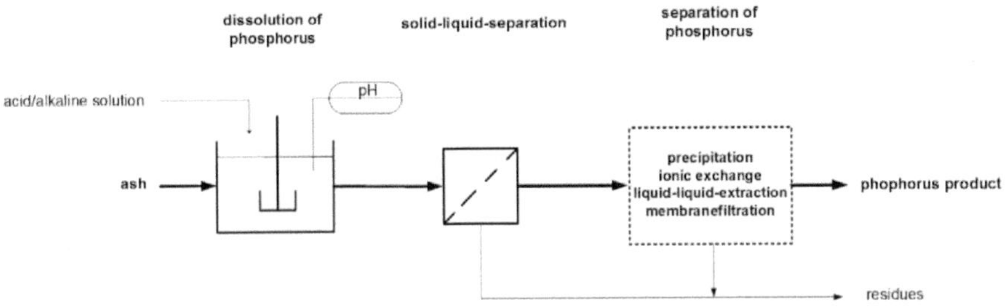

FIGURE 7.9 Simplified flow diagram of a wet chemical process for phosphorus recovery from SSA.

7.5.3.3 THERMO-CHEMICAL PROCESSES FOR SSA

In the field of thermo-chemical processes for the recovery of phosphorus from SSA, two approaches can be distinguished:

- Treatment at T > 1,000°C with the aim of producing magnesium/potassium phosphates; "Rhenania process"; and
- Treatment at T > 1,400/1,500°C with the aim of producing white phosphorus; "electro-thermal process."

At the beginning of the 19th century, the so-called Rhenania process was developed. Core of the process is a thermal treatment of raw phosphate with the addition of soda, lime, and alkali silicates among others, whereby Rhenania phosphate (a calcium silicate phosphate) is recovered (Franck et al., 1936, 1938). Rhenania phosphate is like apatite insoluble in water. However, it can be dissolved by organic acids, and thus be considered as plants available.

Within the AshDec process, ash is thermally treated at a temperature of 1,000°C. By addition of soda/magnesium/potassium chlorides, heavy metals are reduced and separated through the gas-phase as metal-chlorides while the plant availability is positively influenced in analogy to the Rhenania process. The process was investigated on a semi-industrial scale in Leoben (Austria) (Kley et al., 2005; Mattenberger et al., 2008, 2010).

The production of white phosphorus from raw phosphate is carried out in an electric arc furnace. The basic chemical reaction was developed by Wöhler in 1829 (Breil, 1963) and was adapted in the RecoPhos process in order to treat SSA. The ash is treated at approx. 1,400–1,500°C and reduction from phosphate to white phosphorus (P_4) occurs by adding coke and sand.

The phosphorus, which leaves the furnace along with CO via the gas phase, is condensed under exclusion of air as P_4 (Breil, 1963). Iron contained in the raw material forms ferrophosphorus, which reduces the yield of P_4. For this reason, high iron concentrations should be avoided for example by avoiding iron precipitants or other iron containing chemicals. Other precipitants based on aluminum or calcium does not disturb the process (Schipper et al., 2005). The substitution of raw phosphate by SSA using the electro-thermal phosphorus process was practiced by Thermphos in Vlissingen (The Netherlands) (Schipper and Korving, 2009). For financial reasons, production was stopped at the end of 2012. In a research project (RecoPhos-Process), studies on the use of the electro-thermal, method with focus on the design of the reactor was carried out (Schönberg et al., 2012; Langeveld, 2018).

7.6 HOLISTIC BIOSOLIDS MANAGEMENT-OUTLOOK

7.6.1 REALIZATION OF PHOSPHORUS RECOVERY: A PURELY TECHNICAL ISSUE

A motivation for phosphorus recovery cannot be substantiated from an economic point of view. None of the developed technologies, except technologies recovering the phosphorus from sludge liquor, is at this point competitive with regard to price compared to conventional phosphorus fertilizers from rock phosphate.

Table 7.5 shows the specific investment and operational costs of various phosphorus recovery technologies as well as used chemicals, residues, efficiency, and received secondary phosphate. It can be seen that most technologies have production costs above 1,000 €/Mg P_2O_5 (Spörri et al., 2017) whereas industrial fertilizers from raw phosphate have specific costs of less than 500 €/Mg P_2O_5 (Triple superphosphate), cf. Egle (2016). However, these provisional costs should be treated with caution, due to a lack of large-scale implementations. Thus, the positive effects such as the higher operating stability due to lower scaling or the improved dewater ability properties due to the lower P-contents in the sludge cannot be expressed in monetary terms for most processes. In addition, the question of residual waste disposal is still insufficiently clarified.

The operating costs may be offset by the revenue from the marketing of the secondary phosphates. The marketability of the secondary phosphates is highly dependent on the quality (concentration of phosphorus and contaminants) and thus of the chosen process as well as the boundary conditions on

TABLE 7.5 Overview of Costs, Use of Chemicals, P-Product of Some Processes

material flow		technologies	chemicals		residues	products	efficiency	capital costs	operating costs
			extraction	precipitation/ heavy metal separation			[%]	[€/Mg P_2O_5]*	[€/Mg P_2O_5]*
sludge liquor		Crystallactor	none	$Ca(OH)_2$ / $MgCl_2$	residual sludge	struvite/ magnesium phosphate/ calcium phosphate	< 20	> 1,300	1,750 - 4,400
sludge liquor		Pearl		$MgCl_2$		struvite			
sludge	wet-chemical	AirPrex	none	$MgCl_2$	residual sludge	struvite	< 20	> 1,300	< 900
sludge	wet-chemical	Stuttgarter-Process	H_2SO_4	NaOH/ $MgCl_2$	P-poor acidic residual sludge		< 50	> 1,300	> 4,400
sludge	wet-chemical	Gifhorner-Process							
sludge	wet-chemical	ExtraPhos	CO_2	$Ca(OH)_2$	P-poor residual sludge	dicalcium phophate		not specified	> 1,000
sludge	thermal	Kubota	Ca/Si (slag former)	none	heavy metal rich fly ash	Ca-Silico-phosphate	75 - 90**	> 1,300	1,750 - 4,400
sludge	thermal	Mephrec							
ash		LeachPhos	H_2SO_4	NaOH or CaO	(acidic) residual ash, filter cake, gypsum	dicalcium phophate	75 - 90	400 - 900	900 - 1,800
ash		TetraPhos	H_3PO_4	H_2SO_4 (precipitation of gypsym)		phosphoric acid			

*1 € = 1.13 $ (03/15/2019); 1 Mg (megagram) = 1,000 kg (kilogram).

** The efficiency does not take into account the losses in the fly ash and melt.

Source: Modified According to Spörri et al. (2017).

the WRRF. Some secondary phosphates which were recovered technologically from the sludge liquor have achieved market prices. Market prices can also be assumed for phosphoric acid recovery processes.

In addition to investment and operating costs, organizational aspects play an important role in terms of nationwide P-recovery as well. In particular, the question of the commercialization of those secondary phosphates has to be clarified in the near future.

7.6.2 WATER AND RESOURCE PROTECTION

More and more, the question of sustainable management of resources like fossil fuels, metals, water, and last but not least phosphorus is discussed in the broad public (Schaum et al., 2015; Schaum and Cornel, 2016; Schaum, 2018b). For a sustainable society, careful, and efficient use of natural resources is essential (ProgRess, 2012). In this sense, WRRF must adapt to new requirements. In this context, WRRF will have to improve the use of the energy stored in the water on the one side and recover valuable substances contained in the wastewater on the other side in addition to the traditional task of wastewater treatment. Due to its criticality, phosphorus has attracted the public and political interest in recent years, unlike nitrogen and potassium. In this sense, great efforts have been made in the past to close the phosphorus cycle.

A variety of technological methods for phosphorus recovery has been developed and is now ready to be implemented in industrial-scale on WRRF in order to provide a low-polluting and phosphorus-rich fertilizer for application in agriculture. In the sense of sustainable resource protection and to prevent eutrophication, unimpeded discharge of high amounts of phosphorus by effluents of WRRF should be avoided.

In the future, WRRF will not only meet the current requirements of the health and water protection, but increasingly become producers. This also implies that they have to deal with product management and marketing strategies. Adherence to quality standards, product acceptance, availability, delivery guarantees, etc., will be necessary prerequisites in order to commercialize “products” such as industrial or irrigation water, fertilizers, raw phosphate substitutes, but also heat and electricity. In particular, this will challenge small and medium-sized WRRF and will require new business models, e.g., the involvement of private and cooperative marketers (Demmelbauer et al., 2018; Schaum, 2018b).

KEYWORDS

- **adenosine triphosphate**
- **biochemical oxygen demand**
- **deoxyribonucleic acid**
- **enhanced biological phosphorus removal**
- **phosphorus accumulating organisms**
- **population equivalent**

REFERENCES

ATV-DVWK, (2000). *Arbeitsblatt ATV-DVWK A 131-Bemessung von einstufigenBelebungsanlagen, Hennef.*

bis (2017). Abwasserbehandlung: Gewässerbelastung, Bemessungsgrundlagen, Mechanische Verfahren, Biologische Verfahren, Reststoffe aus der Abwasserbehandlung; Weitergehende Abwasserreinigung, Bauhaus-Institut für zukunftsweisende Infrastruktursysteme (b.is), Arbeitsgruppe Weiterbildendes Studium Wasser und Umwelt, 4. überarbeitete Auflage, Weimar: Universitätsverlag.

Barjenbruch, M., & Geyer, P., (2018). Total solids and phosphorus: A cross linked topic? In: Schaum, C., (ed.), *Phosphorus: Polluter and Resource of the Future: Removal and Recovery from Wastewater*. IWA-Publishing, ISBN 9781780408354. London.

Bennett, E., & Carpenter, S., (2002). P soup: Humanity's impact on the phosphorus cycle. *World Watch Magazine, 15*, 24–32.

Böhler, M., & Siegrist, H., (2008). *Möglichkeiten zur Optimierung der chemischen Phosphorfällung an hessischen Kläranlagen; Gutachten der EAWAG im Auftrag der Europa Fachhochschule Fresenius und des Hessischen Landesamtes für Umwelt und Geologie (HLUG).*

Bratby, J., (2018). Phosphorus removal in wastewater treatment plants. In: Schaum, C., (ed.), *Phosphorus: Polluter and Resource of the Future: Removal and Recovery from Wastewater*. IWA-Publishing, ISBN 9781780408354, London.

Breil, G., (1963). Elektrothermische Erzeugung von Phosphor. *Chemie-Ingenieur-Technik, 35*(8).

Bühler, J., & Schlumberger, S., (2014). *Phosphormining aus Klärschlammasche, Nasschemische Extraktion-LEACHPHOS eine Lösung die funktioniert.* http://www.umweltbundesamt.de/sites/default/files/medien/378/dokumente/buehler_bsh_leachphos.pdf (accessed on 10 June 2020).

Calvo, M. S., Moshfegh, A. J., & Tucker, K. L., (2014). Assessing the health impact of phosphorus in the food supply: Issues and considerations. *Advances in Nutrition (Bethesda, Md.), 5*(1), 104–113.

Cornel, P., & Schaum, C., (2009). Phosphorus recovery from wastewater: Needs, technologies and costs. *Water Sci. Technol., 59*(6), 1069–1076.

Demmelbauer, V., Schaum, C., Frank, D., & Kabbe, C., (2018). Anforderungen an Phosphor-Rezyklate und Vermarktungskonzepte: Aspekte für eine zukünftige Phosphorrückgewinnung. GWF Wasser Abwasser.

DWA, (2011). *Arbeitsblatt DWA-A 202-Chemisch-physikalische Verfahren zur Elimination von Phosphor aus Abwasser.* Hennef.

DWA, (2014a). 27. DWA-Leistungsvergleich kommunaler Kläranlagen 2014.

DWA, (2014b). *Arbeitsblatt DWA-A 272-Grundsätze für die Planung und Implementierung Neuartiger Sanitärsysteme (NASS).* Hennef.

DWA, (2016). Bemessung von Kläranlagen in warmen und kalten Klimazonen. DWA-Themen T4/2016. Oktober 2016. Deutsche Vereinigung für Wasserwirtschaft Abwasser und Abfall, Hennef.

Eastham, R. D., Colombo, J. P., & Peheim, E., (1981). Interpretation klinisch-chemischer Laborresultate, 2, neu bearb. und erw. Aufl./übers. und neu bearb. Von E. Peheim und J. P. Colombo. S. Karger, Basel.

Egle, L. R. H., Krampe, J., & Zessner, M., (2016). Phosphorus recovery from municipal wastewater: An integrated comparative technological, environmental, and economic assessment of P recovery technologies. *Science of the Total Environment, 571*, 522–542.

Eicher, N., (2018). The LeachPhos process at the waste-to-energy plant Bern. In: Schaum, C., (ed.), *Phosphorus: Polluter and Resource of the Future* (pp. 411–416). IWA Publishing, London.

Emsley, J., (1980). The phosphorus cycle. In: *The Natural Environment and the Biogeochemical Cycles*. Springer Verlag, Berlin-Heidelberg-New York.

Emsley, J., (2000). *Phosphor-ein Element auf Leben und Tod.* Wiley-VCH Verlag GmbH, Weinheim.

Englmann, E., & Müller, K., (2017). Strategie zur Reduzierung der Phosphoreinträge aus Abwassereinleitungen in Bayern, 45. Abwassertechnisches Seminar, Berichte aus der Siedlungswasserwirtschaft, Technische Universität München, Nr. 215.

Erdmann, L., Behrendt, S., & Feil, M., (2011). *Kritische Rohstoffe für Deutschland-Identifikation aus Sicht deutscher Unternehmen wirtschaftlich bedeutsamer mineralischer Rohstoffe, deren Versorgungslage sich mittel-bis langfristig als kritisch erweisen könnte-Abschlussbericht im Auftrag der KfW Bankengruppe.*

Esemen, T., (2012). *Untersuchungen zur technischen und wirtschaftlichen Optimierung der Nährstoffrückgewinnung aus Klärschlamm.* Ges. zurFörderung des Inst. fürSiedlungswasserwirtschaft, Braunschweig.

EU, (2000). DIRECTIVE 2000/60/EC. In: *The Tepatco and European Union, THE European Parliament and the Council of the European UNION*. Straßbourg.

EU, (2010). *European Commission: Critical Raw Materials for the EU.* Brüssel, Belgien.

EU, (2014). *European Commission: Report on Critical Raw Materials for the EU.* Report of the Ad hoc Working Group on defining critical raw materials.

EU, (2017). *Communication from the Commission to the European Parliament, the Council, the European Economic and Social Committee, and the Committee of the Regions.* In Commission E, *European Commission*. Brüssel.

FAO, (2017).*World Fertilizer Trends and Outlook to 2020*. NATIONS FAAOOTU.

FAO, (2019). Fertilizers archive. *Database with Information about 'Total global Fertilizer Production by Nutrient.* Food and Agriculture Organization of the United Nations (FAO). http://www.fao.org/faostat/en/#data/RA. 22/02/2019 (accessed on 10 June 2020).

Filippeli, G. M., (2002). *The Global Phosphorus Cycle* (p. 48). Mineralogical Society of America. Washington. USA.

Franck, H. H., Bredig, M. A., & Frank, R., (1936). Untersuchungen über Kalk-Alkaliphosphate I-Ein Beitrag zur Kenntnis des Rhenaniaphosphates. *Zeitschrift für Anorganische und Allgemeine Chemie,* 230(1/2).

Franck, H. H., Bredig, M. A., & Kanert, E., (1938). Untersuchungen über kalk-alkali-phosphate II-über calcium-kalium-phosphate. *Zeitschrift für Anorganische und Allgemeine Chemie.,* 237.

Freiberg, T. B., (2017). *Parforce Technologie.* http://www.parforce-technologie.de/unsere-technologie/ (accessed on 10 June 2020).

Fröhlich, P. E., & Martin, G. J., (2018). The PARFORCE-Technology (Germany). In: Schaum, C., (ed.), *Phosphorus: Polluter and Resource of the Future.* IWA Publishing, London.

Günther, L., (2011). *Großtechnische Nährstoffrückgewinnung und Schadstoffausschleusung aus kommunalen Klärschlämmen.* Ges. zur Förderung des Inst. für Siedlungswasserwirtschaft, Braunschweig.

Gysin, A. L. D., & Wirtel, S., (2018). The Pearl and WASSTRIP processes (Canada). In: Schaum, C., (ed.), *Phosphorus: Polluter and Resource of the Future* (pp. 359–366). IWA Publishing, London.

Henze, M. E., (2008). *Biological Wastewater Treatment: Principles, Modeling, and Design.* IWA Pub, London.

Hiller, G. W. J., (2018). Total phosphorus recovery and direct utilization of the sewage sludge ash as a fertilizer at Ulm. In: Schaum, C., (ed.), *Phosphorus: Polluter and Resource of the Future* (pp. 447–454). IWA Publishing, London.

Hosho, F., (2018). Sludge melting system. In: Schaum, C., (ed.), *Phosphorous: Polluter and Resource* (pp. 435–441). IWA Publishing, London.

Janssen, P. M. J., Meinema, K., & Van, D. R. H. F., (2002). Biological phosphorus removal. Manual for design and operation. *Water and Wastewater Practitioner Series: STOWA Report.* IWA-Publishing. London.

Jardin, N., (2002). Phosphorbilanz bei der Abwasser-und Klärschlammbehandlung. In: Rückgewinnung von Phosphor aus Klärschlamm und Klärschlammasche. 66. Darmstädter Seminar-Abwassertechnik ; mit Ausstellung [am 07. November 2002 in Darmstadt]. Schriftenreihe WAR 147, Inst. WAR, Darmstadt, *147*, pp. 17–31.

Jeyanayagam, S., (2018). Phosphorus recovery-North American perspective. In: Schaum, C., (ed.), *Phosphorus: Polluter and Resource of the Future.* London.

Kabbe, C., & Kraus, F., (2017). P recovery: From evolution to revolution. *Fertilizer International,* 479. USA.

Kalantar-Zadeh, K., Gutekunst, L., Mehrotra, R., Kovesdy, C. P., Bross, R., Shinaberger, C. S., Noori, N., et al., (2010). Understanding sources of dietary phosphorus in the treatment of patients with chronic kidney disease. *Clinical Journal of the American Society of Nephrology: CJASN, 5*(3), 519–530.

Kley, G., Adam, C., Brenneis, R., & Simon, F. G., (2005). Thermochemische Aufbereitung von Klärschlammaschen zu Phosphordüngern-Das EU-Projekt SUSAN. In: *75. Darmstädter Seminar-Rückgewinnung von Phosphor aus Abwasser und Klärschlamm.*

Krause, S., (2018). Phosphorus in wastewater. In: Schaum, C., (ed.), *Phosphorus: Polluter and Resource of the Future: Removal and Recovery from Wastewater.* IWA-Publishing, ISBN 9781780408354, London.

Kroiss, H., Rechberger, H., & Egle, L., (2011). *Phosphorus in Water Quality and Waste Management*. INTECH Open Access Publisher.

Langeveld, K., (2018). The RecoPhos/ Inducarb process (the Netherlands). In: Schaum, C., (ed.), *Phosphorus: Polluter and Resource of the Future* (pp. 443–446). IWA Publishing, London.

Lebek, M. R. A., & Hanßen, H., (2018). The REMONIS TertaPhos Process at the WWTP in Hamburg (Germany). In: Schaum, C., (ed.), *Phosphorus: Polluter and Resource of the Future* (pp. 401–410). IWA Publishing, London.

Lebek, M., (2016). P-Rückgewinnung aus Klärschlammaschen am Beispiel des REMONDIS TetraPhos-Verfahren in Hamburg, IWAR-Vortragsreihe, Neues aus der Umwelttechnik und Infrastrukturplanung, Darmstadt.

LeBlanc, R. J., Matthews, P., & Richard, R. P., (2009). *Global Atlas of Excreta, Wastewater Sludge, and Biosolids Management: Moving Forward the Sustainable and Welcome Uses of a Global Resource*. United Nations Human Settlements Programme, Nairobi, Kenya.

Mara, D. D., & Horan, N. J., (2003). *Handbook of Water and Wastewater Microbiology*. Academic Press, Amsterdam.

Mattenberger, H., Fraissler, G., Brunner, T., Herk, P., Hermann, L., & Obernberger, I., (2008). Sewage sludge ash to phosphorus fertilizer: Variables influencing heavy metal removal during thermo chemical treatment. *Waste Manag., 28*(12), 2709–2722.

Mattenberger, H., Fraissler, G., Joller, M., Brunner, T., Obernberger, I., Herk, P., & Hermann, L., (2010). Sewage sludge ash to phosphorus fertilizer (II): Influences of ash and granulate type on heavy metal removal. *Waste Manag., 30*(8/9), 1622–1633.

Mihelcic, J. R., Fry, L. M., & Shaw, R., (2011). Global potential of phosphorus recovery from human urine and feces. *Chemosphere, 84*(6), 832–829.

Ochi, S. M. T., (2018). The PHOSNIX process at the WWTP Lake Shinji East (Japan). In: Schaum, C., (ed.), *Phosphorus: Polluter and Resource of the Future* (pp. 367–374.). IWA Publishing, London.

Opitz, E., (2017). *ExtraPhos-Verfahren zur Rückgewinnung von Phosphor aus Klärschlamm-Erfahreungen aus der Praxis*. Chemische Fabrik Budenheim KG, DWA.

Ortwein, B., (2018). AirPrex sludge optimization and struvite recovery from digested sludge. In: Schaum, C., (ed.), *Phosphorus: Polluter and Resource of the Future* (pp. 343–350). IWA Publishing, London.

Parsonsa, S. A. B. T. A., (2004). Chemical phosphorus removal. In: Valsami-Jones, E., (ed.), *Phosphorus in Environmental Technology-Principles and Applications* (pp. 260–272). IWA Publishing, Padestow, Cornwall, UK.

Petzet, S., (2013). Phosphorrückgewinnung in der Abwassertechnik: Neue Verfahren für Klärschlamm und Klärschlammaschen, Dissertation. Schriftenreihe IWAR220.

Pinnekamp, J., Baumann, P., Cornel, P., Everding, W., Göttlicher-Schmidle, U., Heinzmann, B., Jardin, N., et al., (2013). Stand und Perspektiven der Phosphorrückgewinnung aus Abwasser und Klärschlamm-Teil 1 und Teil 2, 2. Arbeitsbericht der DWA-Arbeitsgruppe KEK-1.1 „Wertstoffrückgewinnung aus Abwasser und Klärschlamm, KA-Korrespondenz Abwasser, Abfall, 60(10) und 60(11).

ProgRess, (2012). Deutsches Ressourceneffizienzprogramm (ProgRess), Beschluss des Bundeskabinetts vom, Bundesministerium für Umwelt, Naturschutz und Reaktorsicherheit (BMU), Berlin.

Schaum, C., & Cornel, P., (2013). Wertstoffrückgewinnung aus Klärschlämmen, DWA-WasserWirtschafts-Kurs O/4-Klärschlammbehandlung, Kassel, ISBN: 978-3-944328-26-3, Hennef.

Schaum, C., & Cornel, P., (2016). Sewage treatment of the future: Health, water and resource protection. Österreichische Wasser-und Abfallwirtschaft, 1–12.

Schaum, C., (2007). Verfahren für eine zukünftige Klärschlammbehandlung-Klärschlammkonditionierung und Rückgewinnung von Phosphor aus Klärschlammasche, Dissertation, IWAR Schriftenreihe 185, Darmstadt.

Schaum, C., (2016). Abwasserbehandlung der Zukunft: Gesundheits-, Gewässer-und Ressourcenschutz, Habilitation. *Schriftenreihe IWAR*233.

Schaum, C., (2018a). In: Christian, S., (ed.), *Phosphorus: Polluter and Resource of the Future: Removal and Recovery from Wastewater*. IWA-Publishing, ISBN 9781780408354. London.

Schaum, C., (2018b). Wastewater treatment of the future: Health, water and resource protection. In: Schaum, C., (ed.), *Phosphorus: Polluter and Resource of the Future: Removal and Recovery from Wastewater*. IWA-Publishing, ISBN 9781780408354. London.

Schaum, C., Fundneider, T., & Cornel, P., (2015). Analysis of methane emissions from digested sludge. *Water Science and Technology*.

Schipper, W., & Korving, L., (2009). Full-scale plant test using sewage sludge ash as raw material for phosphorus production. *International Conference on Nutrient Recovery from Wastewater Streams*.

Schnee, R. O. E., (2018). The ExtraPhos process (Germany). In: Schaum C (ed.), *Phosphorus: Polluter and Resource of the Future* (pp. 391–394). IWA Publishing, London.

Schönberg, A., Raupenstrauch, H., Rapf, M., Cimatoribus, C., & Kranert, M., (2012). Das EU-projekt recophos-phosphorrückgewinnung aus Klärschlamm und Klärschlammasche. In: *Round Table Phosphor-Recycling: Von Der Technologie Bis Zur Erfolgreichen Markteinführung*. Wien, Austria.

Seel, P., (2017). Maßnahmenprogramm zur Reduzierung von Phosphoreinträgen in Hessen, 45. Abwassertechnisches Seminar, Berichte aus der Siedlungswasserwirtschaft, Technische Universität München, Nr. 215.

Sperling, M. V., (2007). *Wastewater Characteristics, Treatment, and Disposal, Biological Wastewater Treatment Series* (Vol. 1). IWA Publishing, London.

Spörri, A., Erny, I., Hermann, L., & Hermann, R., (2017). *Beurteilung von Technolgienzur Phosphor-Rückgewinnung*. Ernst Basler + Partner AG, Bundesamtfür Umwelt BAFU AAuR.

Stumm, W., & Morgan, J. J., (1996). *Aquatic Chemistry: Chemical Equilibria and Rates in Natural Waters*. Wiley, New York.

Szabó, A., Takács, I., Murthy, S., Daigger, G. T., Licskó, I., & Smith, S., (2008). Significance of design and operational variables in chemical phosphorus removal. *Water Environment Research,* 80(5), 407–416.

Takaoka, M., (2018). The current situation regarding phosphorus recovery in Asian countries. In: Schaum, C., (ed.), *Phosphorus: Polluter and Resource of the Future* (pp. 281–304). IWA Publishing, London.

Tchobanoglous, G., (2014). *Wastewater Engineering: Treatment and Resource Recovery* (5th edn.). International Student Edition. McGraw-Hill, New York, NY.

Tchobanoglous, G., Stensel, H. D., Tsuchihashi, R., & Burton, F., (2014). *Wastewater Engineering: Treatment and Resource Recovery*. McGraw-Hill Education, New York.

Udert, K., (2014). *Peak Phosphor-Paralellenzu Peak Oil?* Fachtagung VSA/SVUT. Luzern, Schweiz.

UKWIR, (2009). *Source Control of Phosphorus from Domestic Sources: Options and Impacts.* UKWIR report. London.

UNEP, (2015). *Wastewater Management Platforms for the Wider Caribbean Region: A Situation Analysis.* CEP Technical Report: 89. United Nations Environmental Programme.

USDA, (2017). *Food Composition Databases.* https://ndb.nal.usda.gov/ndb/nutrients/index (accessed on 10 June 2020).

USGS, (2010). *U.S. Geological Survey, Mineral Commodity Summaries-Phosphate Rock.* In.

Weidelener, A., (2010). *Phosphorrückgewinnung aus kommunalem Klärschlamm als Magnesium-Ammonium-Phosphat (MAP).* Oldenbourg, München.

Wentzel, M., Comeau, Y., Ekama, G., Van, L. M., Brdjanovic, D., & Lopez-Vazquez, C., (2008). Enhanced biological phosphorus removal. In: Henze, M., Van, L. M., Ekama, G., & Brdjanovic, D., (eds.), *Biological Wastewater Treatment Principles, Modeling and Design.* IWA Publishing, London, U.K.

CHAPTER 8

Microbial Community Diversity and Monitoring in Anaerobic Digestion

MACARENA MELLADO[1] and OSCAR FRANCHI[2]

[1]*Faculty of Chemistry and Biology, University of Santiago of Chile, Libertador Bernardo O'Higgins Avenue, Santiago, Chile*
Phone: +56962490603, E-mail: macarena.mellado.m@usach.cl

[2]*Faculty of Engineering and Sciences, Adolfo Ibáñez University, Padre Hurtado Avenue 750, Viña del Mar, Chile,*
Phone: +56964695194, E-mail: oscar.franchi@edu.uai.cl

ABSTRACT

Anaerobic digestion is a process carried out by members of *Bacteria* and *Archaea* domains. Anaerobic digesters (ADs) are monitored using physico-chemical parameters. In recent years, microbiological information is being considered as an alternative of monitoring. Studies have been focused on the description of microorganisms based on 16S ribosomal RNA (rRNA) and functional genes, as taxonomic markers. Several bacterial phyla have been detected in ADs, such as *Firmicutes*, and *Bacteoridetes*, among others. Regarding archaea, seven genera have been described, in which the most abundant are *Methanosaeta* and *Methanosarcina*. Aspects that must be considered to analyze these microorganisms are the correct sampling and nucleic acid extraction. Different molecular biology techniques are used in the monitoring, such as the next-generation sequencing (NGS) to describe the 16S rRNA gene and the metagenomes, or real-time PCR to quantify specific microbial groups. However, NGS requires a long time of analysis, which is a disadvantage since decisions about digester management must be made fast. Currently, real-time PCR is the molecular technique more suitable to obtain information in a short time and be used in the monitoring. Finally, new technologies are being developed (nanopore sequencing), opening new horizons to identify warning biomarkers to take better control/operational decisions in ADs.

8.1 INTRODUCTION

The conventional monitoring of anaerobic digesters (ADs) is based on the measurement of several physicochemical parameters, such as pH, alkalinity, chemical oxygen demand (COD), volatile fatty acids (VFAs), methane production, and percentage of methane in the biogas. When ADs present a destabilization event, the value of these parameters could either decrease (pH, alkalinity, methane production, and percentage of methane) or increase (oxygen chemical demand and VFAs). This is the result of both final and intermediate products of an unbalanced microbial community activity that is present in ADs. According to this, the conventional monitoring variables can show us the effects; however, they do not indicate the causes of the instability. For this reason, new monitoring variables are necessary to predict the future performance of ADs. Currently, the studies have been focused on the microbial community, taxonomic groups, and their metabolic functions. In addition, researchers are interested in new biological targets to anticipate the destabilization events which could contribute to take quick operational decisions. This chapter will describe recent advances in knowledge regarding microorganisms detected in the anaerobic digestion process, focusing on the use of this information to monitor the AD stability.

AD is composed of four steps: hydrolysis, acidogenesis, acetogenesis, and methanogenesis. In the first three phases, there are "fermenters" which degrade organic matter to both VFAs and hydrogen, and "syntrophs" which degrade VFAs to both acetate and H_2. All these reactions are performed by bacteria microorganisms and are classified as hydrolytic, acidogenic, and acetogenic steps. In the last phase, "methanogens" convert acetate, CO_2, and H_2 to methane as a final product. All microorganisms that produce methane, described so far, belong to the Archaea domain, and by this way they obtain energy for their metabolic requirements (Liu and Whitman, 2008; Vanwonterghem et al., 2016). Figure 8.1 shows a scheme that represents the steps in anaerobic digestion, products, and group of microorganisms involves in the process. However, it is a representation that summarizes the most important aspects mentioned previously since the process is dynamic and compose for different microorganisms; diverse species or groups degrading organic matter with different metabolic capacities and growth rate. They compete for same substrates or can use the final products of other reactions. As a result, microbial community of ADs can responsed to changes or perturbations as well as pollutants that enter the digester. In the following paragraphs, taxonomic groups will be described and how they have been used to monitor the process and its stability.

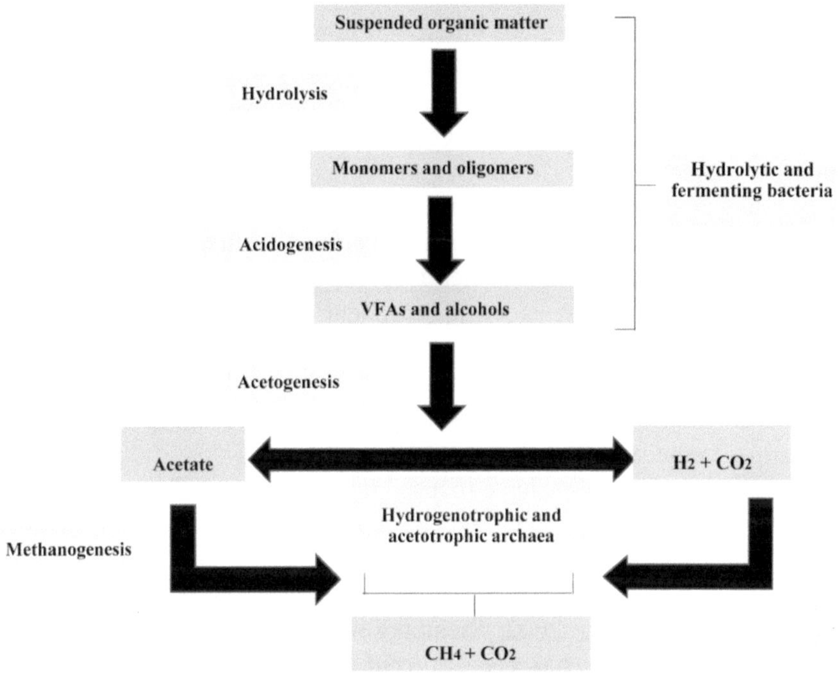

FIGURE 8.1 Anaerobic digestion phases, raw material, products and group of microorganisms involved in each step.

Microbial community in ADs has long been referred as a "black box" due to the uncharacterized and uncultivated microorganisms. However, the information about species involved in the process has increased in the last years, due for the development of next-generation sequencing (NGS) that provides a great amount of data of the taxonomic groups present in digesters (Vanwonterghem et al., 2014, 2016; Cabezas et al., 2015; Treu et al., 2016) as well as the isolation of new species both *Bacteria* and *Archaea* domain (e.g., Köek et al., 2014; Sun and Schnürer et al., 2016; Zhang et al., 2016). Microbial interactions are another topic that researchers have both described and given information to understand the performance of the digesters (Rotaru et al., 2014) as well as metagenomic and metratranscriptome of the microbial community in different stages. However, these last topics will not be covered in this chapter.

8.2 THE RESEARCH OF THE MICROBIAL COMMUNITIES IN ADS

There are two ways to study microbial communities, culture-dependent, and independent methods, and both have been used in ADs. In the first, microorganisms must be both cultured and isolated in a culture medium. It possesses the advantage that gives us information about the unknown microorganisms so far. However, anaerobic microorganisms are fastidious since they possess low growth rates and fastidious nutritional requirements (Leaphart and Lovell, 2001) which make them difficult both to study and monitor by culture methods. Therefore, the most common way to monitor a microbial community is by independent culture (Sanz and Kochling, 2007), such as NGS (Cabezas et al., 2015; Vanwonterghem et al., 2014), real-time PCR (Steinberg and Regan, 2009; Xu et al., 2009; Pereyra et al., 2010), DGGE (Yu et al., 2008; Regueiro et al., 2012), or SSCP (Delbes et al., 2001; Gannoun et al., 2015).

Most research studies have focused on the study of 16S gene as a target to identify species (e.g., De Francisci et al., 2015; Luo et al., 2015). On the other hand, some studies have focused on functional gene to study specific population, e.g., acetogenic bacteria (Xu et al., 2009) or methanogenic archaea (Luton et al., 2002; Steinberg and Regan, 2009; Pereyra et al., 2010). The *16S* gene has been utilized because it is present in bacteria and archaea in high amount in living beings (10^3–10^5 ribosomes/cell) as well as present in both variable and highly conserved regions (Sanz and Kochling, 2007). Some specific primers have been described to be target of the conserved regions, and the variable zones are used to do the taxonomic classification by comparing with sequences of the databases. By amplifying, it is possible to detect all microorganisms of DNA sample obtained from a microbial community. It has been described primers that amplify 16S gene to detect archaea (Yu et al., 2008), bacteria (Delbes et al., 2001) or both (Caporaso et al., 2011). In addition, it is possible to determine microbial community diversity (Cabezas et al., 2015) and to obtain relevant statistic data (Sanz and Kochling, 2007) to compare microbiological and physicochemical results. The importance of the *16S* gene will be discussed later in this chapter.

On the other several primers have been used to detect specific populations, such as cellulose-degrading, fermentative, sulfate-reducing bacteria (Pereyra et al., 2010), acetogenic (Xu et al., 2009), syntrophic β-oxidizing bacteria (Ziels et al., 2015), methanogenic archaea (Luton et al., 2002; Steinberg and Regan, 2009; Pereyra et al., 2010), or acetotrophic methanogens (Smith and Ingram-Smith, 2007; Ince et al., 2011). In this case, the quantification of copy number of these genes by real-time PCR, and the variation of their values are considered as an indicator of the changes in the number of microorganisms. In addition, some researchers have used these primers to determine taxonomic classification (Sun et al., 2013). The use of primers to specific populations is adequate in the case the 16S gene sequence possesses a high similarity and the microorganisms have distinct phenotypes (Achenbach et al., 2010). The importance of functional gene will be discussed later in this chapter.

Finally, several molecular biology techniques have been used in microbiome, and described in detail about their use in ADs (e.g., Vanwonterghem et al., 2014). Previously, techniques such as DGGE, SSCP or 16S gene sequencing by Sanger method were used to determine changes in the microorganisms of AD. However, these required long periods of analysis. Currently, high throughput put sequencing is broadly used to determine the structure of microbial communities since they provide a great quantity of data than the techniques mentioned previously in this paragraph. This technique also needs long periods to perform all assays, including bioinformatic analysis. However, it provides data that supports solid conclusions and to understand the process in depth.

8.3 MICROORGANISMS DETECTED IN ADS

This section will describe both the taxonomic groups detected in ADs and their most important metabolic functions in the degradation of organic matter. First, bacteria microorganisms will be described followed by methanogen archaea.

8.3.1 MICROORGANISMS OF BACTERIA DOMAIN

Kingdom *Bacteria* has two sub-kingdoms: *Posibacteria* (4 phyla) and *Negibacteria* (25 phyla). In ADs is possible to find all members of *Posibacteria* subkingdom and the rest of the members belong to the *Negibacteria* subkingdom (Ruggiero et al., 2015). In general, bacterial taxonomic groups possess higher diversity and abundance as compared to the archaeal group (Sundberg et al., 2013).

Most of the sequences can be classified as either phylum or class, despite few sequences that can be related to lower taxonomic groups, such as family or genus, which suggest that most sequences detected correspond to unknown or unclassified microorganisms so far (Kröber et al., 2009). The isolation and description of new microorganisms from ADs could help increase the information of the database, with the years; the percentage of the unknown sequences could decrease. Some phyla detected in ADs will be described later in this chapter.

Taxonomic classification of *Bacteria* has been done practically with the analysis of the *16S*rRNA gene, while the *Archaea* has been used as both *16S*rRNA gene and a functional gene that will be described later. Regarding phyla, *Firmicutes* has been detected in several studies of ADs. Two classes of this phylum are *Clostridium* and *Bacilli*. *Clostridium* has been related to hydrolysis, acidogenesis, and acetogenesis steps (Kröber et al., 2009; Wirth et al., 2012; Hanreich et al., 2013). Other studies have described microorganisms at the genera level of *Firmicutes*, detecting *Gallicola* and *Fastidiosipila* (Cardinale-Rezende et al., 2016), taxonomic groups related to the conversion of peptone and amino acid (Heider and Fuchs, 1997; Mechini et al., 1999) into acetic and butyric acids without the necessity to ferment carbohydrates. Another microorganism isolated from a thermophilic AD, *Bacillus thermoamylovorans*, could grow in xylane, cellobiose, pectin, and starch which indicates that it is important in the hydrolytic step of polymers. Metabolic pathways related to the production of lactate, acetate, ethanol, and formate that indicate participation in the fermentative step were detected in its genome (Köeck et al., 2014).

Spirochaetes is another phylum detected in ADs and, generally, has been related with hydrogenotrophic methanogens since they can interact by syntrophic associations to produce methane (Morita et al., 2011; Kato et al., 2012). Lee et al (2015) described the methane production by *Spirochaetes* and hydrogenotrophic archaea in the presence of acetate, which is completely different from most studies that have described that hydrogenotrophic archaea cannot produce methane using acetate as a precursor. On the other hand, microorganisms of the *Geobacter* genus (belonging to *Proteobacteria* phylum, another taxonomic group detected in ADs), were able to interact with microorganisms of the genus *Methanosaeta* (methanogen known for its acetoclastic metabolism) by syntrophic association and to produce methane for the reduction of carbon dioxide (Rotaru et al., 2014). This interesting fact described by Lee et al (2015), present both the capacity and versatility of the archaea when interacting with different bacterial taxonomic groups. However, further investigations are required to understand the mechanisms of interaction between bacteria and archaea.

Synergistetes is another phylum related to syntrophic interactions with methanogens. A member of this phylum, *Lactovibrioal coholicus*, in a co-culture with the methanogen *Methanospirillum hungatei* was able to use ethanol, 1-propanol, lactate, glycerol, as well as several amino acids (Qiu et al., 2014). Another member, *Amonibacterium thunnarium* in co-culture with *Methanobacterium formicicum* was also able to oxidize amino acids (Hamdi et al., 2015). Studies have not described so far if these interactions are either among specific species or all methanogens can interact with microorganisms of this phylum to degrade amino acids. Table 8.1 shows phyla and their metabolic functions detected in ADs. Some phyla share similar roles in the degradation of organic matter. *Firmicutes* and *Bacteoridetes* can participate in the hydrolysis of polymers as well as in acidogenic and acetogenic steps. On the other hand, *Thermotogae* has only been involved in cellulose degradation and there is no information about their participation in acidogenic and acetogenic phases. *Chloroflexi* and *Actinobacteria* can perform carbohydrate degradation. This last phylum has also shown syntrophic interactions with members of the phyla *Synergistetes* and *Spirochaetes*. Despite the fact that ADs can possess different taxonomic groups, the microbial community can degrade organic matter and produce methane. Specific microorganisms can be selected depending on the microbe composition in inoculum and feeding as well as operational conditions of the process. It must be considered that microorganisms can compete for raw material and they possess diverse growth rate that can also influence on the microorganisms present in ADs.

Phylum *candidatus* was also detected in ADs; it means microorganisms that have not been both isolated and described so far. However, they are an interesting group since they can present both interactions and functions in the process not described so far (Nobu et al., 2015). One of them is the phylum candidate *Cloacimonetes* that has been detected in several studies of ADs (With et al., 2012; Sundberg et al., 2013; Li et al., 2014; Sun et al., 2015). Studies suggest that these microorganisms are syntrophic bacterium with the capacity of degrading both propionate and amino acids (Pelletier et al., 2008), especially the clade WWE1 associated with the degradation of cellulose (Limam et al., 2014; Lucas et al., 2015). Other phyla *candidatus* detected in ADs are *Atribacteria* (OP9), *Fermentibacteria* (Hyd-24-12),

TABLE 8.1 Phyla Detected in ADs and Their Metabolic Functions

Phylum	Metabolic Function	References
Firmicutes	Participation in hydrolytic step, they possess cellulosomes, supramolecular proteins complexes capable of degrading cellulose and cellulosome- associated proteins. Participation in acetogenic and acidogenic steps.	Schlüter et al., 2008; Kröber et al., 2009; Wirth et al., 2012; Hanreich et al., 2013; Yamazawa et al., 2014
Bacteroidetes	Hydrolysis of polysaccharides and proteins, fermentation of sugars, and production of acids such as butyric, succinic, propionic and acetic acids.	Hahnke et al., 2015; Cardinale-Rezende et al., 2016
	Degradation pathways of amino acids and enzymes related to the production of volatile fatty acids.	
Proteobacteria	Oxidize acid and ethanol possibly produced by fermentative degradation of sugar and polyethylene glycol in association with hydrogenotrophic methanogens.	aNarihiro et al., 2015
Chloroflexi	Carbohydrate degradation. Microorganisms of this phylum may participates in syntrophic interactions with methanogens.	Ariesyady et al., 2007
Actinobacteria	Carbohydrate degradation.	Ariesyady et al., 2007
Spirochaetes	Syntrophic interactions with methanogens, in metabolic activities of acetate, ethanol, lactate fermentations from glucose. Syntrophic interaction only with hydrogenotrophic methanogenic.	Godon et al., 1997; Graber et al., 2004; Lee et al., 2015
Synergistetes	Syntrophic interactions with methanogens, related to proteolytic activity and could ferment amino acids to acetic, propionic and butyric acid, formate, valeric acid ornithine and H2. Degradation of specific compounds; benzoato, terephtalate, phenol and to different petrochemicals.	Vartoukian et al., 2007; Chen et al., 2009; Perkins et al., 2011; aNarihiro et al., 2015a
Thermothogae	Degradation of cellulose	Yamazawa et al., 2014

Aminicenantes (OP8) (Kirkegaard et al., 2017), *Parcubacteria, Gracilibacteria*, and *Microgenomates* (Narihiro et al., 2015a), OD1, SR1, TM7 (Ziganshin et al., 2013), KSB3 (Sekiguchi et al., 2015), GN04 (Narihiro et al., 2015b), or *Marinimicrobia* (SAR406) (Nobu et al., 2015). In general, they have been detected either as high or low percentage of abundance. In addition, these phyla must present both physiology aspects and a role in the degradation of organic matter unknown so far (Narihiro et al., 2015b). On the other hand, some studies have been focused on the description of their metabolic function; despite they have not been isolated so far. For example, Sekiguchi et al (2015) described the presence of the *candidatus* KSB3 in a UASB system. Results suggested the ability both to ferment sugars and amino acids, the capacity to respond chemotactically to glucose and maltose as well as a role in the wastewater bulking.

8.3.2 MICROORGANISMS OF ARCHAEA DOMAIN

Kingdom *Archaea* is formed by the phyla *Crenarchaeota* and *Euryarchaeota*; however, the methanogenic microorganisms only belong to *Euryarchaeota*, which 4 out of 9 classes correspond to methanogenic archaea. There are seven orders: *Methanopyrales*, *Methanococcales*, *Methanobacteriales*, *Methanomicrobiales*, *Methanosarcinales*, *Methanocellales*, and *Methanomassiliicoccales* (Ruggiero et al., 2015). Methanogens are lesser diverse than bacteria in ADs. They are described using their both family and genus name. In this section, methanogens were described considering their genus name.

Methanogenic species have been classified using both *16S* and *mcrA* genes as a marker. All methanogens share the *mcrA* gene which encodes the alpha subunit of methyl-coenzyme M reductase that is the last enzyme of the pathway to produce methane. This gene is highly conserved in the *microperin* which is shared by all methanogens known (Liu and Whitman, 2008), with the exception of the methane-oxidizing archaea that possess this gene in their genomes (Friedich, 2005). Other enzymes that are part of the methanogenic pathway (e.g., methylene tetrahydromethanopterin dehydrogenase or methenyltetrahydromethanopterincyclohydrolase) can be found in other microorganisms, such as aerobic methanotrophic bacteria (Chistoserdova et al., 1998). Therefore, the *mcrA* gene is considered an adequate taxonomic marker. Phylogenetic studies about methanogens using both *16S* and *mcrA* genes showed similar taxonomic classification (Luton et al., 2002; Friedrich, 2005). However, other authors described that the *mcrA* gene showed the presence of different taxonomic groups in relation to the 16S gene (Wilkins et al., 2015).

On the other hand, taxonomic classification using the *mcrA* gene has provided information about either the unknown species or new phylogenetic orders of methanogens (Fry et al., 2009). Some taxonomic groups unknown detected are WCHA1-57 (Saito et al., 2015), Arc-7 (Steinberg and Regan,

2008), Fen Cluster, MCR-7, MCR-2b and MCR-2a (Steinberg and Regan, 2009). Studies using the 16S gene as a genetic marker have also detected uncultured groups such as WSA2 (Wilkins et al., 2015). This information could be used to find new methanogenic archaeal groups. One example is the study by Saito et al (2015) that found new phylotypes in relation to known methanogens. This conclusion was obtained by sequencing both the *16S* and the *mcrA* genes of granular sludge samples. The sequences were classified into WCHA1-57, an uncultured archaeal group which could be considered a putative new order of methanogens.

Methanogen genera found more abundantly in ADs are *Methanosarcina* (FitzGerald et al., 2015), *Methanosaeta* (Sundberg et al., 2013; DeVrieze et al., 2015b; Wilkins et al., 2015), or both at the same time (Li et al., 2013; Abendroth et al., 2015). They belong to the same family, *Methanosarcinaceae*. These studies have been performed with different designs of reactors (e.g., continue, and semi-continue stirrer tank reactor, leach-bed batch), scales (e.g., industry, pilot, laboratory), and feeding (e.g., pig and cow manure, silage, straw, sludge, kitchen waste or co-digestion of macroalgae, dairy slurry) (Goberna et al., 2015; Sun et al., 2015).

Methanosarcina is a special group among methanogens, because it is the only archaeal genus able to use both hydrogenotrophic and acetoclastic pathways (2008; Liu and Whitman, 2008) as well as pathways that use either the methyl reduction with H_2 or methylotrophic catabolism of methanol, methylated amines, and dimethylsulfide (Galagan et al., 2002; Welander and Metclaf, 2005). This metabolic characteristic gives them an advantage if you compare with other methanogens, the possibility to use several substrates as precursors.

On the other hand, *Methanosaeta* genus was considered an acetoclastic methanogen for several years (Liu and Whitman, 2008). However, Rotaru, and cols. (2014) described that the methanogen *Methanosaeta harundinacea* could produce methane using CO_2 as a precursor, a metabolic characteristic of the hydrogenotrophic methanogens. This was possible by the syntrophic interaction with *Geobacter metallireducens* (*Proteobacteria*) that was the donor of electrons. This suggests that the metabolic capabilities of this *Methanosaeta* have not been totally described so far. In addition, further studies are necessary to understand if this ability is present only in *M. harundinacea*, or other species of this genus can produce methane using CO_2 like a precursor, as well as to emphasize the importance of the microbial community in ADs in detail. In addition, *Methanosaeta*, and *Methanosarcina* have been related to high and low yields of biogas, respectively (Abendroth et al., 2015). The first uses between 5–20 μM of acetate and the second needs about 1 mM. Probably this situation is due to the use of high or low-affinity enzymes for acetate, respectively (Teh and Zinder, 1992; Singh-Wissmann and Ferry, 1995). The mechanisms of competition between these taxonomic groups are not yet clear, consequently, more studies are necessary to understand them (Li et al., 2015).

On the other hand, it has also been described about the prevalence of hydrogenotrophic methanogens in ADs over acetoclastic methanogens using the 16S gene as a genetic marker. Morris et al (2015) showed a similar situation studying the *mcrA* gene in industrial/municipal full-scale digesters by real-time PCR. In addition, specific methanogenic activity with propionate, H_2 y CO_2, but not with acetate would indicate the prevalence of hydrogenotrophic methanogens. Several hydrogenotrophic genera have been detected as the most abundant methanogens in several studies, where ADs were operated in different conditions, such as *Methanoculleus* (Kröber et al., 2009; Wirth et al., 2012; Ziganshin et al., 2013; Abendroth et al., 2015), *Methanolinea* (Maspolim et al., 2015), *Methanobacterium* (Sun et al., 2015; Zhang et al., 2015), or *Methanocalculus* (Nolla-Ardevol et al., 2015). In addition, other studies have described similar abundance both acetoclastic and hydrogenotrophic methanogens (DeVrieze et al., 2015a).

Acetoclastic methanogens were considered the microorganisms most abundant in ADs for a long time; this was perceived as an indicator that they were the largest producer of methane. However, the role of hydrogenotrophic microorganisms was not considered as an important factor for their low abundance. Studies that have detected these microorganisms as the most abundance in some ADs suggest that it is necessary to understand all factors involved in the process, such as operational conditions (e.g., feeding, the design of the reactor, temperature, hydraulic retention time) as well as both interactions and activity of microorganisms to define the dominant pathway in the methane production in ADs.

Figure 8.2 shows a diagram of the principal pathways involved in methane production and taxonomic groups that participle in the process. Regarding the hydrogenotrophic pathway, only order classification is represented. However, the acetoclastic pathway is divided into two genera described so far that can use acetate to produce methane. Unclassified microorganisms were not mentioned because their metabolic capacities are still unknown. It is also shown that the interaction of *Methanosaeta harundinacea* and *Geobacter metallireducens* can produce methane from H_2 and CO_2. This situation emphasizes the importance of microbial interactions in the community as well as those microorganisms that work together for mutual benefits.

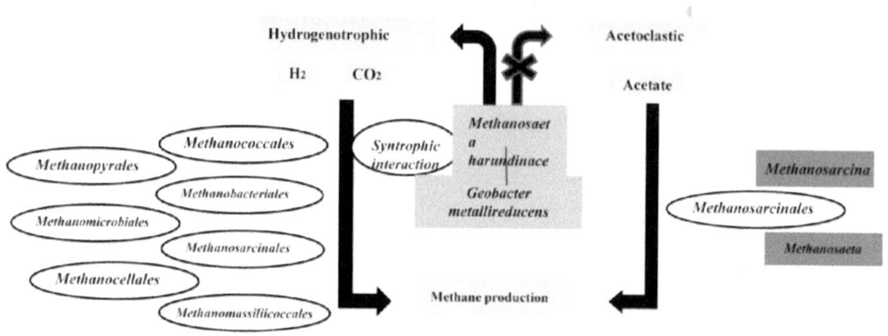

FIGURE 8.2 Scheme of the principal pathways involved in methane production and taxonomic groups that participate in the process.

8.4 MONITORING OF MICROBIAL COMMUNITY OF ADS

Monitoring of ADs has been performed using physicochemical parameters, such as pH, alkalinity, VFA, and OCD. However, they cannot be used to predict destabilizations since they are the final products or chemical indicators. As mentioned previously, the microbial community is dynamic and can change depending on chemical (toxic or inhibitor compounds) or physical perturbations (temperature). Microorganisms are responsible for organic matter degradation and to find a way of following their changes could support a better strategy of monitoring. In addition, it could give information to predict perturbations and take decisions to avoid destabilizations.

On the other hand, ADs possess a microbial community with diverse populations. However, there is no standard way of monitoring them. In this part of the chapter, several procedures involved in the analysis of microbial communities, such as sampling or DNA extract methods, as well as different techniques used in the detection of microorganisms in ADs are discussed.

8.4.1 PROCEDURES FOR THE ANALYSIS OF MICROBIAL COMMUNITIES IN ADS

8.4.1.1 SAMPLING

Sampling involves several aspects. First, to select the digester zone from which the samples will be taken. This decision is important since the samples must be representative of the microbial community. Then, the procedures to take the samples, such as tools or sterile conditions. Finally, if they will be both shipped and stored in another location, or if the samples will be analyzed immediately after the sampling.

Regarding the sampling zone, the recirculation loop seems to be the most common point to take samples (Gao et al., 2016; Hao et al., 2016). Dead volumes of the outlet pipelines are suggested to be discarded before sampling to eliminate a source of microorganisms not representative of the process. It is also suggested that the sample should be stored during transport in a gas purged container to avoid the growth either aerobic or facultative microorganisms that could change the original microbial community composition (Stolze et al., 2016). Other authors reported to consider multiple sampling strategies to obtain a representative sample (Wilkins et al., 2015; Cai et al., 2016), as well as both the homogenization of the sample and cold shipping (0°C–20°C) (Hao et al., 2016; Luo et al., 2016; Smith et al., 2017). Finally, Albertsen et al. (2015) suggested avoiding the shipping sludge samples at room temperature with overnight delivery, if a weekly resolution of the microbial community is required.

8.4.1.2 NUCLEIC ACID EXTRACTION

Before starting the analysis workflow, it is necessary to provide a DNA of good quality that contains the genomic material from the entire community present in the system. For that purpose, it is mandatory to choose the best DNA extraction protocol available.

First, it is necessary to choose an adequate kit for DNA extraction of anaerobic sludge. There is no specific kit for sludge; however, researches have used kits for soils of different providers since it is possible to obtain DNA of good quality than other kits or methodologies. Lebuhn et al. (2016) evaluated the efficiencies obtained in parallel of DNA extractions after *Escherichia coli* spiking with the FastDNA™ SPIN Kit for Soil, the UltraClean® Soil DNA Isolation Kit and the QIAamp DNA Stool Mini Kit applying *E. coli* specific qPCR. Results indicated that the FastDNA™ SPIN Kit for Soil is the most efficient kit with a recovery of 88.24%, followed by UltraClean® Soil DNA Isolation Kit (60.58%) and QIAamp DNA Stool Mini Kit (37.48%). For comparison, conventional phenol-chloroform-isoamylalcohol extraction of *E. coli* spiked cattle manure did not yield PCR amplifiable DNA.

After the extraction, it is necessary to quantify the DNA samples. There are several alternatives both to quantify and check its purity. One of them is Nanodrop, the simplest and cheapest method. However, Nanodrop quantification results are usually higher than those obtained using Qubit or Picogreen kit (fluorescence-based methods), especially for low-quality extracts. Quantification results based on fluorescence methods may be more reliable because the impurities in the extract (phenols, oligonucleotides, humic acids, etc) could also result in UV absorbance, whereas fluorescence-based quantification is more specific (Guo et al., 2013).

8.4.2 SEQUENCING METHOD SELECTION

In general, the 16S gene is the most common gene to be amplified and sequence in the microbial communities of ADs. In this section, methods used in the sequencing of 16S gene will be discussed.

8.4.2.1 16S RRNA GENE AMPLICON SEQUENCING AND TARGETED REGIONS

The most common sequencing approach to analyze the microbiome is amplicon analysis of the 16S ribosomal RNA (rRNA) gene (e.g., Regueiro et al., 2012; De Francisci et al., 2015; Luo et al., 2015) since databases possess a great number of the 16S gene sequences of microorganisms isolated from different environments, and it is possible to compare with the amplicons obtained from DNA samples of ADs. If a functional gene is both amplified and sequenced, there is a high probability that the databases do not possess huge sequences to compare with the amplicons obtained. As a consequence, results could show that several amplicons correspond to 'unknown microorganisms' when the sequences correspond to microorganisms that have been both isolated and described in the laboratory. However, the microorganism detected has been both isolated and described in the laboratory. The problem is that the databases do not possess enough information.

The 16S gene possesses several regions that have been well-characterized. The hyper variable region V4 is the most commonly used for amplicon sequencing using the primer set F515 and R806 (for bacteria and archaea). The better accuracy of V4 region to describe the microbial population was confirmed by Tremblay et al. (2015) who reported that V4 target region sequenced using Illumina Miseq showed the highest similarity toward the expected taxonomic distribution of a mock community containing a known number of species. However, other authors have chosen other regions to study the microbial community of ADs, such as segments that includes

V1–V2 (Wilkins et al., 2015), V1–V3 (Jang et al., 2016; Albertensen et al., 2015) or V3–V5 (Fykse et al., 2016).

The increasing power of sequencing technologies, and the decision to perform 16S amplicon sequencing should be balanced by its limitations, which include PCR primer selection and amplification bias (Xie et al., 2016). It is known that PCR primers able to amplify all known bacterial taxonomic groups efficiently (Tremblay et al., 2015) are not available leading to biased rRNA profiling analysis. If the decision is to apply the 16S rRNA gene amplicon sequencing method, the recommendation is to use the primer set targeting the V4 region for the analysis of the anaerobic sludge community.

8.4.2.2 SHOTGUN METAGENOMIC SEQUENCING (SMS)

The major limitation of the 16S amplicon method is that sequences provide only information of a single region of the bacterial genome, whereas the advantage of SMS method can provide broad regions of the genome allowing an accurate taxonomic and functional analysis defined at the species level, even in bacteria of low abundance (Ranjan et al., 2016). Even though WGS is usually more expensive than 16S amplicon sequencing, it offers increased resolution, enabling a more specific taxonomic and functional classification of sequences as well as the discovery of new bacterial genes and genomes (Jovel et al., 2016). SMS method also has some constraints regarding the number of sequenced genomes for the taxonomical assignment of genes. However, this limitation was overcome by bioinformatics tools that allow the analysis of 16S rRNA gene sequences based on shotgun sequencing (Logares et al., 2014; Guo et al., 2016; Xie et al., 2016) which gives the advantage of retrieving simultaneously both taxonomic (bacteria and archaea) and functional information from the same microbial community (Logares et al., 2014). Based on the advantages that the SMS method has this approach is recommended over 16S rRNA amplicon sequencing.

8.4.2.3 SMS TECHNOLOGY COMPARISON

Among sequencing technologies available in the market, Illumina (Miseq and Hiseq), 454, Ion Torrent, and PacBio are the most commercialized. All these technologies possess advantages and disadvantages, and every year kits and protocols for sequencing are updated to fully exploit the potential of each technology. Regarding the error rate of sequencing, D'Amore et al.

(2016) compared the performance of these four technologies and concluded that both the PacBio and IonTorrent possess error rates (1.9% and 1.47%, respectively) that were significantly higher than the Miseq (0.92%) and 454 (1.06%). The non-matching rate was similar for Miseq and 454. However, PacBio showed a higher value (21.58%) as well as IonTorrent (43.77%). This agrees with the conclusions of Loman et al. (2012) who found that the Illumina technology produced the highest quality reads, due to a low substitution error rate and the near absence of indel errors compared to the other technologies.

If we compare the time per run, IonTorrent technology is superior (3 hrs) than 454 (8 hrs), and Illumina (24 h). However, Illumina has a better throughput (1500 Mb) than IonTorrent (1000 Mb) and 454 (35 Mb) which possess lower prices per megabase (Liu et al., 2012). Currently, due to the advantages that Illumina offers (less error rate and high-throughput), it is recommended to use this technology instead of others.

8.4.2.4 THIRD-GENERATION SEQUENCING

The third-generation sequencing is the future trend in genome sequencing. It appeared after the NGS and possesses several advantages over other techniques of sequencing. The process is fast, performed in real-time, does not require PCR amplification, and it is possible to obtain good assembled genomes as this technology produces long reads, and the elimination of PCR step avoids several mistakes produced in NGS that can influence the results and their interpretation. Only Illumina platform uses PCR amplification. For this reason, it is not considered third-generation sequencing; however, it is an improvement of the NGS and is being developed in parallel with the other platforms (Vandijk et al., 2018).

Currently, there are three platforms; PacBio (SMRT), ONT, and Illumina/10X Genomics SLR. PacBio provides long reads, and does not possess problems with repeats or low/high %GC. The disadvantages are that the sequencer is expensive; it requires a high amount of genomic sample for preparation of the library with potential high error rate (Quail et al., 2012). Nanopore also provides long reads, libraries can be prepared quickly, and it is capable of the direct sequencing of RNA. The disadvantages are a high overall error rate, the necessity of a great amount of sample for the preparation of library, and versions of software are constantly changing (Weirather et al., 2017; Garalde et al., 2018). Finally, Illumina possesses low error rates, relatively cheaper, and small amounts of sample are necessary for the preparation of libraries. The disadvantages are that it does not provide long reads, and it requires PCR amplification for the preparation of the library (Vandijk et al., 2018).

Unfortunately, these technologies are still being improved due to their disadvantages. Despite this they will contribute to obtain data of the microbial community, and it could be possible to monitor microorganisms in real-time, with the possibility to obtain fast data in a short time to make decisions over the management of the ADs.

8.4.3 REAL-TIME PCR TO QUANTIFY MICROBIAL POPULATIONS

NGS has increased the data about the microbial community of ADs (Cabezas et al., 2015). However, these analyses need time to be performed, including the bioinformatic analysis. If the overall goal is to monitor the microorganisms to make decisions about the management of ADs, NGS is not adequate since in a wastewater treatment plant, decision about digester must be made fast. In this case, real-time PCR is a technique that could give information in a short period of time.

Real-time PCR has been used to study both 16S and functional genes in ADs. In this case, the 16S gene can provide information about the presence of both bacteria and archaea (Aydin et al., 2015). However, it is not possible to detect differences in specific taxonomic groups or populations. On the other hand, the use of functional genes for a specific population is suitable to simplify the study of the microorganisms in ADs since it is difficult to design primers or probes to detect specific taxonomic groups by 16S gene (Leaphart and Lovell, 2001; Xu et al., 2009).

Functional genes can provide information about the changes of different populations since the copy number of genes is related to the abundance of microorganisms (Sharma et al., 2007). In addition to that new primers could be designed using updated information from the databases which are continually expanding, as well as the designed primers could be improved (Xu et al., 2009).

The hydrolytic population has been studied using glycoside hydrolases as a target. They are involved in cellulose degradation. Among all hydrolase families described so far, anaerobic bacteria possess members of the families both *cel5* and *cel48*, genes that have been used to characterize as well as quantify hydrolytic microorganisms in ADs (Pereyra et al., 2009; Sun et al.,

2013). Sun et al (2013) used hydrolytic genes to do a taxonomic classification of bacteria in ADs. They detected members of the phyla *Firmicutes* and *Bacteroidetes* as well as microorganisms uncultured. As mentioned previously, the detection of uncultivated microorganisms so far is an important topic to develop comprehensive microbial community in ADs.

Acetogens, another population that is part of ADs, can produce acetate by the reduction of both CO_2/CO and H_2. They have been isolated from different environmentals and detected by the functional gene formyltetrahydrofolate synthetase (FTHFS) as a target. This gene is highly conserved in the acetyl-CoA pathway. Some members of this group belong to the phyla *Firmicutes* and *Spirochaetes* (Xu et al., 2009), and some previous studies have detected them by real-time PCR of the FTHFS (Aydin etal., 2015).

Some studies focused on the number of both archaea and bacteria using the 16S gene (Zhang et al., 2015). Regarding archaea, primers have been designed to detect order (*Methanomicrobiales*, *Methanosarcinales*, *Methanobacteriales*, *Methanococcales*) or family (*Methanosarcinaceae*, *Methanosaetaceae*) (May et al., 2015). The monitoring can provide information about specific taxonomic groups depending on the primer used. Steinberg et al (2011) determined that the Fen Cluster (methanogens not isolated so far) was present in digester after organic loading rates as well as in neutral conditions. On the other hand, the monitoring of transcripts is more adequate to predict the metabolic activity of microorganisms and their response to the changes in the digester. For example, Wilkins et al (2015) described a linear correlation between the abundance of *mcrA* transcripts and methane production, and concluded that the gene transcripts could be a good marker to monitor methanogenesis in ADs.

Currently, real-time PCR offers a fast and accurate alternative to monitor microorganisms in ADs. Figure 8.3 shows a scheme of genes used in the study of microbial community populations in anaerobic digestion. Five functional genes and 16S, both archaea and bacteria, were considered in the figure and related to each population. Some authors correlate the copy number of a gene to the number of microorganisms of a population and have tried to establish correlations with the physicochemical parameters of the process. In addition, it is possible to obtain results in a shorter time than other techniques mentioned previously. Therefore, real-time PCR could be considered as the best alternative to be used in the monitoring of microorganisms, and results could be used to take decisions in the management of ADs.

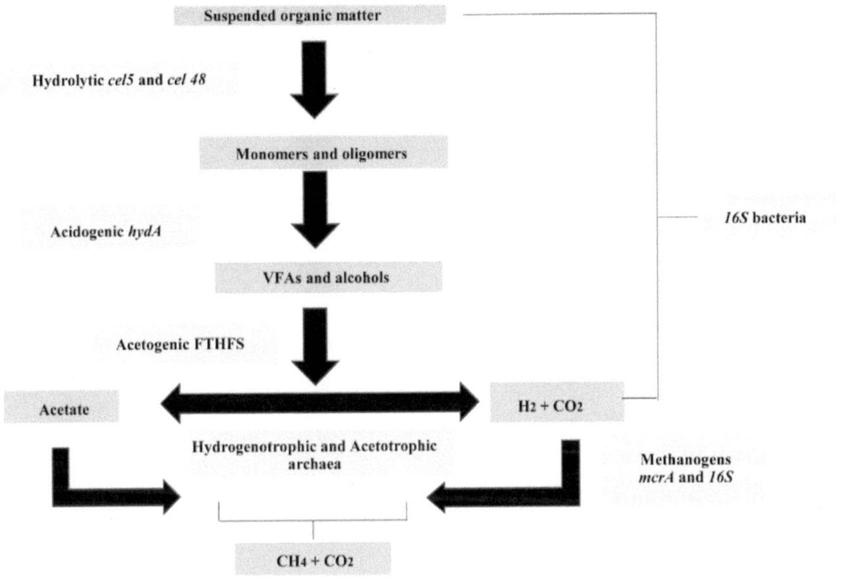

FIGURE 8.3 Scheme of functional gene used in the quantification of the populations that participle in steps of anaerobic digestion.

8.5 CONCLUDING REMARKS

The microbial community of ADs has been studied by the sequencing of both 16S and functional genes; detecting bacteria, archaea or specific populations.

NGS has contributed to increasing the amount of microbial community data, such as 16S gene, metagenomic or functional gene, in ADs operated under different conditions. Several bacterial phyla have been detected, such as *Firmicutes* and *Bacteroidetes* as the most abundant. Regarding archaea, *Methanosaeta*, and *Methanosarcina* were both considered as the most abundant genera for a long time. However, other genera have been reported with a high abundance, suggesting the possibility that other methanogens could possess a role in methane production.

Several techniques have been used to monitor the microbial community of ADs. However, they require a long time of analysis that is a disadvantage since it is necessary to detect imbalances and make decisions quickly.

Currently, real-time PCR is the most affordable technique to monitor the dynamics of microbial populations, because it is fast and to permit the quantification of microorganisms. It is necessary to detect the best microbial targets to establish adequate microbial parameters to monitor the process. Therefore, further studies are required to determine the best microbial parameters to quantify.

KEYWORDS

- **16S rRNA**
- **anaerobic digesters**
- **archaea**
- **bacteria**
- **functional gene**
- **microbial community**
- **microbial population**

REFERENCES

Abendroth, C., Vilanova, C., Günther, T., Luschning, O., &Porcar, M., (2015). Eubacteria and archaea communities in seven mesophile anaerobic digester plants in Germany. *Biotechnol. Biofuels., 8*, 87.

Achenbach, L. A., & Coates, J. D., (2000). Disparity between bacterial phylogeny and physiology-comparing 16S rRNA sequences to assess relationships can be a powerful tool, but its limitations need to be considered. *AMS News., 66*, 714–715.

Albertsen, M., Karst, S. M., Ziegler, A. S., Kirkegaard, R. H., & Nielsen, P. H., (2015). Back to basics-the influence of DNA extraction and primer choice on phylogenetic analysis of activated sludge communities. *PloS One, 10*, e0132783.

Ariesyady, H. D., Ito, T., & Okabe, S., (2007). Functional bacterial and archaeal community structures of major trophic groups in a full-scale anaerobic sludge digester. *Water Res., 41*, 1554–1568.

Aydin, S., Ince, B., & Ince, O., (2015). Application of real-time PCR for determination of combined effect of antibiotics on Bacteria, Methanogenic Archaea, Archaea in anaerobic sequencing batch reactors. *Water Research, 76*, 88–98.

Cabezas, A., Calabria, D. A. J., Callejas, C., Gales, A., Hamelin, J., Marone, A., Sousa, D., et al., (2015). How to use molecular biology tools for the study of the anaerobic digestion process. *Rev. Environ. Sci. Biotecnol., 14*, 555–593.

Cai, M., Wilkins, D., Chen, J., Ng, S. K., Lu, H., Jia, Y., & Lee, P. K., (2016). Metagenomic reconstruction of key anaerobic digestion pathways in municipal sludge and industrial wastewater biogas-producing systems. *FrontMicrobiol.*, 7.

Caporaso, J., Lauber, C., Walters, W., Berg-Lyons, D., Lozupone, C., Turnbaugh, P., Fierer, N., & Knight, R., (2011). Global patterns of 16S rRNA diversity at a depth of millions of sequences per sample. *PNAS, 108*, 4516–4522.

Cardinale-Rezende, J., Rojas-Ojeda, P., Nascimiento, A., & Sanz, J. I., (2016). Proteolytic bacterial dominance in a full-scale municipal solid waste anaerobic reactor assessed by 454 pyrosequencing technology. *Chemosphere, 146*, 519–525.

Chen, C. L., Wu, J. H., Tseng, I. C., Liang, T. M., & Liu, W. T., (2009). Characterization of active microbes in a full-scale anaerobic fluidized bed reactor treating phenolic wastewater. *Microbes. Environ., 24*, 144–153.

Chistoserdova, L., Vorholt, J., Thauer, R. K., &Lidstro, M. E., (1998). C1 Transfer enzymes and coenzymes linking methylotrophic bacteria and methanogenic archaea. *Science, 281*, 99–102.

D'Amore, R., Ijaz, U. Z., Schirmer, M., Kenny, J. G., Gregory, R., Darby, A. C., & Hall, N., (2016). A comprehensive benchmarking study of protocols and sequencing platforms for 16S rRNA community profiling. *BMC Genomic, 17*, 55.

Daelman, M. R. J., Sorokin, D., Kruse, O., Van, L. M. C. M., &Strous, M., (2016). Haloalkaline bioconversions for methane production from microalgae grown on sunlight. *Trends Biotechnol., 34*, 450–457.

De Francisci, D., Kougias, P. G., Treu, L., Campanaro, S., &Angelidaki, I., (2015). Microbial diversity and dynamicity of biogas reactors due to radical changes of feedstock composition. *Bioresource Technol., 176*, 56–64.

Delbes, C., Moletta, R., &Godon, J., (2001). Bacterial and archaeal 16S rDNA and 16S rRNA dynamics during an acetate crisis in an anaerobic digestor ecosystem. *FEMS Microbiol. Ecol., 35*, 19–26.

DeVrieze, J., Gildemyn, S., Vilchez-Vargas, R., Jáuregui, R., Pieper, D., Verstraete, W., & Boon, N., (2015a). Inoculum selection is crucial to ensure operational stability in anaerobic digestion. *Appl. Microbiol Biotechnol., 99*, 189–199.

DeVrieze, J., Saunders, A. M., He, Y., Fang, J., Nielsen, P. H., Verstraete, W., & Boon, N., (2015b). Ammonia and temperature determine potential clustering in the anaerobic digestion microbiome. *Water Res., 75*, 312–323.

Friedrich, M. W., (2005). Methyl-coenzyme M reductase genes: Unique functional markers for methanogenic and anaerobic methane-oxidizing Archaea. *Methods Enzymol., 397*, 428–442.

Fry, J. C., Horsfield, B., Sykes, R., Cragg, B. A., Heywood, C., Kim, G. T., Mangelsdorf, K., et al., (2009). Prokaryotic populations and activities in an interbedded coal deposit, including a previously deeply buried section (1.6–2.3 km) above ~ 150 Ma basement rock. *Geomicrobiol. J., 26*, 163–178.

Galagan, J. E., Nusbaum, C., Roy, A., Endrizzi, M. G., Macdonald, P., FitzHugh, W., Calvo, S., et al., (2002). The genome of *Methanosarcina acetivorans* reveals extensive metabolic and physiological diversity. *Genome Res., 12*, 532–542.

Gannoun, H., Omri, I., Chouari, R., Khelifi, E., Keskes, S., Godon, J. J., Hamdi, M., Sghir, A., & Bouallagui, H., (2015). Microbial community structure associated with the high loading anaerobic co-digestion of olive mill and abattoir wastewaters. *Bioresource Technol., 201*, 337–346.

Gao, J., Liu, G., Li, H., Xu, L., Du, L., & Yang, B., (2016). Predictive functional profiling using marker gene sequences and community diversity analyses of microbes in full-scale anaerobic sludge digesters. *Bioprocess. Biosyst. Eng., 39*, 1115–1127.

Garalde, D. R., Snell, E. A., Jachimowicz, D., Sipos, B., Lloyd, J. H., Bruce, M., Pantic, N., et al., (2018). Highly parallel direct RNA sequencing on an-array of nanopores. *Nat. Methods., 15*, 201–206.

Goberna, M., Gadermaier, M., Franke-Whittle, L. H., García, C., Wett, B., & Insam, H., (2015). Start-up strategies in manure-fed biogas reactors: Process parameters and methanogenic communities. *Biomass. Bioenergy, 75*, 46–56.

Godon, J. J., Zumstein, E., Dabert, P., Habouzit, F., & Moletta, R., (1997). Molecular microbial diversity of an anaerobic digestor as determined by small-subunit rDNA sequence analysis. *Appl. Microbiol. Biotechnol., 63*, 2802–2813.

Graber, J. R., & Breznak, J. A., (2004). Physiology and nutrition of *Treponemaprimitia*, an H_2/CO_2-acetogenic spirochete from termite hindguts. *Appl. Environ. Microbiol., 70*, 1307–1314.

Guo, F., & Zhang, T., (2013). Biases during DNA extraction of activated sludge samples revealed by high throughput sequencing. *Appl. Microbiol. Biotechnol., 97*, 4607–4616.

Guo, J., Cole, J. R., Zhang, Q., Brown, C. T., & Tiedje, J. M., (2016). Microbial community analysis with ribosomal gene fragments from shotgun metagenomes. *Appl. Environ. Microbiol., 82*, 157–166.

Hanreich, A., Schimpf, U., Zakrzewski, M., Schlüter, A., Benndorf, D., Heyer, R., Rapp, E., Pühler, A., Reichl, U., & Klocke, M., (2013). Metagenome and metaproteome analyses of microbial communities in mesophilic biogas-producing anaerobic batch fermentations indicate concerted plant carbohydrate degradation. *Syst. Appl. Microbiol., 36*, 330–338.

Hao, L., Bize, A., Conteau, D., Chapleur, O., Courtois, S., Kroff, P., Desmond-Le, Q. E., Bouchez, T., & Mazéas, L., (2016). New insights into the key microbial phylotypes of anaerobic sludge digesters under different operational conditions. *Water. Res., 102*, 158–169.

Heider, J., & Fuchs, G., (1997). Microbial anaerobic aromatic metabolism. *Anaerobe., 3*, 1–22.

Ince, B., Koksel, G., Cetecioglu, Z., Oz, N. A., Coban, H., & Ince, O., (2011). Inhibition effect of isopropanol on acetyl-CoA synthetase expression level of acetoclastic methanogen, *Methanosaeta concilii. J. Biotechnol., 156*, 95–99.

Ito, T., Yoshiguchi, K., Ariesyady, H. D., & Okabe, S., (2012). Identification and quantification of key microbial trophic groups of methanogenic glucose degradation in an anaerobic digester sludge. *Bioresour. Technol., 123*, 599–607.

Jovel, J., Patterson, J., Wang, W., Hotte, N., O'Keefe, S., Mitche, T., Perry, T., et al., (2016). Characterization of the gut microbiome using 16S or shotgun metagenomics. *Front Microbiol.*, 7.

Kirkegaard, R. H., McIlroy, S. J., Kristensen, J. M., Nierychlo, M., Karst, S. M., Dueholm, M. S., Albertsen, M., & Nielsen, P. H., (2017). The impact of immigration on microbial community composition in full-scale anaerobic digesters. *Nature, 7*, 9343.

Köek, D. E., Wibberg, D., Maus, I., Winkler, A., Albersmeier, A., Zverlov, V. V., Pühler, A., et al., (2014). First draft genome sequence of the amylolytic *Bacillus thermoamyovorans* wild-type strain 1A1 isolated from a thermophilic biogas plant. *J. Biotechnol., 192*, 154–155.

Leaphart, A. B., & Lovell, C. R., (2001). Recovery and analysis of formyltetrahydrofolate synthetase gene sequences from natural populations of acetogenic bacteria. *Appl. Environ. Microbiol., 65*, 5117–5123.

Lebuhn, M., Derenkó, J., Rademacher, A., Helbig, S., Munk, B., Pechtl, A., Stolze, Y., Prowe, S., et al., (2016). DNA and RNA extraction and quantitative real-time PCR-based assays for Biogas biocenoses in an interlaboratory comparison. *Bioengineering, 3*, 7.

Lee, S. H., Park, J. H., Kang, H. J., Lee, Y. H., Lee, T. J., & Park, H. D., (2013). Distribution and abundance of *Spirochaetes* in full-scale anaerobic digesters. *Bioresour. Technol., 145*, 25–32.

Lee, S. H., Park, J. H., Kim, S. H., Yu, B. J., Yoon, J. J., & Park, H. D., (2015). Evidence of syntrophic acetate oxidation by *Spirochaetes* during anaerobic methane production. *Bioresour. Technol., 190*, 543–549.

Li, J., Rui, J., Pei, Z., Sun, X., Zhang, S., Yan, Z., Wang, Y., et al., (2014). Straw-and slurry-associated prokaryotic communities differ during co-fermentation of straw and swine manure. *Appl. Microbiol. Biotechnol., 98*, 4771–4780.

Li, L., He, Q., Ma, Y., Wang, X., & Peng, X., (2015). Dynamics of microbial community in a mesophilic anaerobic digester treating food waste: Relationship between community structure and process stability. *Bioresour. Technol., 189*, 113–120.

Limam, R. D., Chouari, R., Mazéas, L., Wu, T. D., Li, T., Grossin-Debattista, J., Guerquin-Kern, J. L., et al., (2014). Members of the uncultured bacterial candidate division WWE1 are implicated in anaerobic digestion of cellulose. *Microbiologyopen, 3*, 157–167.

Liu, L., Li, Y., Li, S., Hum, N., He, Y., Pong, R., Lin, D., Lu, L., & Law, M., (2016). Comparison of next-generation sequencing systems. *J. Biomed. Biotechnol., 2012*, 11, 251364, .

Liu, Y., & Whitman, W., (2008). Metabolic, phylogenetic, and ecological diversity of the methanogenic archaea. *Ann. New York. Acad. Sci., 1125*, 171–189.

Logare, S. R., Sunagawa, S., Salazar, G., Cornejo-Castillo, F. M., Ferrera, I., Sarmento, H., Hingamp, P., et al., (2014). Metagenomic, 16S rDNA Illumina tags are a powerful alternative to amplicon sequencing to explore diversity and structure of microbial communities. *Environ. Microbiol., 16*, 2659–2671.

Lucas, R., Kuchenbuch, A., Fetzer, I., Harms, H., & Kleinsteuber, S., (2015). Long-term monitoring reveals stable and remarkably similar microbial communities in parallel full-scale biogas reactors digesting energy crops. *FEMS Microbiol. Ecol.*, 91.

Luo, G., De Francisci, D., Kougias, P. G., Laura, T., Zhu, X., & Angelidaki, I., (2015). New steady-state microbial community compositions and process performances in biogas reactors induced by temperature disturbances. *Biotechnol. Biofuels., 8*, 3.

Luo, G., Fotidis, I. A., & Angelidaki, I., (2016). Comparative analysis of taxonomic, functional, and metabolic patterns of micro biomes from 14 full-scale biogas reactors by metagenomic sequencing and radio isotopic analysis. *Biotechnol. Biofuels., 9*, 51.

Luton, P., Wayne, J., Sharp, R., & Riley, P., (2002). The *mcrA* gene as an alternative to 16S rRNA in the phylogenetic analysis of methanogen populations in landfill. *Microbiology, 148*, 3521–3530.

Maspolim, Y., Zhou, Y., Guo, C., Xiao, K., & Ng, W. J., (2015). Determination of the archaeal and bacterial communities in two-phase and single-stage anaerobic systems by 454 pyrosequencinng. *J. Environ. Sci., 36*, 121–129.

May, T., Koch-Singenstreu, M., Ebling, J., Stantscheff, R., Müller, L., Jacobi, F., Polag, D., et al., (2015). Design and application of a synthetic DNA standard for real-time PCR analysis of microbial communities in a biogas digester. *Appl. Microbiol. Biotechnol., 99*, 6855–6863.

Mechichi, T., Labat, M., Patel, B. K., Woo, T. H., Thomas, P., & Garcia, J. L., (1999). *Clostridium methoxybenzovorans* sp. nov., a new aromatic o-demethylating homoacetogen from an olive mill wastewater treatment reactor. *Int. J. Syst. Evol. Microb., 49*, 1201–1209.

Militon, C., Hamdi, O., Michotey, V., Fardeau, M. L., Ollivier, B., Bouallagui, H., Hamdi, M., & Bonin, P., (2015). Ecological significance of *Synergistetes* in the biological treatment of tuna cooking wastewater by an anaerobic sequencing batch reactor. *Environ. Sci. Pollut. Res., 22*, 18230–18238.

Narihiro, T., Kim, N. K., Mei, R., Nobu, M., & Liu, W. T., (2015b). Microbial community analysis of anaerobic reactors treating soft drink wastewater. *PLoS One,* 10.

Narihiro, T., Nobu, M. K., Kim, N. K., Kamagata, Y., & Liu, W. T., (2015a). The nexus of syntrophy-associated microbiota in anaerobic digestion revealed by long-term enrichment and community survey. *Environ. Microbiol., 17*, 1707–1720.

Nobu, M., Narihiro, T., Rinke, C., Kamagata, Y., Tringe, S., Woyke, T., & Liu, W., (2015). Microbial dark matter ecogenomics reveals complex synergistic networks in a methanogenic bioreactor. *The ISME Journal, 9*, 1710–1722.

Nolla-Ardevol, V., Strous, M., & Tegetmeyer, H. E., (2015). Anaerobic digestion of the microalga *Spirulina* at extreme alkaline conditions: Biogas production, metagenome and metatranscriptome. *FrontMicrobiol., 6*, 597.

Pelletier, E., Kreimeyer, A., Bocs, S., Rouy, Z., Gyapay, G., Chouari, R., Rivière, D., et al., (2008). '*Candidatus cloaca monas acid aminovorans*': Genome sequence reconstruction provides a first glimpse of a new bacterial division. *J. Bacteriol., 190*, 2572–2579.

Pereyra, L. P., Hiibel, S. R., Prieto, R. M. V., Reardon, K. F., & Pruden, A., (2010). Detection and quantification of functional genes of cellulose-degrading, fermentative, and sulfate-reducing bacteria and methanogenic archaea. *Appl. Environ. Microbiol., 76*, 2192–2202.

Perkins, S. D., Scalfone, N. B., & Angenent, L. T., (2011). Comparative 16S rRNA gene surveys of granular sludge from three upflow anaerobic bioreactors treating purified terephthalic acid (PTA) wastewater. *Water Sci. Technol., 64*, 1406–1412.

Qiu, Y. L., Hanada, S., Kamagata, Y., Guo, R. B., & Sekiguchi, Y., (2014). *Lactivibrioal coholicus* gen. nov., sp. nov., an anaerobic, mesophilic, lactate-, alcohol-, carbohydrate-and amino-acid-degrading bacterium in the phylum *Synergistetes. Int. J. Syst. Evol. Microbiol., 64*, 2137–2145.

Quail, M. A., Smith, M., Coupland, P., Otto, T. D., Harris, S. R., Connor, T. R., Bertoni, A., et al., (2012). A tale of three next generation sequencing platforms: Comparison of ion torrent, pacific biosciences and Illumina MiSeq sequencers. *BMC Genomics, 13*, 341.

Ranjan, R., Rani, A., Metwally, A., McGee, H. S., & Perkins, D. L., (2016). Analysis of the micro biome: Advantages of whole genome shotgun versus 16S amplicon sequencing. *Biochem. Biophys. Res. Commun., 469*, 967–977.

Regueiro, L., Veiga, P., Figueroa, M., Alonso-Gutierrez, J., Stams, A., Lema, J. M., & Carballa, M., (2012). Relationship between microbial activity and microbial community structure in six full-scale anaerobic digesters. *Microbiol. Res., 167*, 581–589.

Robin, J. D., Ludlow, A. T., LaRanger, R., Wright, W. E., & Shay, J. W., (2016). Comparison of DNA quantification methods for next generation sequencing. *SCI REP-UK.*, 6.

Rotaru, A. E., Shrestha, P. M., Liu, F. H., Shrestha, M., Shrestha, D., Embree, M., Zengler, K., et al., (2014). A new model for electron flow during anaerobic digestion: Direct interspecies electron transfer for *Methanosaeta* for the reduction of carbon dioxide to methane. *Energy. Environ. Sci., 7*, 408–415.

Ruggiero, M. A., Gordon, D. P., Orrel, T. M., Bailly, N., Bourgoin, T., Brusca, R., Cavalier-Smith, T., Guiry, M., & Kirk, P., (2015). A higher-level classification of all living organisms. *PLoS One*, 10.

Saito, Y., Aoki, M., Hatamoto, M., Yamaguchi, T., Takai, K., & Imachi, H., (2015). Presence of a novel methanogenic archaeal lineage in anaerobic digesters inferred from mcrA and 16S rRNA gene phylogenetic analyses. *J. Water. Environ. Technol., 13*, 279–289.

Sanz, J., & Kochling, T., (2007). Molecular biology techniques used in wastewater treatment: An overview. *ProcessBiochem., 42*, 119–133.

Schlüter, A., Bekel, T., Diaz, N., Dondrup, M., Eichenlaub, R., Gartemann, K. H., Krahn, I., et al., (2008). The metagenome of a biogas-producing microbial community of a production-scale biogas plant fermenter analyzed by the 454-pyrosequencing technology. *J. Biotechnology., 136*, 77–90.

Sekiguchi, Y., Ohashi, A., Parks, D. H., Yamauchi, T., Tyson, G. W., & Hugenholtz, P., (2015). First genomic insights into members of a candidate bacterial phylum responsible for wastewater bulking. *Peer J.*, 3.

Sharma, S., Radl, V., Hai, B., Kloos, K., Fuka, M. M., Engel, M., Schauss, K., & Schloter, M., (2007). Quantification of functional genes from procaryotes in soil by PCR. *J. Microbial. Methods, 68*, 445–452.

Singh-Wissmann, K., & Ferry, J. G., (1995). Transcriptional regulation of the phosphotransacetylase-encoding and acetate kinase-encoding genes (*pta* and *ack*) from *Methanosarcina thermophila. J. Bacteriol., 177*, 1699–1702.

Smith, A. L., Shimada, T., & Raskin, L., (2017). A comparative evaluation of community structure in full-scale digesters indicates that two-phase digesters exhibit greater microbial diversity than single-phase digesters. *Environ. Sci. Water. Res. Technol., 3*, 304–311.

Smith, K. S., & Ingram-Smith, C., (2007). Methanosaeta, the forgotten methanogen? *Trends Microbiol., 15*, 150–155.

Steinberg, L. M., & Regan, J. M., (2011). Response of lab-scale methanogenic reactors inoculated from different sources to organic loading rate shocks. *Bioresour. Technol., 102*, 8790–8798.

Steinberg, L., & Regan, J., (2008). Phylogenetic comparison of the methanogenic communities from an acidic, oligotrophic fen and an anaerobic digester treating municipal wastewater sludge. *Appl. Environ. Microbiol., 74*, 6663–6671.

Steinberg, L., & Regan, J., (2009). *McrA*-Targeted real-time quantitative PCR method to examine methanogen communities. *Appl. Environ. Microbiol., 75*, 4435–4442.

Stolze, Y., Bremges, A., Rumming, M., Henke, C., Maus, I., Pühler, A., Sczyrba, A., & Schlüter, A., (2016). Identification and genome reconstruction of abundant distinct taxa in microbiomes from one thermophilic and three mesophilic production-scale biogas plants. *Biotechnol. Biofuels., 9*, 156.

Sun, L., & Schnürer, A., (2016). Draft genome sequence of the cellulolytic strain *Clostridium* sp. Bc-iso-3 isolated from an industrial-scale anaerobic digester. *Genome. Announc., 4*, e01188–16.

Sun, L., Müller, B., & Schnürer, A., (2013). Biogas production from wheat straw: Community structure of cellulose-degrading bacteria. *Energy. Sustain. Soc., 3*, 15.

Sun, L., Pope, P., Eijsink, V., & Schnürer, A., (2015). Characterization of microbial community structure during continuous anaerobic digestion of straw and cow manure. *Microb. Biotechnol., 8*, 815–827.

Sundberg, C., Al-Soud, W., Larsson, M., Alm, E., Yekta, S., Svensson, B., Sorensen, S., & Karlsson, A., (2013). 454 pyrosequencing analyses of bacterial and archaeal richness in 21 full-scale biogas digesters. *FEMS Microbiol. Ecol., 85*, 612–626.

Teh, Y. L., & Zinder, S., (1992). Acetyl-coenzyme a synthetase in the thermophilic, acetate-utilizing methanogen *Methanothrix* sp. strain CALS-1. *FEMS Microbiol. Lett., 98*, 1–7.

Tremblay, J., Singh, K., Fern, A., Kirton, E. S., He, S., Woyke, T., Lee, J., et al., (2015). Primer and platform effects on 16S rRNA tag sequencing. *FrontMicrobiol., 6*, 771.

Treu, L., Kougias, P., Campanaro, S., Bassani, I., & Angelidaki, I., (2016). Deeper insight into the structure of the anaerobic digestion microbial community; the biogas microbiome database is expanded with 157 new genomes. *Bioresource Technol., 216*, 260–266.

Van, D. E. L., Jaszczyszyn, Y., Naquin, D., & Thermes, C., (2018). The third revolution in sequencing technology. *Trends Genet., 34*, 666–681.

Vanwonterghem, I., Jensen, P. D., Ho, D. P., Batstone, D. J., & Tyson, G. W., (2014). Linking microbial community structure, interactions and function in anaerobic digesters using new molecular techniques. *Curr. Opin. Biotechnol., 27*, 55–64.

Vanwonterghem, I., Jensen, P. D., Rabaey, K., & Tyson, G., (2016). Genome-centric resolution of microbial diversity, metabolism, and interactions in anaerobic digestion. *Environ Microbial., 18*, 3144–3158.

Vartoukian, S. R., Palmer, R. M., & Wade, W. G., (2007). The division "Synergistes." *Anaerobe., 13*, 99–106.

Weirather, J. L., De Cesare, M., Wang, Y., Piazza, P., Sebastiano, V., Wang, X. J., Buck, D., & Au, K. F., (2017). Comprehensive comparison of pacific biosciences and Oxford nanopore technologies and their applications to transcriptome analysis. *F1000Res., 6*, 100.

Welander, P. V., & Metcalf, W. W., (2005). Loss of the *mtr* operon in *Methanosarcina* blocks growth on methanol, but not methanogenesis, and reveals an unknown methanogenic pathway. *Proc. Natl. Acad. Sci.USA., 102*, 10664–10669.

Wilkins, D., Lu, X. T., Shen, Z., Chen, J., & Lee, P. K. H., (2015). Pyrosequencing of *mcrA* and archaeal *16S*rRNA genes reveals diversity and substrate preferences of methanogen communities in anaerobic digesters. *Appl. Environ. Microbiol., 81*, 604–613.

Wirth, R., Kovács, E., Maróti, G., Bagi, Z., Rákhely, G., & Kovács, K., (2012). Characterization of a biogas-producing microbial community by short-read next generation DNA sequencing. *Biotechnol. Biofuels., 5*, 41.

Xie, C., Goi, C. L. W., Huson, D. H., Little, P. F., & Williams, R. B., (2016). Ribo tagger: Fast and unbiased 16S/18S profiling using whole community shotgun metagenomic or metatranscriptome surveys. *BMC Bioinformatics, 17*, 277.

Xu, K., Liu, H., Du, G., & Chen, J., (2009). Real-time PCR assays targeting formyltetrahydrofolate synthetase gene to enumerate acetogens in natural and engineered environments. *Anaerobe, 15*, 204–213.

Yamazawa, A., Iikura, T., Morioka, Y., Shino, A., Ogata, Y., Date, Y., & Kikuchi, J., (2014). Cellulose digestion and metabolism induced biocatalytic transitions in anaerobic microbial ecosystems. *Metabolites, 4*, 36–52.

Yu, Z., García-González, R., Schanbacher, F., & Morrison, M., (2008). Evaluations of different hypervariable regions of archaeal 16S rRNA genes in profiling of methanogens by archaea-specific PCR and denaturing gradient gel electrophoresis. *Appl. Environ. Microbiol., 74*, 889–893.

Zhang, J., Zhang, Y., Quan, X., & Chen, S., (2015). Enhacement of anaerobic acidogenesis by integrating an electrochemical system into an acidogenic reactor: Effect of hydraulic retention times (HRT) and role of bacteria and acidophilic methanogenic Archaea. *Bioresour. Technol., 179*, 43–49.

Zhang, W., Ge, X., Li, Y., Yu, Z., Yu, Z., & Li, Y., (2016). Isolation of a methanotroph from a hydrogen sulfide-rich anaerobic digester for methanol production from biogas. *Process. Biochem., 51*, 838–844.

Ziganshin, A., Liebetrau, J., Pröter, J., & Kleinsteuber, S., (2013). Microbial community structure and dynamics during anaerobic digestion of various agricultural waste materials. *Appl. Microbiol. Biotechnol., 97*, 5161–5174.

CHAPTER 9

Decentralized Systems for Wastewater Treatment and Resource Recovery

ANDREA ARIAS, GUMERSINDO FEIJOO, and MARIA TERESA MOREIRA

Department of Chemical Engineering, CRETUS institute.
University of Santiago de Compostela, E-15782, Santiago de Compostela, Galicia, Spain, E-mail: maite.moreira@usc.es (M. T. Moreira)

ABSTRACT

Conventional centralized treatment of wastewater cannot be the only option to ensure adequate treatment of polluted water in future with growing population. The adoption of a new paradigm based on a decentralized approach has been proposed to accomplish present and future sustainability goals. Decentralized management in wastewater treatment has been defined as the collection, treatment, and reuse of wastewater at or near the point of generation. Initially considered for isolated residential areas, its implementation at the service of new housing developments such as suburban areas, industrial parks, and small communities is becoming increasingly common. The reason behind this alternative is the possibility of avoiding wastewater pumping over large distances and overloading existing treatment facilities.

It is important to establish the importance and current status of decentralized treatment systems, especially in the European context, where the approach of conventional centralized treatment is predominant. Thus, this chapter aims to provide an analysis of different scenarios for the collection and treatment of wastewater from individual households and small communities, moving from centralized to decentralized approaches. Moreover, different treatment schemes are compared from a life-cycle assessment perspective. Finally, implications for the reuse of treated effluents and the recovery of resources such as biofertilizers are estimated and discussed.

9.1 SEARCHING A SUSTAINABLE WASTEWATER SCHEME: A PARADIGM SHIFT

Wastewater treatment has traditionally been one of the most important topics in the development of environmental engineering. In recent years, the consumption of water and energy resources has been increased significantly, suggesting that this trend will continue in the coming decades (Larsen and Gujer, 2013). The conventional approach of wastewater management is characterized by being supply-driven, centralized, and developed on a large-scale. This model has led to overexploitation or depletion of water resources and deterioration of water quality. In addition, conventional systems present high capital and operating costs associated with the sewerage network and the treatment plant. This fact implies a large number of people living in developing countries do not have access to safe sanitation (Remy and Jekel, 2008). Keeping sustainability criteria in mind, wastewater treatment technologies should be geared towards the development of technically viable, environmentally friendly, low-cost, and socially accepted solutions (Massoud et al., 2009; Meuler et al., 2008).

In the context of a constantly growing urban population and water scarcity, a shift towards decentralization and separation of domestic wastewater at source is an increasingly relevant option for the treatment of much more concentrated wastewaters, with a fit-for-purpose technology (Lam et al., 2015; Ng et al., 2014). Decentralized systems, defined as wastewater treatment processes located near the point of discharge, arise as a potential alternative. Historically, this approach was applied to on-site wastewater solutions adopted in small communities where the connection to centralized sewer systems was not feasible due to technical or economic limitations. At present, these systems are not only characteristic of peri-urban, rural, and remote areas, but are also implemented in new housing developments within or at the boundaries of large cities, with the aim of avoiding the overload of existing sewer network and centralized facilities. In a prospective scenario in which decentralized wastewater systems are widespread, we will witness significant changes in the current "urban wastewater paradigm" centered on centralized systems (Libralato et al., 2012; Massoud et al., 2009; Opher and Friedler, 2016; Venhuizen, 1997).

However, a change of approach needs to demonstrate significant advantages over well-established technologies. The aim of this chapter is to identify technologies for decentralized wastewater treatment and to discuss the main benefits, drivers, and barriers that influence the transition from one model to another. On the road to a paradigm shift, the awareness of the vulnerability and scarcity of water resources plays a major role. Understanding decentralized systems are important for greater acceptance by end-users and policy-makers, recognizing the role of these solutions as a counterpart to existing centralized systems.

9.2 MOVING FROM CENTRALIZED TO DECENTRALIZED SYSTEMS

As society faces a changing world, the current concept of wastewater treatment needs to adapt flexibly to changing scenarios. Centralized wastewater treatment will not be able to cope with the future challenges associated with lower environmental impacts, cost-effective solutions, and resource recovery. Having in mind the need of guaranteed treatment for the world's population, the adoption of alternative schemes of water management, such as source separation, arises (Hophmayer-Tokich, 2006).

Trade-off analysis between the advantages and disadvantages of centralized and decentralized systems has been investigated. Those associated with the centralized approach are mainly due to economies of scale in terms of resource consumption. Starting from the premise that facilities are well designed and constructed, they can develop water leaks, which can contaminate groundwater (Remy and Jekel, 2008). The advantages of the decentralized approach derive mainly from the potential reduction in the demand of drinking water due to the potential of water reuse for irrigation (Nogueira et al., 2009; Shehabi et al., 2012). It is generally agreed that the preference for one over the other depends to a large extent on the specific circumstances, and that energy consumption, nutrient emissions, and water sources are among the decisive factors in the comparison (Table 9.1).

Decentralized systems can be classified according to different scales: (i) on-site: treatment technologies and/or management systems at the scale of an individual property owned and managed by the owners; (ii) cluster or development, where a common ownership model serves two or more dwellings with close sewage treatment (these approaches include wastewater recycling and stormwater collection); (iii) distributed systems for large housing developments (over 100 properties), with outsourced services operated by water utility companies (Bakir, 2001; Sharma et al., 2013; Tchobanoglous et al., 1998).

In the context of decentralized systems, several important issues are under consideration: (i) technical and economic feasibility; (ii) proximity to source; (iii) compliance with discharge limits of reclaimed water (RW); (iv) risk analysis focused on the management and consequences of failures; (v) independence, from an autonomous option to a semi-integrated system and (vi) social perception and acceptance of decentralized systems (Sharma et al., 2013; Wilderer and Schreff, 2000).

TABLE 9.1 Main Differences Between Decentralized and Centralized Wastewater Treatment Systems

	Decentralized	**Centralized**
Disadvantages	Lack of management	Advanced collection and process that treat and discharge a large quantity of wastewater.
	Social acceptance	
	Odor nuisance	High construction costs.
	Noise disturbance	Higher personnel costs
		More robust
		Based on the end of line treatments
Advantages	Low construction costs.	More knowledge about the technologies that formed the plant.
	Low environmental costs	
	Increase of wastewater reuse	Far to the houses
	Proximity to source	Wasted resources
	More flexible	Food wastes can be incorporated into AD for obtaining for biogas
	Based on circular economic	

Because of the potential of decentralized solutions to contribute to sustainable development, it may be useful to synthesize the results of existing studies on decentralized wastewater treatment under a holistic approach. In decreasing order, the countries with more than 30 publications according to the SCOPUS database are the United States (US) (333), China (127), Germany (127), Australia (97), The Netherlands (55), India (40), and Spain (47). Given the growing number of publications in recent years, there is an undeniable interest in the subject. Most publications in the literature have investigated the viability of different technologies at laboratory and pilotscale.

In some countries such as the US, the decentralized approach to water supply and wastewater treatment has traditionally been considered for urban developments in dispersed populations, where centralized collection and treatment are not economically viable. Conversely, communities with higher population density tend to have common collection systems that transport wastewater to a centralized treatment plant (Gikas and Tchobanoglous, 2009; Glick and Guggemos, 2013; Shehabi et al., 2012). Beyond proximity to centralized facilities, other additional causes behind the studies in China, Australia, and India can be explained by the limited water resources associated with severe drought, as well as the lack of infrastructure for centralized wastewater collection (Lam et al., 2015; Massoud et al., 2009; Ng et al., 2014; Singh et al., 2016).

The European countries with the largest number of related studies are Germany and The Netherlands, especially linked to the concept of New Sanitation, which allows source separation, reduction of water consumption and reuse, a system that has been applied in several residential areas of Europe (Igos et al., 2012, 2017). In Spain and other Mediterranean countries, the frequency of droughts has favored the proliferation of alternative wastewater systems (Bdour et al., 2009); however, the full development of this strategy is far from being widely implemented in the short term.

9.3 DRIVERS AND BARRIERS IN THE IMPLEMENTATION OF DECENTRALIZED SYSTEMS

Several factors can be identified that drive the implementation of decentralized systems, mainly associated with environmental and social pressures such as water scarcity (Bakir, 2001), climate change (CC), population growth (Singh et al., 2016), water loss associated with long-distance pumping, time needed for infrastructure construction and local community empowerment (Wilderer, 2004).

Decentralized approaches can offer many benefits compared to centralized systems, such as preservation of water resources, increased independence from centralized systems, economic development of rural and water-stressed regions and public health benefits for socially vulnerable populations, especially in poor and rural regions (Parkinson and Tayler, 2003). The environmental, social, and economic benefits of alternative supply schemes also require site-specific analysis to avoid under- or over-estimation of real costs and how they can be comparable and competitive (Chen and Wang, 2009).

Beyond drivers and benefits, there are significant barriers to the full implementation of decentralized treatment. The most notable is the public reluctance to change the wastewater treatment model and their perception and uncertainty about the consequences of this change for future sales decisions. Other factors are frequently mentioned as they can influence the implementation and success of decentralized systems: risk perception, lack of professional skills and knowledge, limited regulatory incentives and perceived threat of potential failure or water shortage (Brown et al., 2009; Mankad and Tapsuwan, 2011). Therefore, while progress has been made towards sustainable urban water systems, especially in terms of technological advances, more research is needed on how to overcome the obstacles mentioned above. As a concluding remark, not only the ease of use of the technology, but also its environmental sustainability and social acceptance are major factors influencing the implementation of decentralized systems. In this context, end-users need to be aware about the resilience of the decentralized technologies for wastewater treatment and the possibilities they offer.

9.4 DESCRIPTION OF DECENTRALIZED TECHNOLOGIES

Starting with the collection and transport of wastewater, there are different sewer options for wastewater collection: gravity, pressurized, and vacuum. The most common are small diameter gravity (SDGS) and pressurized sewers, combined with septic tanks. Both alternatives provide economic advantages due to the use of low-cost materials such as PVC and there is no need of pumping, resulting in energy savings. The vacuum sewer system consists of an interceptor tank when the wastewater from each household is collected and then discharged to main sewage through a negative pressure valve. In vacuum toilets, the flow of water used in these toilets is lower than in conventional toilets.

Once the wastewater has been collected, it is important to choose the best configuration for its treatment. There are several configurations; the most common is septic tank (ST) with a soil absorption system (SAS). The combination of ST and SAS accomplishes the removal of particles from wastewater. Due to the lack of aeration, it is a gravity-driven anaerobic system, which implies no electricity consumption. The effluent is discharged directly into the soil, so it can reach the groundwater courses. Considering that the sludge is accumulated in the bottom of the tank, the system needs periodical removal of the sludge that must be transported to a conventional wastewater treatment plant (WWTPs). Over time, it may encounter water-proofing problems and sewage leaks.

Another technology used in small communities is constructed wetlands (CWs). These technologies have advantages such as low cost, ease of operation and maintenance (O&M) and have a strong potential in developing countries, but it is currently difficult to ensure that such systems are built to an adequate standard and that the effluent quality is acceptable (Kivaisi, 2001). Wetlands can be classified into two types: free water surface (FWS) and subsurface flow (SSF) depending on the wetland hydraulics. FWS systems are similar to natural wetlands, with the shallow flow of wastewater over-saturated substrate. In SSF systems, wastewater influent can be horizontal flows (HF) or vertical flows (VF) through the substrate which supports the growth of plants. Wetlands are very effective in removing suspended solids, organic matter, and microbial pollutants (Wu et al., 2015). However, the removal of nitrogen or phosphorus in these systems is not effective. For nitrogen removal, it will be necessary to add an external source of organic matter to sustain denitrification while the oxygen supply rate can be insufficient for nitrification (Babatunde et al., 2010). Waste stabilization ponds (WSP) is another alternative that has been developed according to two different types: facultative or aerobic ponds. Often, WSP systems are combined with wetlands. This is due to the high concentration of suspended solids that increase biochemical oxygen demand (BOD) concentration. Moreover, WSP may have problems such as large land occupation and odor development, among others (Senzia et al., 2003). Other configurations such as advanced integrated wastewater pond systems (AIWPS) include four or more ponds where primary sedimentation, flotation, fermentation, aeration, secondary sedimentation, nutrient removal, and effluent storage are performed similarly to a conventional treatment plant. Land occupation is the major problem of these systems, but operating costs are low (Green et al., 1996).

In recent years, more advanced and sophisticated systems for nitrogen removal have been developed such as the sequencing batch reactors (SBR), membrane bioreactors (MBRs), or rotating biological contactor (RBC). The advantages are that depending on the wastewater composition, an external organic matter source maybe not required. Moreover, the quality of the effluent from the MBR is suitable for its reuse in the irrigation of green areas. The use of conventional activated sludge (CAS) for nitrogen removal leads to lower efficiency, with the requirement of an external carbon source and high energy consumption. An up-flow anaerobic sludge blanket (UASB) reactor has been used in decentralized systems for organic matter removal. Moreover, the UASB reactor produces biogas that can be used as energy, which can lead to significant savings in the energy consumption of the plant.

Also, the treated effluent from the UASB ensures the removal of pathogens, which makes it suitable for irrigation. The struvite crystallization unit aims to recover nitrogen and phosphorus from wastewater and produce a fertilizer (Rodriguez-Garcia et al., 2014). Struvite production has positive effects such as reducing eutrophication and mineral fertilizer requirements, with potential savings due to nutrient recovery. Conversely, struvite production can lead to operational problems such as pipe blockage, which may increase energy consumption for pumping (Jaffer et al., 2002).

Finally, decentralized systems are usually linked to the separation of all domestic flows. This approach ranges from the differentiated collection of rainwater to the segregation of the main streams in the household. The segregation of the main streams is known as the source separation term which consists of the separation of BW - Black Water (toilets) and GW - Grey Water (wastewater from sinks, showers, bathtubs, kitchen sinks, and washing machines). The different types of wastewater are collected in different pipelines and treated in separate units. GW can be treated to an acceptable quality level for non-potable reuse (NPR) by extensive biological treatment systems, due to its relatively low organic load (Penn et al., 2012). Another potential advantage of separate collection is that volumes of wastewater requiring treatment or disposal can be substantially lower when vacuum toilets are used, reducing water use at each flush from 26 to 6 L. Moreover, the higher concentration of pollutants (organic load and nutrients) in the stream is favorable for its treatment (Gual et al., 2008).

Rainwater (RW) can be combined with GW and used for irrigation of green areas, washing machines, or toilets. The separation of rainwater allows the system to be more ecological. In addition, it can also offer benefits to other parts of the water cycle, such as alleviating peak stormwater runoff and reducing the influent flow into wastewater facilities. This alternative can be used to relieve water stress by reducing the amount of water demands and providing resilience to the water supply system (Partzsch, 2009; Sharma et al., 2013).

Choosing the most appropriate technology is not an easy task. In fact, there is a wide variety of investments and programs that are implemented focusing only on the construction of new facilities and technologies. However, it is also necessary to implement appropriate management models to operate and maintain the wastewater system, considering social, technical, financial, and institutional issues (Meleg, 2012). Beyond demonstrating technical feasibility and economic costs, the selection process for appropriate technologies should also consider the life cycle impacts of the system, including construction, operation, and maintenance. Understanding the environment in which the solution will be implemented is also crucial and should be achieved through a comprehensive site assessment process (Massoud et al., 2009). Furthermore, for the implementation and promotion of a new technology, a risk assessment should be conducted, including health risks and quality restrictions (Bdour et al., 2009; Massoud et al., 2009).

9.5 LIFE CYCLE ASSESSMENT (LCA) OF DIFFERENT DECENTRALIZED SCHEMES

Life cycle assessment (LCA) of centralized wastewater systems have been extensively studied. When only focusing on the LCA studies of conventional WWTPs, Gallego et al. (2008) analyzed 13 WWTPs for small populations (less than 20,000 inhabitants) located in Galicia. They concluded that the most important category in WWTPs is eutrophication due to the presence of residual nitrogen, phosphorus, and chemical oxygen demand (COD) in the treated effluent. Electricity consumption during the operational phase causes an important impact, which is evidenced for the better performance of anoxic/anaerobic treatments compared to extended aeration (EA). Foley et al. (2010) considered ten different WWTP scenarios. The effects related to infrastructure, energy consumption, chemical consumption, or direct emissions are remarkable when standards are stricter for nitrogen discharge. Lorenzo-Toja et al. (2016b) evaluated 22 conventional WWTPs in Spain under the eco-efficiency perspective using the methodologies of LCA and life cycle costing (LCC). Facilities with anaerobic digestion have the best environmental impact, while medium-sized plants (24,000 to 29,400 equivalent inhabitants) presented the worst results, potentially attributed to the lack of anaerobic digestion. Finally, several authors applied the LCA methodology to assess the environmental impacts of a specific treatment technology. Høibe et al. (2008) studied five advanced technologies: (i) sand filtration, (ii) ozone, (iii) disinfection ultraviolet (UV), (iv) MBR and v) UV+ advanced oxidation. They concluded that the best technology from a technical and environmental perspective is sand filtration due to low energy consumption and extensive removal of heavy metals. Hospido et al. (2012) analyzed four different MBR configurations. The main contributing factors to the impact were energy consumption and soil application. Rodriguez-Garcia et al. (2014) evaluated three different technologies to treat the anaerobic supernatant in the sludge line. These technologies are: (i) partial nitrification-anammox (CANON), (ii) nitrate short-cut (SCN) and, (iii) struvite crystallization process (SCP). They modeled systems to pilot-plant scale before being integrated into a full-scale WWTP. These technologies reduce the impacts on the WWTP especially for eutrophication potential (EP). More recently, Longo et al. (2017) evaluated the integration of an emerging technology (short cut enhanced nutrient abatement (SCENA)) into a real WWTP. They concluded that SCENA has economic and environmental advantages over conventional WWTPs due to the reduction of chemical and energy requirements. Casas et al. (2017) analyzed greenhouse gas (GHG) emissions associated with horizontal subsurface flow construction wetland (HSSF) technologies. They concluded that emissions (12–22 kg CO_2eq/pe·y) and energy demand in these systems were lower than conventional wastewater treatment plants (WWTPs) (67 kg CO_2 eq/pe·y). Abou-Elela et al. (2017) evaluated a new passively aerated biological filter (PABF) with low energy consumption. The results showed that this technology can be applied to water quality and decrease the production of sludge and energy. A novel three-stage filler-based water aeration grooves biofilm reactor (WAGBR) for decentralized wastewater treatment was analyzed by Zhang et al. (2018). The results in terms of COD and NH_3 removal were satisfactory (80%), which could help the implementation of decentralized systems in rural areas. Kaetzl et al. (2018) reported the use of anaerobic biochar filters as a decentralized technology, with reasonable performance for treating wastewater at a low cost. A similar approach was adopted by Mattos et al. (2019), who developed an anaerobic filter filled with green coconut husk for wastewater treatment in a rural area of Brazil.

When opening the scope to include the water cycle, Amores et al. (2013) studied the urban water cycle in Tarragona (Spain). Moreover, the study proposed two scenarios to improve the environmental impact and water stress. These scenarios are: (i) the reuse of treated wastewater and, (ii) the use of desalination plants and RW during the drought period. Slagstad and Brattebø (2014) evaluated the impacts associated to the water cycle in Trondheim (Norway). Energy and chemical consumption, as well as freshwater eutrophication were the main impacts. Finally, Risch et al. (2015) evaluated the construction and operation phases of the sewer and WWTPs and concluded that the impacts associated with the construction phase are relevant in all impact categories.

Evaluations of decentralized and/or source separation systems are less common, but have become more frequent in recent years. Shehabi et al. (2012) compared the environmental impacts of decentralized and centralized systems in California. They concluded that the centralized system helps to reduce the environmental impact and the decentralized system should be carefully evaluated as part of the design process. Lehtoranta et al. (2014) analyzed six decentralized wastewater scenarios in Finland and concluded that the optimal solution depends on local conditions, so it is important to have appropriate guidance in such systems. Thibodeau et al. (2014) compared conventional and decentralized systems. Impacts on CC, human health, and ecosystem quality and resource depletion are higher in the centralized system. However, when focusing on the decentralized system, the main impacts are attributed to ammonia emissions and digestate management. Matos et al. (2014) compared a decentralized system that considered the reuse of treated effluent with a centralized system with no water reuse under the perspective of CO_2 emissions and energy consumption. The lower impacts of the decentralized system make it a better candidate for wastewater treatment from an environmental point of view. Lam et al. (2015) studied four scenarios: (i) off-site treatment; (ii) onsite treatment; (iii) source-separation system; and (iv) a water flushing toilet in Tiajin (China). They concluded that the scenario with a source-separation system had better environmental profile in terms of environmental impacts and water use. The interest in the development of decentralized systems for the treatment of urban wastewater is evident in the growing number of publications analyzing different options for the segregation of flows and treatment technologies. Opher and Friedler (2016) studied four different alternatives for treating wastewater. Two options are based on centralized treatment in conventional WWTP, considering the possibility of using RW for irrigation, while the other two are based on alternatives of decentralized treatment with separation of GW and BW, one of them implemented at cluster level (320 households) whereas the second was applied at building level (40 households). They concluded that the decentralized alternative at cluster level presented better environmental results in terms of energy consumption and infrastructure than the other options. Kavvada et al. (2017) analyzed a new system for nitrogen recovery from a decentralized life cycle perspective. The method consisted of separating the urine fraction at households for its further treatment in resin systems through an ion exchange. The study concluded that this decentralized approach has a lower cost, lower GHG emissions, and energy consumption, but at this time, these systems are in a preliminary phase of research, so their use as fertilizers is still a pending task. In addition, Garfí et al. (2017) evaluated three wastewater treatment alternatives for small communities. These alternatives were: (i) a conventional WWTP;(ii) hybrid CW; and (iii) a high rate algal pond (HRAP) system. They concluded that the eco-friendliest technologies were the two-nature technologies (hybrid CW and HRAP system). In terms of economic indicators, the best option was the high algal pond system due to the fact that wetlands require a large land occupation. Schwaickhardt et al. (2017) analyzed the hospital laundry wastewater from an environmental point of view using photochemical VUV and UVC reactors. They concluded that the VUV process is better due to lower energy consumption. Kabdaşlı and Tünay (2018) studied the recovery of phosphorus in human urine in decentralized systems. Three phosphorus recovery processes were evaluated: (i) struvite precipitation;(ii) ion exchange with zeolites; and (iii) adsorption processes. They concluded that adsorption processes such as membrane methods, biochemical, or electrochemical processes can help improve phosphorus recovery in decentralized wastewater systems. Jeong et al. (2018) analyzed a hybrid system that combines a conventional centralized system and a decentralized system for GW recovery. They studied this system in nine residential zones: five single-family house zones (SFZs) and four apartment zones (MFZs). The decentralization system for the GW stream allowed greater benefits in terms of water and energy demand from an economic and environmental point of view. Dominguez et al. (2018) analyzed the GW separation in a hotel in Cantabria (Spain). Three GW reuse technologies were implemented: (i) photocatalysis, (ii) solar-driven photocatalysis and, (iii) an MBR. The best option from an environmental point of view was the solar-driven photocatalysis because it has lower energy consumption. Singh et al. (2018) performed the techno-economic analysis of 16 decentralized WWTPs. Different technologies were assessed: (i) moving bed bioreactor (MBBR), (ii) RBC, (iii) combination of aerobic and/or anaerobic units as a fixed film aeration tank and a septic tank, and finally (iv) MBR. MBR was the best solution in terms of effluent quality, while aerobic systems were the most efficient in removing organic matter. Finally, in the recent years, there are a few publications related to the decentralized topic. Hazarika and Pandit (2019) studied the sustainability of three decentralized WWTPs in Ahmadabad. The technologies evaluated in these plants were: (i) fluidized bioreactor, (ii) activated sludge process, and (iii) decentralized wastewater treatment installed at the households. They concluded that the small-scale decentralized system was not economical due to high energy consumption. Another alternative relied on a SSF artificial wetland to treat wastewater from a hotel located on the Central Pacific coast of Costa Rica (Pérez-Salazar et al., 2019). The results showed that these systems are efficient in terms of energy consumption, but the main problem is land occupation. The most recent work includes an analysis of costs as well as environmental impacts (Yerri and Piratla, 2019). This model can assist water utilities in planning water supplies.

In this book chapter, we have considered four different scenarios for the comparison of centralized and decentralized systems under the LCA methodology (Figure 9.1). Scenarios based on centralized systems differ in the partial reuse (with or without) of the treated water. Decentralized treatment scenarios consider source separation, on-site treatment, with the possibility of greywater (GW) reuse and nutrient recovery in a system with higher complexity. In all scenarios, the sludge will be treated in a composting plant before its use as fertilizer for agricultural soils. In terms of system boundaries, the urban water lifecycle consists of a comprehensive network of co-dependent services, including WW collection, treatment, RW distribution, and urban irrigation, for a fair comparison between alternatives, in its various phases and quality levels.

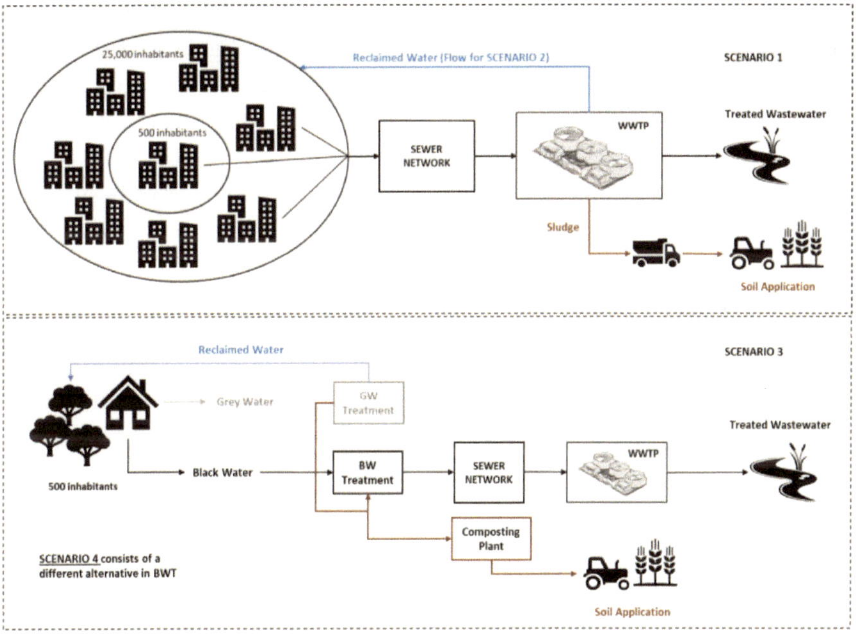

FIGURE 9.1 Scheme of the centralized (1 and 2) and decentralized (3 and 4) scenarios evaluated for urban wastewater treatment.

9.5.1 DEFINITION OF SCENARIOS

Scenario 1 consists of a sewerage network that transports all the domestic wastewater to a conventional WWTP designed for a population equivalent

(PE) of 25,000 inhabitants. The system includes pretreatment, secondary treatment in the water line while a thickener, a homogenizer, and a band filter in the sludge line. The effluent meets quality standards and is released to a nearby watercourse. For comparison purposes, a neighborhood of 500 inhabitants was considered.

Scenario 2 is the same as Scenario 1 regarding WW collection and treatment, but differs in reuse. The non-potable RW is used for private gardens and cluster parks, and can be supplied from the central WWTP through a dual municipal water distribution system. The WWTP effluent is discharged to a natural watercourse. The RW represents 40% of the treated flow, while the remaining fraction (60%) is discharged as treated effluent (Opher and Friedler, 2016).

Scenario 3 implements the separation of domestic wastewater at source into GW and BW. The GW system consists of a physical-chemical treatment (bioflocculation) and a secondary treatment based on activated sludge. The water is collected in a tank and reused for irrigation of green areas. The BW system consists of a UASB reactor coupled to an RBC. The excess sludge from the GW treatment facilities is sent to the UASB reactor. The excess sludge of BW is collected and treated in a composting plant before its use as fertilizer. The treated BW is discharged to the main sewer network and treated in a conventional WWTP.

Scenario 4 also implements source separation, as in Scenario 3, but differs in the final scheme of BW treatment. The BW scheme consists of an UASB, RBC, and a struvite unit for phosphorus recovery. As in Scenario 3, biogas is used in the WWTP and its valorization in a CHP unit may offset the requirements of energy from the grid. Moreover, the possibility of not valorizing the biogas stream was also considered in a sensitivity analysis.

9.5.2 FUNCTIONAL UNIT (FU), SYSTEM BOUNDARIES AND LIFE CYCLE INVENTORY (LCI)

The functional unit (FU) represents the quantification of the system. The main purpose of a FU is to provide a reference to which inputs and outputs are related (ISO 14040, 2006). Based on other LCA works on WWTP (Hospido et al., 2012; Lorenzo-Toja et al., 2016a; Pasqualino et al., 2011), the FU selected is 1 m^3 of treated wastewater.

The selection of system boundaries is another important step in the LCA methodology (Lundin et al., 2000). In conventional WWTPs, the main impacts occur during the operation phase (Corominas et al., 2013; Lorenzo-Toja et al., 2016b), while the construction and decommissioning phases have less significant impacts, around 25–35% (Lorenzo-Toja et al., 2016a) or even negligible (Foley et al., 2010). This study takes into account the construction phase of the plant and the sewer network, as well as the operation phase and irrigation of green areas. In the construction phase, considering that the domestic water use is identical in all scenarios, all processes related to domestic water use were excluded. It has been considered that the wastewater collection pipelines from the household to the point of connection with the municipal collection network have not been affected by the reduction in wastewater flow (due to separation at source).

Inventory analysis involves data collection and quantification of inputs and outputs. The quality of data use is a major issue in any LCA study as it determines the validity of the environmental results. Therefore, to the extent possible, it is recommended to handle actual data from the system being studied or to consider simulating the system with a critical review by technology experts (ISO 14040, 2006).

Primary data were obtained from academic publications, reports, catalogs, and personal communications. Secondary data were obtained from the Ecoinvent v3.2. Database (Weidema et al., 2013). Electricity production and the import/export mix in Spain were updated for 2016. The medium voltage used in WWTP was modeled, including atmospheric emissions and transport losses (Dones et al., 2007). Euro 4 trucks (16–32t) were selected as transport vehicles for construction materials during the construction phase. At the operational phase, the trucks selected for the transport of chemicals were smaller (3.5–7.5t). Emissions to air (N_2O and NO) were calculated from Kampschreur et al. (2008). The inventory data for each scenario are shown in Tables 9.2–9.5.

TABLE 9.2 Main Life Cycle Inventory (LCI) Data of Scenario 1 (FU: 1 m^3 of Treated Wastewater)

Inputs		**Outputs**	
From the Technosphere		**To the Environment**	
Materials and Fuel		**Emissions to Water**	
Influent			
		COD (g)	0.76
COD (g)	7.96	TN (g)	0.22
TP (g)	0.09	TP (g)	0.04
TN (g)	0.38	Al (mg)	2.28
Al (mg)	58.1	Fe (mg)	1.93

TABLE 9.2 *(Continued)*

Inputs		Outputs	
From the Technosphere		**To the Environment**	
Materials and Fuel		**Emissions to Water**	
Fe (mg)	48.08	Co (mg)	0.01
Co (mg)	0.03	Ni (mg)	0.04
Ni (mg)	0.19	Cu (mg)	0.13
Cu (mg)	7.40	Zn (mg)	0.38
Zn (mg)	3.07	As (mg)	0.06
As (mg)	0.10	Pb (mg)	0.01
Pb (mg)	0.17	**To the Technosphere**	
Electricity Consumption		**Products and Co-Products**	
Energy consumption (kWh)	0.51	Inert waste	1.00×10^{-2}
Chemical Consumption		Solid waste	4.42×10^{-2}
Polyelectrolyte (kg)	9.72×10^{-4}	Composting production (kg)	0.03
Infrastructure		Transport (kg·km)	0.05
Civil infrastructure	0.0159	Avoided Products	
Sewer	4.44×10^{-4}	N based fertilizer (g)	0.07
		P_2O_5 based fertilizer (g)	0.08

TABLE 9.3 Main Life Cycle Inventory (LCI) Data of Scenario 2 (FU: 1 m³ of Treated Wastewater)

Inputs		Outputs	
From the Technosphere		**To the Environment**	
Materials and Fuel		**Emissions to Water**	
Influent		COD (g)	0.45
COD (g)	7.96	TN (g)	0.13
TP (g)	0.09	TP (g)	0.02
TN (g)	0.38	Al (mg)	1.37
Al (mg)	58.1	Fe (mg)	1.19
Fe (mg)	48.08	Co (mg)	0.05
Co (mg)	0.03	Ni (mg)	0.02
Ni (mg)	0.19	Cu (mg)	0.08
Cu (mg)	7.40	Zn (mg)	0.23
Zn (mg)	3.07	As (mg)	0.04
As (mg)	0.10	Pb (mg)	0.01

TABLE 9.3 *(Continued)*

Inputs		**Outputs**	
From the Technosphere		**To the Environment**	
Materials and Fuel		**Emissions to Water**	
Pb (mg)	0.17	**To the Technosphere**	
Electricity Consumption		**Products and Co-Products**	
Energy consumption (kWh)	0.51	Inert waste	1.00×10^{-2}
Chemical Consumption		Solid waste	4.42×10^{-2}
Polyelectrolyte (kg)	9.72×10^{-4}	Composting production (kg)	0.03
Infrastructure		Transport (kg·km)	0.05
Civil infrastructure	0.0159	Avoided Products	
Sewer	4.44×10^{-4}	N based fertilizer (g)	0.07
Reclaimed water supply	7.00×10^{-4}	P_2O_5 based fertilizer (g)	0.08

TABLE 9.4 Main Life Cycle Inventory (LCI) Data of Scenario 3 (FU: 1 m^3 of Treated Wastewater)

Inputs		**Outputs**	
From the Technosphere		**To the Environment**	
Materials and Fuel		**Emissions to Water**	
Influent		COD (kg)	0.53
COD (kg)	10.9	TN (kg)	0.59
TN (kg)	1.31	NH_4 (kg)	0.53
TP (kg)	0.18	TP (kg)	0.08
NH_4 (kg)	0.79	Emissions to Air	
PO_4^{3-} (kg)	0.06	CH_4 (mg)	20
Electricity Consumption		N_2O (g)	6.01
Energy consumption (kWh)	4.31	**To the Technosphere**	
Transport		**Products and Co-Products**	
Polyelectrolyte (kg·km)	0.23	Composting production (g)	1.86
Chemical Consumption		Transport (kg·km)	3.50×10^{-3}
Polyelectrolyte (g)	9.53	Biogas production (m^3)	2.15
Infrastructure		**Avoided Products**	
Civil infrastructure	5.36×10^{-5}	N based fertilizer (g)	4.62×10^{-3}
Sewage	6.26×10^{-4}	P_2O_5 based fertilizer (g)	6.13×10^{-3}
Reclaimed water supply	4.71×10^{-5}		

TABLE 9.5 Main Life Cycle Inventory (LCI) data of Scenario 4 (FU: 1 m^3 of Treated Wastewater)

Inputs		Outputs	
From the Technosphere		**To the Environment**	
Materials and Fuel		**Emissions to Water**	
Influent		COD (kg)	0.12
COD (kg)	11.8	TN (kg)	0.02
TN (kg)	1.31	NH_4 (kg)	0.01
TP (kg)	0.18	TP (g)	1.74
NH_4 (kg)	0.79	**Emissions to Air**	
PO_4^{3-} (kg)	0.05	CH_4 (mg)	20
Electricity Consumption		N_2O (g)	6.01
Energy consumption (kWh)	5.44	**To the Technosphere**	
Transport		**Products and Co-Products**	
Polyelectrolyte (kg·km)	0.20	Composting production (g)	1.47
Struvite chemicals (kg*km)	6.3	Transport (kg·km)	2.84×10^{-3}
Chemical Consumption		Biogas production (m^3)	2.18
Polyelectrolyte (g)	9.53	**Avoided Products**	
$MgCl_2$ (g)	222.8	N based fertilizer (g)	0.41
NaOH (g)	27.57	P_2O_5 based fertilizer (g)	2.12
Infrastructure			
Civil infrastructure	6.27×10^{-5}		
Sewage	6.26×10^{-4}		
Reclaimed water supply	4.71×10^{-5}		

9.5.3 ASSESSMENT METHODOLOGY AND IMPACT CATEGORIES

SimaPro v.8.2 software was used for calculating the impact assessment. Two methodologies were selected the ReCiPe Midpoint (H), version 1.12 and the CML 2001. Potential Eutrophication (EP) was calculated to the CML 2001 methodology (Guinée, 2002). CC, ozone depletion (OD), terrestrial acidification (TA), photochemical oxidation formation (POF), particulate matter formation (PMF), human toxicity (HT), terrestrial ecotoxicity (TE), freshwater ecotoxicity (FE), marine ecotoxicity (ME), water depletion (WD) and fossil depletion (FD) were calculated using the ReCiPe midpoint method (Goedkoop et al., 2009). The selection of two methodologies is largely justified by the need to include as an essential parameter in the operation of the system: COD. It is well known that the WWTP operation must meet COD discharge requirements according to the legislation (91/271/EEC). However, this variable does not have a characterization factor in the ReCiPe methodology but is included in the CML 2001 method. With this premise, both evaluation methodologies were combined.

9.5.4 *LIFE CYCLE IMPACT ASSESSMENT (LCA)*

Plant infrastructure, sewer, and reclaimed wastewater were considered in all scenarios. In Scenarios 3 and 4, it was considered that biogas is valorized to produce energy. After the analysis of each individual scenario, a comparison between the different alternatives was conducted.

In Scenario 1, the main contributor to the impact in all categories is the sewer construction except for TA where plant infrastructure is more relevant and for TE and EP which depend on the use of sludge as fertilizer (Figure 9.2). The impact of the sewer varies from 3% in TE to 84% in WD. The high impact on WD is due to drinking water consumption during the construction phase. Land application has a negative effect on TE (96%) and EP (45%) categories. This is due to the content of heavy metals in the wastewater and leachate from soil application, respectively. Infrastructure causes the main impact in TA category (61%), which is attributed to the indirect emissions related to the electricity consumption in the construction phase. Electricity consumption ranged from 4% in FE and ME to 39% in OD. The disposal of the products generated in the WWTPs has more relevance in ME (22%) and FE (24%). This impact is associated with landfill leachate. As for composting sludge, the most important impact is on CC (20%), the consumption of electricity in the plant is responsible for this impact. Finally, the consumption of polyelectrolyte can be considered negligible in all categories except EP. In this category, the impact is around 40% which is associated with indirect emissions related to the production of polyelectrolyte.

As explained in Section 5.1, Scenario 2 is similar to Scenario 1, but the treated effluent (40%) is reused for irrigation of green areas. Thus, the distribution network of RW is taken into account. Based on this difference, sewer, and RW distribution networks are the main generators of impacts in all categories except in EP, WD, and TE (Figure 9.3). In TE (95%) and EP (78%) categories, as in Scenario 1, the most important contributor is land application. Again, this is attributed to the content of heavy metals in wastewater and leachate from soil application. As in scenario 1, infrastructure causes the main impact in TA category (77%). Irrigation has a positive

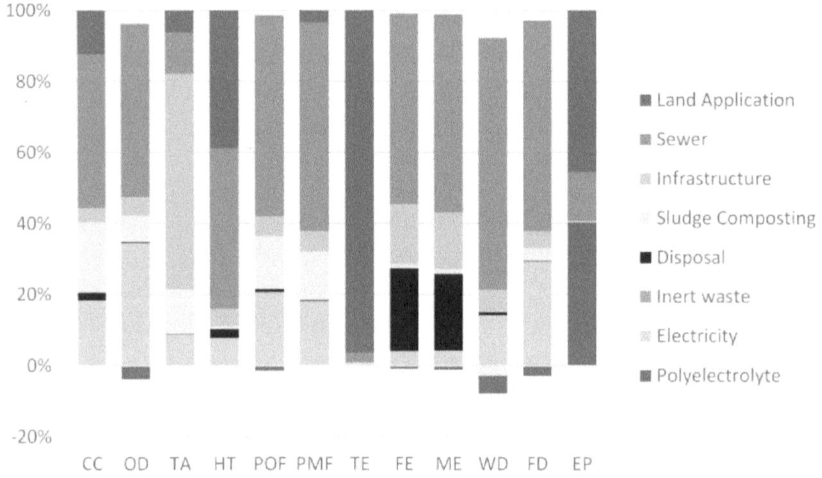

FIGURE 9.2 Environmental impacts for Scenario 1 (centralized system without reuse). FU: 1 m^3 of treated wastewater.

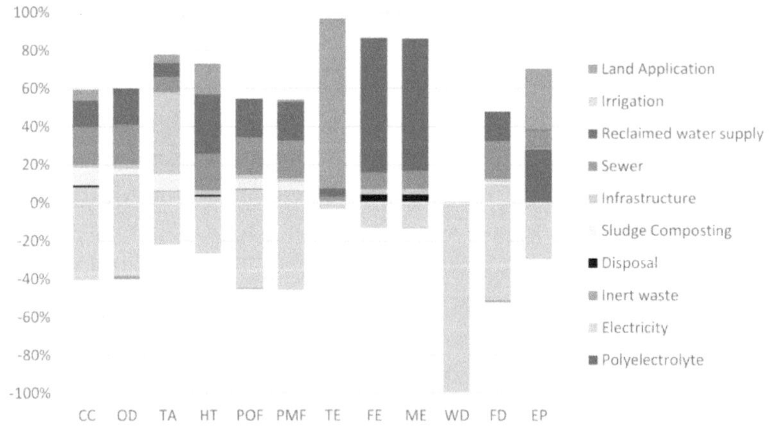

FIGURE 9.3 Environmental impact for Scenario 2 (centralized system with water reuse in irrigation): FU: 1 m^3 of treated wastewater.

effect on all categories, especially on WD, because if RW is reused, it is not necessary to use fresh drinking water from the network. In this case, the impact related to the disposal of products generated in the WWTP can be considered negligible. Polyelectrolyte has a significant impact on the eutrophication category (69%) due to the indirect emissions associated with its production. As for compost, as in Scenario 1, the main impact is associated with electricity consumption, which represents about 48% in CC category.

In Scenario 3, the BW and GW configurations were analyzed together. Moreover, consideration was given to whether or not the biogas produced is reused in the plant. When biogas is valorized in the plant, the main contributors to the impact are sewer and RW distribution in all categories except FD, TA, and TE (Figure 9.4). In TE, infrastructure causes the main impact due to land use. In FD and TA, the consumption of polyelectrolyte presents a greater relevance due to the indirect emissions associated with the production of this chemical. Both electricity and irrigation have a positive effect on the system. The electricity produced has been considered sufficient to supply the system for the treatment of BW (UASB + OLAND (oxygen limited autotrophic nitrification-denitrification)) and GW (bioflocculation + activated sludge) (Figure 9.4. Scenario a).

The possibility of recovering water for irrigation has a beneficial effect on the WD indicator because it avoids using fresh drinking water. The composting of sludge and the impacts of terrestrial applications can be considered negligible, since the amount of sludge generated in these systems is much lower than that of conventional WWTPs, which also has a positive impact on the lower contribution of transport. If biogas will not be reused in the BW plant, it can be flared. In this context, electricity consumption accounts for the largest impacts in all categories except for EP and WD. The impact of electricity is associated with indirect emissions associated with the capture of electricity from the grid and CO_2emissions associated with biogas combustion (Figure 9.4. Scenario b). In EP category, the main impact is caused by the sewer construction, whereas in the WD category, irrigation has a positive effect. In Scenario 3, both transport and plant infrastructure can be considered negligible.

Scenario 4 implies a significant change from the previous scenario, as a struvite precipitation unit is included. In this case, the infrastructure has more impact on all categories except EP and WD because a new unit is added to the system. The networks of sewer and RW remains important, but the impact is not as high as in Scenario 3. In EP category, the impact (41%) is attributed to chemical consumption to produce struvite. Irrigation and electricity consumption, as in Scenario 3, have a positive effect. The plant is energy sufficient and no need of extra electricity or drinking water for irrigation is required (Figure 9.5. Scenario a).

However, the results vary substantially if biogas is not used (Figure 9.5. Scenario b). As in Scenario 3, network energy consumption has a

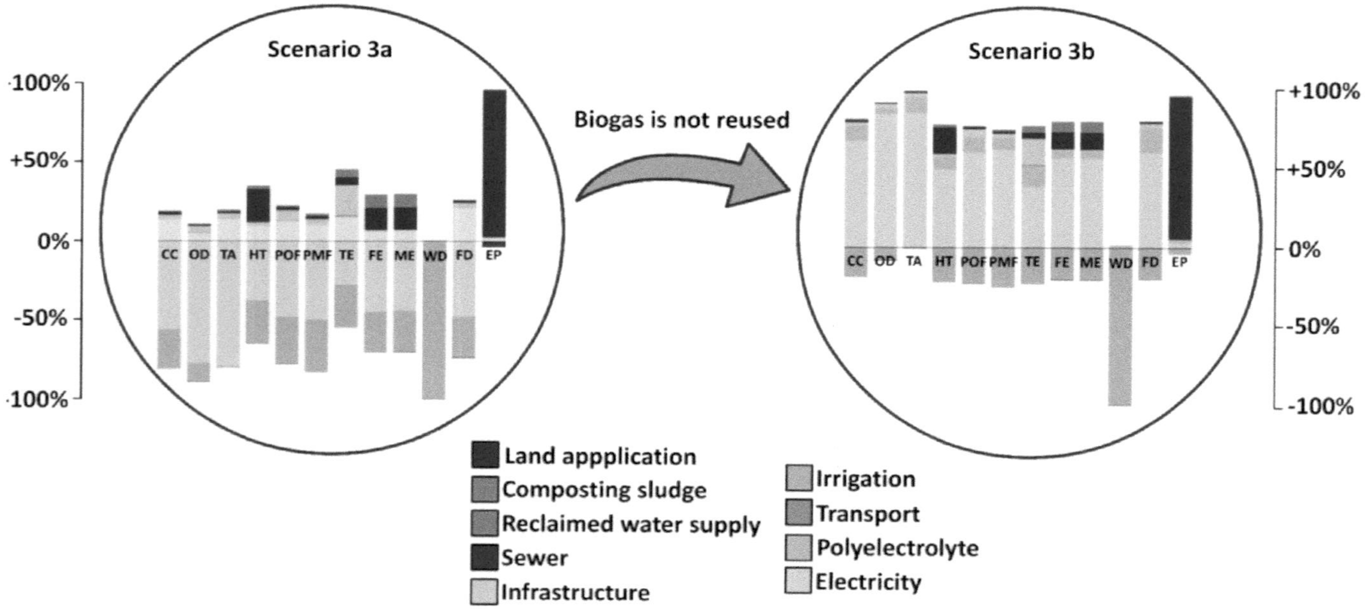

FIGURE 9.4 Environmental impacts for Scenarios 3a (scenario 3 consists in a UASB + RBC for BW treatment while the GW line is formed by a bioflocculation unit + activated sludge unit. The biogas in this scenario is valorized in the decentralized plant): biogas reuse and 3b (scenario 3b is the same that scenario 3a, but in this case, the biogas is not recovery in the plant, thus, the electricity is provided to the grid): no biogas valorization. FU: 1 m^3 of treated wastewater.

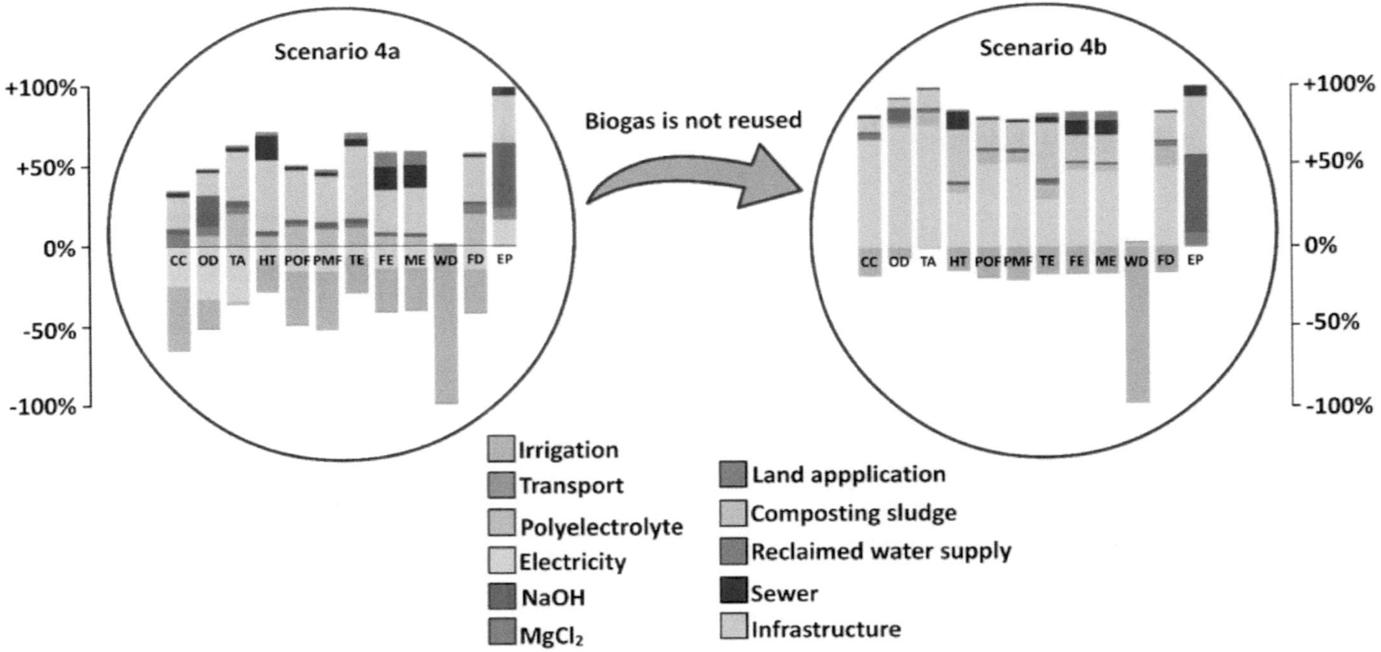

FIGURE 9.5 Environmental impacts for Scenarios 4a (this scenario is similar to scenario 3, but in the BW line incorporates a struvite unit for recovering phosphorus. In this case, of study, the biogas is reused in the plant as energy): biogas reuse and 4b (in the same configuration that scenario 4a, but the biogas is not recovery in the plant and electricity is provided to the grid): no biogas valorization. FU: 1 m^3 of treated wastewater.

negative effect and is the main contributor to all impact categories except EP and WD. For the WD indicator alone, the observed trend is maintained as electricity consumption does not affect and irrigation continues to have a positive effect. The impact on EP is due to the consumption of chemicals to produce struvite. By contrast, other subsystems such as sludge composting, transport, and soil application can be considered negligible. Furthermore, in a scenario where biogas is not reused, impacts related to sewer and water supply distribution are not of significant importance, as they are hidden by direct and indirect emissions related to electricity consumption.

Having studied the systems separately, the next step was to study them from an integrated perspective. Figure 9.6 shows the results for the different scenarios with biogas recovery.

In this perspective, it is observed that conventional systems have more impacts than decentralized systems in all categories except TE, ME, and FE. In these categories, the worst-case scenario is 4 due to indirect emissions associated with the production of chemicals necessary for struvite precipitation. Scenario 1 presents a better environmental profile than Scenario 2, except in eutrophication, as a fraction of treated water is reused in the latter scenario.

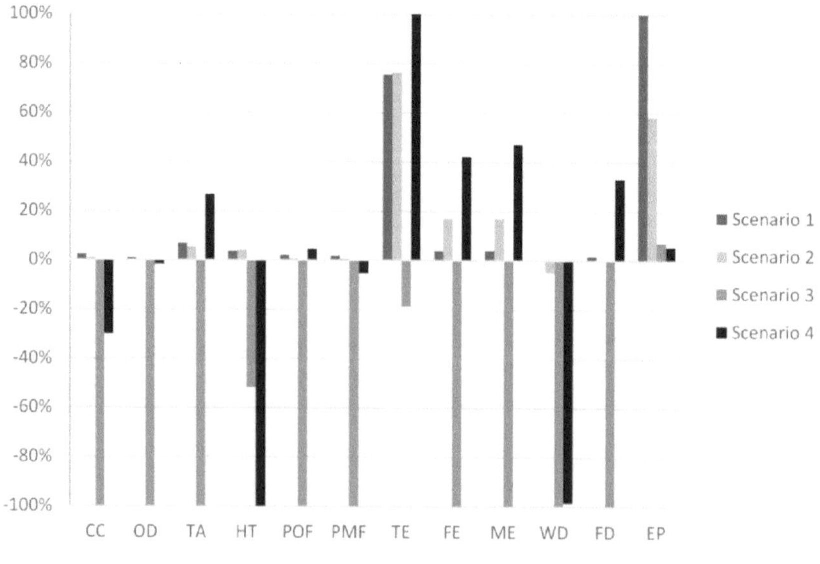

FIGURE 9.6 Comparison between the different scenarios (construction phase included). FU: 1 m^3 of wastewater.

If biogas is not recovered, impacts change considerably (Figure 9.7). Scenarios 3 and 4 now have greater impacts in all categories except EP and WD, mainly due to the use of RW. Impacts related to infrastructure and sewer is lower in Scenarios 3 and 4, but impacts associated with the operation are greater than in Scenarios 1 and 2. These impacts are greater because the flows of BW and GW are treated separately, with their corresponding energy consumption for each current. Assuming that the sum of both flows is equivalent to the centralized system, we must be aware that it is necessary to consider the operational impacts of both configurations, which lead to significant impacts. Therefore, the valorization of biogas is very important in decentralized systems as it can improve impacts in all categories and be more efficient than centralized systems.

In different works in which centralized and decentralized systems were compared, the construction phase was not taken into account, but only the operational phase (Matos et al., 2014; Thibodeau et al., 2014). As presented in the previous results, the construction phase affects all scenarios, so an additional analysis has been carried out in which this subsystem is obvious. The results of the different scenarios are shown in Figure 9.8.

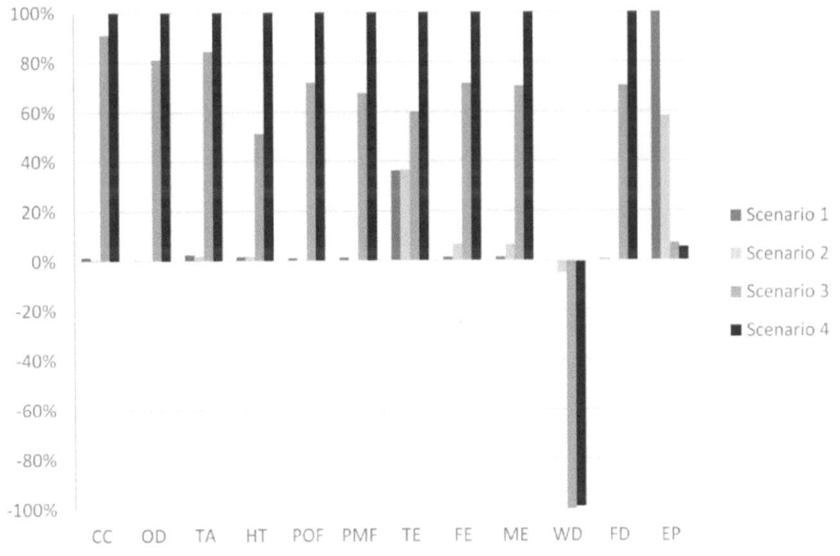

FIGURE 9.7 Comparison between the different scenarios (neglecting biogas production). FU: 1 m^3 of wastewater.

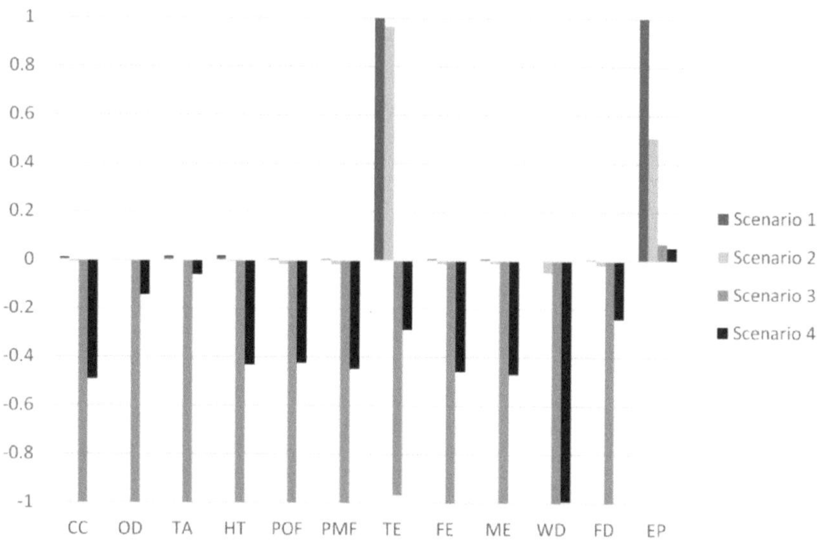

FIGURE 9.8 Environmental results of each scenario for the impact categories under assessment without construction phase. FU: 1 m^3 of treated wastewater.

Results without construction show that the worst-case scenarios are conventional systems in all categories. The most important impacts occur in EP and TE categories, which is associated with the discharge of treated effluents and the presence of heavy metals in sludge when used as fertilizers in agricultural soils.

In this study, the net environmental benefit (NEB) methodology was considered since it is a methodology widely used in the study of eutrophication of WWTPs (Lorenzo-Toja et al., 2016b; Rodriguez-Garcia et al., 2011). The NEB indicator is the difference between the potential environmental impacts (PEI) caused and avoided by the WWTP and is represented by the following Eqn. (9.1) (Godin et al., 2012).

$$NEB = [PI_{NO} - PI_{TW}] - PI_{SLC} \tag{9.1}$$

PI_{NO} is the scenario with no treatment, PI_{TW} corresponds to the treatment scenario, and PI_{SLC} corresponds to the impacts produced by the WWTP during its life cycle. The results of the NEB indicator are shown in Figure 9.9.

Decentralized systems have worse values than centralized systems. In particular, Scenario 4, as the treated BW stream still has significant concentrations of nutrients and organic matter. Therefore, it is necessary to treat this water in a conventional system or to increase the efficiency of the operation.

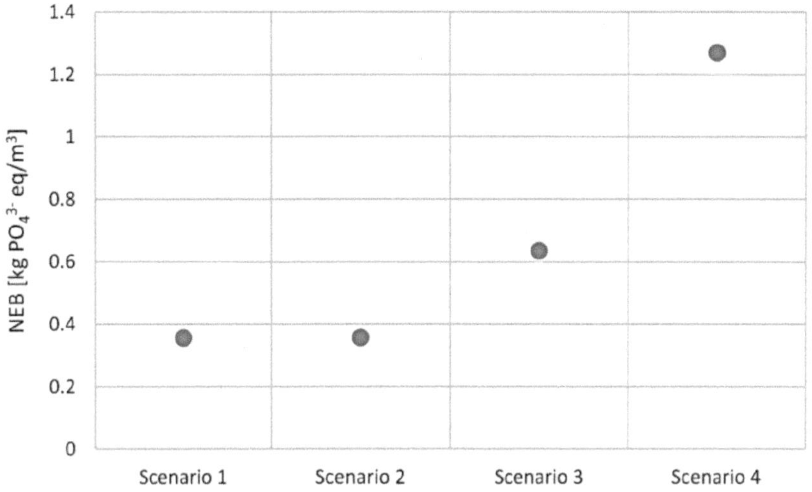

FIGURE 9.9 Net environmental benefit (NEB) for different scenarios. FU: 1 m^3 of treated wastewater.

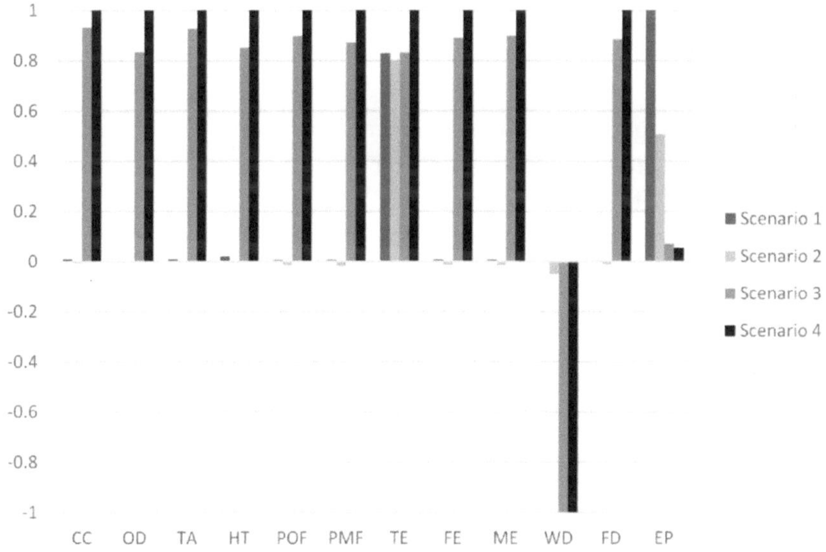

FIGURE 9.10 Environmental results of each scenario for the impact categories under assessment without construction phase and without biogas recovery. FU: 1 m^3 of treated wastewater.

Finally, it is important to analyze the system without infrastructure and without biogas recovery (Figure 9.10). Scenario 4 presents the main impacts in all categories except EP, which is mainly due to energy and chemical consumption in the operation phase. Irrigation has a positive effect on WD in decentralized systems (Scenarios 3 and 4). In the EP category, the amount of sludge generated in the conventional system is the main factor contributing to the impact of agricultural land use. Again, the importance of energy and nutrient recovery in decentralized systems is highlighted. These systems may be more efficient and sustainable, but if the recovered flows are not valued, impacts may be even greater than those of centralized systems.

9.6 ENVISIONING SOCIAL ASPECTS

The social aspect is associated with local factors that can affect the functioning and maintenance of a given system. These included parameters such as community habits, lifestyle, health protection, or social acceptance (Massoud et al., 2009; Saurí and Del Moral, 2001). Decentralized systems have less sewer infrastructure and fewer leaks. In addition, the role of individuals may involve a higher level of local governance. In fact, the implementation of new water schemes requires public acceptance and political support (Hurlimann and Dolnicar, 2010). End-users are responsible for the O&M of decentralized systems, as they own the system and need to develop new capabilities. More empirical studies are needed to explore the process of social learning and socio-technical transition involving institutional and socio-cultural settings. However, so far only a limited number of studies have been identified (Domènech et al., 2015) that empirically address the dynamics of learning. Previous research on the acceptance of decentralized systems has shown that the general public has a very limited understanding of these approaches and how they can be used. People, in general, do not yet have committed attitudes on the decentralized issue of water and, therefore, their attitudes are subject to change.

9.7 IDENTIFYING HOTSPOTS: PROCESS EFFICIENCY AND ENERGY PROFILE

The main impacts in the selected impact categories are dominated by land application of sludge and electricity consumption in the operational phase and the activities related to the construction of sewer, RW supply, and the infrastructure. It is important to take into account the valorization of biogas in decentralized systems. The emissions associated with energy use and biogas combustion are quite significant (Figures 9.4 and 9.5, Scenario b). In Scenario 1, 9 out of 12 categories are affected by the sewer. In TA, the main impact is produced by the plant infrastructure while in two impact categories (TE and EP) the sludge used in agriculture is the main contributor. Scenario 2 presents a wider variety of impacts: 6 categories are affected by sewer, 2 by land application and 1 by the construction of the plant. Finally, irrigation had a positive impact on the WD category. In Scenarios 3 and 4, the benefit of energy recovery affects all categories. In conventional WWTPs, the main impact contributor is usually associated with energy consumption (Longo et al., 2017; Rodriguez-Garcia et al., 2011). In this case, if construction was not taken into account, electricity would gain importance in all categories, being the main driver of the impact in 6 categories. Therefore, trying to recover this resource is very important from an environmental and economic point of view.

The energy profile in this study is based on Spanish country mix, which accounts for 59.2% of non-renewable energies and 40.8% of renewable energies. Non-renewable energies (coal, combined cycle, and nuclear energy) are responsible for indirect emissions associated with the use of fossil fuels. If these systems are studied in other countries with a lower proportion of non-renewable energy sources, such as Norway (18 g CO_2 eq/kWh) or Sweden (58 g CO_2 eq/kWh), indirect emissions would be lower than in Spain (300 g CO_2 eq/kWh). Other countries are more dependent on non-renewable energies, such as Turkey (519 g CO_2 eq/kWh) or Poland (734 g CO_2 eq/kWh), which would lead to greater environmental impacts. It is important to consider the energy profile associated with developing countries such as Brazil, Mexico, Tanzania, Taiwan, or Bangladesh, among others. For example, when the electricity profile relies heavily on non-renewable energy as in Bangladesh, the indirect emissions related to electricity consumption are 681 g CO_2/eq. The energy mix of this country depends largely on natural gas (63%) followed by fuel oil (20%). Tanzania, where the mixture consists of natural gas energy (49%) followed by fuel oil (20%), has emissions in the order of 547 g CO_2/eq. In Taiwan, emissions increase by 709 g CO_2/eq. This is because the main source of energy production is coal (40%), followed by natural gas (25%). Mexico where 86% of energy is produced from hydrocarbons such as fuel oil or natural gas produces emissions into the atmosphere of 568 g CO_2/eq. Finally, Brazil has the best combination of developing countries, even better than Spain. Emissions involve 241 g CO_2/eq. and renewable energies account for 84% of total energy demand (Goedkoop et al., 2009). The problem in these countries is the lack of funding for wastewater treatment. Therefore, decentralized systems may be a good option because they may be cheaper than conventional systems.

In addition, electricity consumption decreases, implying lower atmospheric emissions related to fossil carbon dioxide (Massoud et al., 2009).

9.8 CONCLUSIONS AND FUTURE PERSPECTIVES

Water scarcity, climate change and population growth were identified as the main drivers of the implementation of decentralized systems. Forecasts of population growth and increasing urbanization predict that water use will increase by 2050, leading to a shortage of water resources. This study compared the environmental impacts of four alternative approaches to wastewater treatment using LCA. Proper management of the decentralized system is vital if these systems are to function properly and resources such as biogas and struvite are to be exploited in such a way as to reduce operational costs and environmental impact, including economic benefits.

While the direct and indirect impacts of fossil fuel electric power use dominate most impact categories in all scenarios, the impact of the system can be reduced in most impact categories (18% on average) by applying urban GW reclamation in either of the two suggested decentralized approaches. Further analysis of the interests and preferences of the various stakeholders involved in water management is also needed to increase the potential for success of decentralized wastewater treatment.

ACKNOWLEDGMENTS

This research was supported by the UE project Run4Life (730285-1). The authors belong to the Galician Competitive Research Group (GRC ED431C 2017/29) and to the CRETUS Strategic Partnership (AGRUP2015/02). The authors gratefully acknowledge the constructive comments of Dr. Miriamva-nEekertfrom Wageningen University and Research.

KEYWORDS

- **blackwater (BW)**
- **decentralization**
- **greywater (GW)**
- **life cycle assessment (LCA)**
- **resource management**
- **wastewater reuse**

REFERENCES

Abou-Elela, S. I., Hellal, M. S., Aly, O. H., & Abo-Elenin, S. A., (2017). Decentralized wastewater treatment using passively aerated biological filter. *Environ. Technol., 0*, 1–11. https://doi.org/10.1080/09593330.2017.1385648 (accessed on 11 June 2020).

Amores, M. J., Meneses, M., Pasqualino, J., Antón, A., & Castells, F., (2013). Environmental assessment of urban water cycle on Mediterranean conditions by LCA approach. *J. Clean. Prod., 43*, 84–92. https://doi.org/10.1016/j.jclepro.2012.12.033 (accessed on 11 June 2020).

Babatunde, A. O., Zhao, Y. Q., & Zhao, X. H., (2010). Alum sludge-based constructed wetland system for enhanced removal of P and OM from wastewater: Concept, design and performance analysis. *Bioresour. Technol., 101*, 6576–6579. https://doi.org/10.1016/j.biortech.2010.03.066 (accessed on 11 June 2020).

Bakir, H., (2001). Water demand management and pollution control: Key to securing and safeguarding the water supplies of MENA in the 21. *Proc. 4th Gulf Water Conf. Water Sci. Technol. Assoc.* (pp. 1–20).

Bdour, A. N., Hamdi, M. R., & Tarawneh, Z., (2009). Perspectives on sustainable wastewater treatment technologies and reuse options in the urban areas of the Mediterranean region. *Desalination, 237*, 162–174. https://doi.org/10.1016/j.desal.2007.12.030.

Casas, Y., Rivas, A., López, D., & Vidal, G., (2017). Life-cycle greenhouse gas emissions assessment and extended exergy accounting of a horizontal-flow constructed wetland for municipal wastewater treatment : A case study in Chile. *Ecol. Indic., 74*, 130–139. https://doi.org/10.1016/j.ecolind.2016.11.014 (accessed on 11 June 2020).

Chen, R., & Wang, X. C., (2009). Cost-benefit evaluation of a decentralized water system for wastewater reuse and environmental protection. *Water Sci. Technol., 59*, 1515–1522. https://doi.org/10.2166/wst.2009.156 (accessed on 11 June 2020).

Corominas, L., Foley, J., Guest, J. S., Hospido, A., Larsen, H. F., Morera, S., & Shaw, A., (2013). Life cycle assessment applied to wastewater treatment: State of the art. *Water Res., 47*, 5480–5492. https://doi.org/10.1016/j.watres.2013.06.049 (accessed on 11 June 2020).

Domènech, L., (2011). Rethinking water management: From centralized to decentralized water supply and sanitation models. *Doc. d'AnàlisiGeogràfica, 57*(2), 293–310.

Dominguez, S., Laso, J., Margallo, M., Aldaco, R., Rivero, M. J., Irabien, Á., & Ortiz, I., (2018). LCA of greywater management within a water circular economy restorative thinking framework. *Sci. Total Environ., 621*, 1047–1056. https://doi.org/10.1016/j.scitotenv.2017.10.122 (accessed on 11 June 2020).

Dones, R., Bauer, C., Bolliger, R., Burger, B., Faist, E. M., Frischknecht, R., Heck, T., Jungbluth, N., & Röder, A. T. M., (2007). Life cycle inventories of energy systems: Results of current systems in Switzerland and other UCTE countries. *Ecoinvent Rep.,* 5. https://doi.org/10.1007/s11367-014-0838-7 (accessed on 11 June 2020).

Eriksson, O. N., & Bisaillon, M., (2004). How model-based systems analysis can be improved for waste management planning. *Waste Manag. Res., 21*, 488–500.

Ferguson, B. C., Brown, R. R., Frantzeskaki, N., De Haan, F. J., & Deletic, A., (2013). The enabling institutional context for integrated water management: Lessons from Melbourne. *Water Res., 47*, 7300–7314. https://doi.org/10.1016/j.watres.2013.09.045 (accessed on 11 June 2020).

Foley, J., De Haas, D., Hartley, K., & Lant, P., (2010b). Comprehensive life cycle inventories of alternative wastewater treatment systems. *Water Res., 44*, 1654–1666. https://doi.org/10.1016/j.watres.2009.11.031 (accessed on 11 June 2020).

Foley, J., De Haas, D., Hartley, K., & Lant, P., (2010a). Comprehensive life cycle inventories of alternative wastewater treatment systems. *Water Res., 44*, 1654–1666. https://doi.org/10.1016/j.watres.2009.11.031 (accessed on 11 June 2020).

Gallego, A., Hospido, A., Moreira, M. T., & Feijoo, G., (2008). Environmental performance of wastewater treatment plants for small populations. *Resour. Conserv. Recycl., 52*, 931–940. https://doi.org/10.1016/j.resconrec.2008.02.001 (accessed on 11 June 2020).

Garfí, M., Flores, L., & Ferrer, I., (2017). Life cycle assessment of wastewater treatment systems for small communities : Activated sludge, constructed wetlands, and high rate algal ponds. *J. Clean. Prod., 161*, 211–219. https://doi.org/10.1016/j.jclepro.2017.05.116 (accessed on 11 June 2020).

Gikas, P., & Tchobanoglous, G., (2009). The role of satellite and decentralized strategies in water resources management. *J. Environ. Manage., 90*, 144–152. https://doi.org/10.1016/j.jenvman.2007.08.016 (accessed on 11 June 2020).

Glick, S., & Guggemos, A., (2013). Rethinking wastewater-treatment infrastructure: Case study using life-cycle cost and life-cycle assessment to highlight sustainability considerations. *J. Constr. Eng. Manag., 139*(12).

Godin, D., Bouchard, C., & Vanrolleghem, P. A., (2012). Net environmental benefit: Introducing a new LCA approach on wastewater treatment systems. *Water Sci. Technol., 65*, 1624–1631. https://doi.org/10.2166/wst.2012.056 (accessed on 11 June 2020).

Goedkoop, M., Heijungs, R., Huijbregts, M., Schryver, A. De, Struijs, J., & Zelm, R. Van, 2009. ReCiPe 2008, a Life Cycle Impact Assessment Method Which Comprises Harmonised Category Indicators at the Midpoint and the Endpoint Level. Uni- versity of Leiden, Radboud University Nijmegen, RIVM, Bilthoven, Amersfoort, Netherlands.

Green, F. B., Bernstone, L. S., Lundquist, T. J., & Oswald, W. J., (1996). Advanced integrated wastewater pond systems for nitrogen removal. *Water Sci. Technol., 33*, 207–217. https://doi.org/10.1016/0273-1223(96)00356-3 (accessed on 11 June 2020).

Gual, M., Moià, A., & March, J. G., (2008). Monitoring of an indoor pilot plant for osmosis rejection and greywater reuse to flush toilets in a hotel. *Desalination, 219*, 81–88. https://doi.org/10.1016/j.desal.2007.02.033 (accessed on 11 June 2020).

Guinée, J., (2002). Handbook on life cycle assessment operational guide to the ISO standards. *Environ. Impact Assess. Rev., 23*, 129–130. https://https://doi.org/10.1016/S0195-9255(02)00101-4 (accessed on 11 June 2020).

Hazarika, S., & Pandit, D., (2019). *Feasibility Analysis of Decentralized Wastewater Treatment Systems.* Springer, Singapore. https://doi.org/10.1007/978-981-13-1966-2 (accessed on 11 June 2020).

Høibe, L., Clauson-Kaas, J., Wenzel, H., Larsen, H. F., Jacobsen, B. N., & Dalgaard, O., (2008). Sustainability assessment of advanced wastewater treatment technologies. *Water Sci. Technol., 58*, 963–968.

Hophmayer-Tokich, S., (2006). Wastewater management strategy: Centralized versus decentralized technologies for small communities. *CSTM-Reeks, 271*, 27.

Hospido, A., Sanchez, I., Rodriguez-Garcia, G., Iglesias, A., Buntner, D., Reif, R., Moreira, M. T., & Feijoo, G., (2012). Are all membrane reactors equal from an environmental point of view? *Desalination, 285*, 263–270. https://doi.org/10.1016/j.desal.2011.10.011 (accessed on 11 June 2020).

Igos, E., Benetto, E., Venditti, S., Kohler, C., Cornelissen, A., Moeller, R., & Biwer, A., (2012). Is it better to remove pharmaceuticals in decentralized or conventional wastewater treatment plants? A life cycle assessment comparison. *Sci. Total Environ., 438*, 533–540. https://doi.org/10.1016/j.scitotenv.2012.08.096 (accessed on 11 June 2020).

Igos, E., Besson, M., Navarrete, G. T., Bisinella, D. F. A. B., Benetto, E., Barna, L., Ahmadi, A., & Spérandio, M., (2017). Assessment of environmental impacts and operational costs of the implementation of an innovative source-separated urine treatment. *Water Res., 126*, 50–59. https://doi.org/10.1016/j.watres.2017.09.016 (accessed on 11 June 2020).

ISO 14040, (2006). *International Organization for Standardization, ISO 14040*. Environmental Management-Life Cycle Assessment-Principles and Framework, Geneve.

Jeong, H., Broesicke, O. A., Drew, B., & Crittenden, J. C., (2018). Life cycle assessment of small-scale grey water reclamation systems combined with conventional centralized water systems for the city of Atlanta, Georgia. *J. Clean. Prod., 174*, 333–342. https://doi.org/10.1016/j.jclepro.2017.10.193 (accessed on 11 June 2020).

Kabdaşlı, I., & Tünay, O., (2018). *Nutrient Recovery by Struvite Precipitation, Ion Exchange, and Adsorption from Source-Separated Human Urine: A Review 2515*. https://doi.org/10.1080/21622515.2018.1473504 (accessed on 11 June 2020).

Kaetzl, K., Lübken, M., Gehring, T., & Wichern, M., (2018). *Efficient Low-Cost Anaerobic Treatment of Wastewater Using Biochar and Woodchip Filters*. https://doi.org/10.3390/w10070818 (accessed on 11 June 2020).

Kampschreur, M. J., Van, D. S. W. R. L., Wielders, H. A., Mulder, J. W., Jetten, M. S. M., & Van, L. M. C. M., (2008). Dynamics of nitric oxide and nitrous oxide emission during full-scale reject water treatment. *Water Res., 42*, 812–826. https://doi.org/10.1016/j.watres.2007.08.022 (accessed on 11 June 2020).

Kavvada, O., Tarpeh, W. A., Horvath, A., Nelson, K. L., Kavvada, O., Tarpeh, W. A., Horvath, A., & Nelson, K. L., (2017). *Life-Cycle Cost and Environmental Assessment of Decentralized Nitrogen Recovery Using Ion Exchange from Source-Separated Urine Through Spatial Modeling*. https://doi.org/10.1021/acs.est.7b02244 (accessed on 11 June 2020).

Kivaisi, A. K., (2001). The potential for constructed wetlands for wastewater treatment and reuse in developing countries: A review. *Ecol. Eng., 16*, 545–560. https://doi.org/10.1016/S0925-8574(00)00113-0 (accessed on 11 June 2020).

Lam, L., Kurisu, K., & Hanaki, K., (2015). Comparative environmental impacts of source-separation systems for domestic wastewater management in rural China. *J. Clean. Prod., 104*, 185–198. https://doi.org/10.1016/j.jclepro.2015.04.126 (accessed on 11 June 2020).

Larsen, T. A., & Gujer, W., (2013). Implementation of source separation and decentralization in cities. In: *Source Separation and Decentralization for Wastewater Management* (pp. 135–150). IWA Publishing. London.

Lehtoranta, S., Vilpas, R., & Mattila, T. J., (2014). Comparison of carbon footprints and eutrophication impacts of rural on-site wastewater treatment plants in Finland. *J. Clean. Prod., 65*, 439–446. https://doi.org/10.1016/j.jclepro.2013.08.024 (accessed on 11 June 2020).

Libralato, G., Volpi, G. A., & Avezzù, F., (2012). To centralize or to decentralize: An overview of the most recent trends in wastewater treatment management. *J. Environ. Manage., 94*, 61–68. https://doi.org/10.1016/j.jenvman.2011.07.010 (accessed on 11 June 2020).

Longo, S., Frison, N., Renzi, D., Fatone, F., & Hospido, A., (2017a). Is SCENA a good approach for side-stream integrated treatment from an environmental and economic point of view? *Water Res., 125*, 478–489. https://doi.org/10.1016/j.watres.2017.09.006 (accessed on 11 June 2020).

Longo, S., Frison, N., Renzi, D., Fatone, F., & Hospido, A., (2017b). Is SCENA a good approach for side-stream integrated treatment from an environmental and economic point of view? *Water Res., 125*, 478–489. https://doi.org/10.1016/j.watres.2017.09.006 (accessed on 11 June 2020).

Lorenzo-Toja, Y., Alfonsín, C., Amores, M. J., Aldea, X., Marin, D., Moreira, M. T., & Feijoo, G., (2016a). Beyond the conventional life cycle inventory in wastewater treatment plants. *Sci. Total Environ., 553*, 71–82. https://doi.org/10.1016/j.scitotenv.2016.02.073 (accessed on 11 June 2020).

Lorenzo-Toja, Y., Vázquez-Rowe, I., Amores, M. J., Termes-Rifé, M., Marín-Navarro, D., Moreira, M. T., & Feijoo, G., (2016b). Benchmarking wastewater treatment plants under an eco-efficiency perspective. *Sci. Total Environ., 566, 567*, 468–479. https://doi.org/10.1016/j.scitotenv.2016.05.110 (accessed on 11 June 2020).

Lundin, M., Bengtsson, M., & Molander, S., (2000). Life cycle assessment of wastewater systems: Influence of system boundaries and scale on calculated environmental loads. *Environ. Sci. Technol., 34*, 180–186. https://doi.org/10.1021/es990003f (accessed on 11 June 2020).

Makropoulos, C. K., & Butler, D., (2010). Distributed water infrastructure for sustainable communities. *Water Resour. Manag., 24*, 2795–2816. https://doi.org/10.1007/s11269-010-9580-5 (accessed on 11 June 2020).

Mankad, A., & Tapsuwan, S., (2011). Review of socio-economic drivers of community acceptance and adoption of decentralized water systems. *J. Environ. Manage, 92*, 380–391. https://doi.org/10.1016/j.jenvman.2010.10.037 (accessed on 11 June 2020).

Massoud, M. A., Tarhini, A., & Nasr, J. A., (2009). Decentralized approaches to wastewater treatment and management: Applicability in developing countries. *J. Environ. Manage., 90*, 652–659. https://doi.org/10.1016/j.jenvman.2008.07.001 (accessed on 11 June 2020).

Matos, C., Pereira, S., Amorim, E. V., Bentes, I., & Briga-Sa, A., (2014). Wastewater and greywater reuse on irrigation in centralized and decentralized systems-An integrated approach on water quality, energy consumption, and CO_2 emissions. *Sci. Total Environ., 493*, 463–471. https://doi.org/10.1016/j.scitotenv.2014.05.129 (accessed on 11 June 2020).

Mattos, L., Cruz, D. O., Graziella, B., Araujo, L., & Tonetti, A. L., (2019). *Using Coconut Husks in a Full-Scale Decentralized Wastewater Treatment System : The Influence of an Anaerobic Filter on Maintenance and Operational Conditions of a Sand Filter, 127*, 454–459. https://doi.org/10.1016/j.ecoleng.2018.12.021 (accessed on 11 June 2020).

Meleg, A., (2012). SISAR: A sustainable management model for small rural decentralized water and wastewater systems in developing countries. *J. Water, Sanit. Hyg. Dev., 2*, 291. https://doi.org/10.2166/washdev.2012.007 (accessed on 11 June 2020).

Meuler, S., Paris, S., & Hackner, T., (2008). Membrane bio-reactors for decentralized wastewater treatment and reuse. *Water Sci. Technol., 58*, 285–294. https://doi.org/10.2166/wst.2008.356 (accessed on 11 June 2020).

Ng, B. J. H., Zhou, J., Giannis, A., Chang, V. W. C., & Wang, J. Y., (2014). Environmental life cycle assessment of different domestic wastewater streams: Policy effectiveness in a tropical urban environment. *J. Environ. Manage., 140*, 60–68. https://doi.org/10.1016/j.jenvman.2014.01.052 (accessed on 11 June 2020).

Nogueira, R., Brito, A. G., Machado, A. P., Janknecht, P., Salas, J. J., Vera, L., & Martel, G., (2009). Economic and environmental assessment of small and decentralized wastewater treatment systems. *Desalin. Water Treat, 4*, 16–21. https://doi.org/10.5004/dwt.2009.349 (accessed on 11 June 2020).

Opher, T., & Friedler, E., (2016). Comparative LCA of decentralized wastewater treatment alternatives for non-potable urban reuse. *J. Environ. Manage., 182*, 464–476. https://doi.org/10.1016/j.jenvman.2016.07.080 (accessed on 11 June 2020).

Parkinson, J., & Tayler, K., (2003). Decentralized wastewater management in peri-urban areas in low-income countries. *Environ. Urban, 15*, 75–90. https://doi.org/10.1177/095624780301500119 (accessed on 11 June 2020).

Partzsch, L., (2009). Smart regulation for water innovation-the case of decentralized rainwater technology. *J. Clean. Prod., 17*, 985–991. https://doi.org/10.1016/j.jclepro.2009.01.009 (accessed on 11 June 2020).

Pasqualino, J. C., Meneses, M., & Castells, F., (2011). Life cycle assessment of urban wastewater reclamation and reuse alternatives. *J. Ind. Ecol., 15*, 49–63. https://doi.org/10.1111/j.1530-9290.2010.00293.x (accessed on 11 June 2020).

Penn, R., Hadari, M., & Friedler, E., (2012). Evaluation of the effects of greywater reuse on domestic wastewater quality and quantity. *Urban Water J., 9*, 137–148. https://doi.org/10.1080/1573062X.2011.652132 (accessed on 11 June 2020).

Pérez-salazar, R., Mora-aparicio, C., Alfaro-chinchilla, C., Sasa-marín, J., Scholz, C., & Rodríguez-corrales, J. Á., (2019). Science of the total environment biogardens as constructed wetlands in tropical climate : A case study in the central Pacific coast of Costa Rica. *Sci. Total Environ., 658*, 1023–1028. https://doi.org/10.1016/j.scitotenv.2018.12.259 (accessed on 11 June 2020).

Remy, C., & Jekel, M., (2008). Sustainable wastewater management: Life cycle assessment of conventional and source-separating urban sanitation systems. *Water Sci. Technol., 58*, 1555–1562. https://doi.org/10.2166/wst.2008.533 (accessed on 11 June 2020).

Risch, E., Gutierrez, O., Roux, P., Boutin, C., & Corominas, L., (2015). Life cycle assessment of urban wastewater systems: Quantifying the relative contribution of sewer systems. *Water Res. 77*, 35–48. https://doi.org/10.1016/j.watres.2015.03.006 (accessed on 11 June 2020).

Rodriguez-Garcia, G., Frison, N., Vázquez-Padín, J. R., Hospido, A., Garrido, J. M., Fatone, F., Bolzonella, D., et al., (2014). Life cycle assessment of nutrient removal technologies for the treatment of anaerobic digestion supernatant and its integration in a wastewater treatment plant. *Sci. Total Environ., 490*, 871–879. https://doi.org/10.1016/j.scitotenv.2014.05.077 (accessed on 11 June 2020).

Rodriguez-Garcia, G., Molinos-Senante, M., Hospido, A., Hernández-Sancho, F., Moreira, M. T., & Feijoo, G., (2011). Environmental and economic profile of six typologies of wastewater treatment plants. *Water Res., 45*, 5997–6010. https://doi.org/10.1016/j.watres.2011.08.053 (accessed on 11 June 2020).

Schwaickhardt, R. D. O., Machado, Ê. L., & Lutterbeck, C. A., (2017). Science of the total environment combined use of VUV and UVC photoreactors for the treatment of hospital laundry wastewaters : Reduction of load parameters, detoxification, and life cycle assessment of different con figurations. *Sci. Total Environ., 590, 591*, 233–241. https://doi.org/10.1016/j.scitotenv.2017.02.218 (accessed on 11 June 2020).

Senzia, M. A., Mashauri, D. A., & Mayo, A. W., (2003). Suitability of constructed wetlands and waste stabilization ponds in wastewater treatment: Nitrogen transformation and removal. *Phys. Chem. Earth, 28*, 1117–1124. https://doi.org/10.1016/j.pce.2003.08.033 (accessed on 11 June 2020).

Sharma, A. K., Tjandraatmadja, G., Cook, S., & Gardner, T., (2013). Decentralized systems-Definition and drivers in the current context. *Water Sci. Technol., 67*, 2091–2101. https://doi.org/10.2166/wst.2013.093 (accessed on 11 June 2020).

Shehabi, A., Stokes, J. R., & Horvath, A., (2012). Energy and air emission implications of a decentralized wastewater system. *Environ. Res. Lett.*, 7. https://doi.org/10.1088/1748-9326/8/1/019001 (accessed on 11 June 2020).

Singh, N. K., Asce, S. M., & Kazmi, A. A., (2018). *Performance and Cost Analysis of Decentralized Wastewater Treatment Plants in Northern India : Case Study, 144*, 1–9. https://doi.org/10.1061/(ASCE)WR.1943-5452.0000886 (accessed on 11 June 2020).

Singh, P., Kansal, A., & Carliell-Marquet, C., (2016). Energy and carbon footprints of sewage treatment methods. *J. Environ. Manage., 165*, 22–30. https://doi.org/10.1016/j.jenvman.2015.09.017 (accessed on 11 June 2020).

Slagstad, H., & Brattebø, H., (2014). Life cycle assessment of the water and wastewater system in Trondheim, Norway: A case study. *Urban Water J., 11*, 323–334. https://doi.org/10.1080/1573062X.2013.795232 (accessed on 11 June 2020).

Tchobanoglous, G., Burton, F., & Stensel, H. D., (1998). Wastewater engineering: An overview. In: *Wastewater Engineering Treatment and Reuse* (pp. 1–24). https://doi.org/10.1016/0309-1708(80)90067-6 (accessed on 11 June 2020).

Thibodeau, C., Monette, F., Bulle, C., & Glaus, M., (2014). Comparison of black water source-separation and conventional sanitation systems using life cycle assessment. *J. Clean. Prod., 67*, 45–57. https://doi.org/10.1016/j.jclepro.2013.12.012 (accessed on 11 June 2020).

Venhuizen, D., (1997). Paradigm shift. *Water Environ. Technol., 9*, 49–54.

Weidema, B. P., Bauer, C., Hischier, R., Nemecek, T., Reinhard, J., Vadenbo, C. O., & Wernet, G., (2013). *Overview and Methodology: Data Quality Guideline for the Ecoinvent Database version 3*. Swiss Cent. Life Cycle Invent. St. Gall.

Wilderer, P. A., & Schreff, D., (2000). Decentralized and centralized wastewater management: A challenge for technology developers. *Water Sci. Technol., 41*(1), 1–8.

Wilderer, P. A., (2004). Applying sustainable water management concepts in rural and urban areas: Some thoughts about reasons, means, and needs. *Water Sci. Technol., 49*(7), 7–16.

Wu, H., Zhang, J., Ngo, H. H., Guo, W., Hu, Z., Liang, S., Fan, J., & Liu, H., (2015). A review on the sustainability of constructed wetlands for wastewater treatment: Design and operation. *Bioresour. Technol., 175*, 594–601. https://doi.org/10.1016/j.biortech.2014.10.068 (accessed on 11 June 2020).

Yerri, S., & Piratla, K. R., (2019). *Resources, Conservation, and Recycling Decentralized Water Reuse Planning : Evaluation of Life Cycle Costs and Benefits, 141*, 339–346. https://doi.org/10.1016/j.resconrec.2018.05.016 (accessed on 11 June 2020).

Zhang, R., Tang, M., & Pu, Y., (2018). *Performance and Microbial Community of Novel Three-Stage Waterfall Aeration Grooves Biofilm Reactor for Treating Decentralized Wastewater in Rural Areas, 00*, 1–8. https://doi.org/10.1089/ees.2018.0018 (accessed on 11 June 2020).

CHAPTER 10

Environmental Impacts of Wastewater Treatment Alternatives for Small Communities

MUSTAFA YILDIRIM[1] and BÜLENT TOPKAYA[2]

[1]*Antalya Water and Wastewater Administration General Directorate (ASAT), Antalya, Turkey, E-mail: yildirim.mustafa@gmail.com*

[2]*Akdeniz University, Department of Environmental Engineering, Antalya, Turkey, E-mail: btopkaya@akdeniz.edu.tr*

10.1 INTRODUCTION

The main purpose of the wastewater treatment systems is to minimize the adverse impacts of the wastewater effluent on the receiving environment. Wastewater treatment systems have advanced significantly within the last 150 years since the introduction of the first cesspits and filtration in the mid-18th century. Scientific discoveries parallel with increasing environmental awareness as well as social, economic, and legislative interests have led to enhanced solutions. For instance, the clean water act (CWA) in 1972 mandated secondary treatment in addition to the primary treatment step. Following the establishment of primary and secondary treatment steps, advanced treatment technologies were developed as the prevention of eutrophication became the next goal for wastewater treatment. Depending on the quality of the receiving waters, many treatment plants are required to remove nitrogen, phosphorous, or both. The first studies carried out by Downing et al. in 1964 are incorporated into design methods of biological nitrification (Lofrano et al., 2010).

Environmental regulations are becoming more stringent in recent years. On the other side, each measure brings additional environmental burdens in the form of resource consumption. The treatment options developed for each need, have different performance characteristics and environmental impacts during construction, operation, or maintenance phases.

Pollution load of the effluent and the discharge restrictions of the receiving water body play an important role on the design of the wastewater treatment plants (WWTP). The Urban Wastewater Treatment Directive (UWWTD) of the European Union (EU) of 1991 differentiates the receiving environment as "sensitive" and "less sensitive" which increases the responsibility of the respective plant. Accordance with the existing regulations, the specific costs of the process are also important.

Due to the technical developments, there are several different options for the solution to pollution. In addition to these technical/financial parameters, environmental impacts of the processes with similar performance can also be incorporated into the decision making process. Life cycle assessment (LCA) methodology offers an opportunity to evaluate the environmental performances of the respective treatment options (Jensen et al., 1997; Machado et al., 2007).

LCA methodology was first applied in the 1960s. The establishment took place in the years 1980–1990. The first publications on the evaluation of the environmental impacts of a different configuration of WWTP using LCA methodology were published in the mid-1990s (Emmerson et al., 1995; Ødegaard, 1995; Dalemo, 1995; Tillman et al., 1996; Roeleveld et al., 1997; Sonesson et al., 1997; Mels et al., 1998).

A summary of studies conducted on the environmental impacts of WWTP is conducted by Yildirim et al. (2012). According to this study, Hospido investigated the environmental performance of the municipal WWTP in dry and humid season from a life cycle aspect (Hospido et al., 2004). Zhang analyzed the environmental performance of a conventional activated sludge treatment (AST) plant considering the construction, operational, and demolition phases (Zhang et al., 2010).

Foley, in his review, assessed 10 different treatment systems using LCA-Impact 2002+ (Foley et al., 2010). Pasqualino investigated the operational improvement of a municipal WWTPs in an environmental perspective by using LCA as a decision support tool (Pasqualino et al., 2009).

Sludge treatment is an important part of the WWTPs and meanwhile there are several treatment options of the produced sludge. The alternatives were also evaluated using LCA procedure (Neumayr et al., 1997; Suh et al., 2002; Houillion et al., 2005; Hospido et al., 2005). Munoz et al. in 2005 investigated specific treatment options.

The common results of the LCA studies on the WWTP show that the environment impacts of the treatment mostly stem from indirect emissions that releasing during production of consumed energy and chemicals

(Zhang et al., 2000; Hospido et al., 2008; Gallego et al., 2008; Zhang et al., 2010;Pasqualino et al., 2010; Li et al., 2013; Ontiveros et al., 2013).

In the framework of the UWWTD (91/271/EEC), for the wastewater produced in communities of more than 2000 p.e., the secondary treatment or an equivalent treatment options should be applied. On the other hand, for communities with less than 2000 p.e. only "appropriate treatment" is foreseen in case of discharges to sensitive water environments.

As in recent years, due to economic concerns, the municipalities tend to low-cost and energy saving systems, eco-technological, and economical solutions, such as natural systems, gained increasing interest. Vegetated land treatment (VLT), constructed wetland (CW), and stabilization ponds (SP) are among these systems (Yildirim et al., 2012).

There are several studies discussing the environmental performance of natural systems (Dixon et al., 2003; Sovik et al., 2006; Ström et al., 2007; Machado et al., 2007; Yildirim et al., 2012; Molinos-Senante et al., 2014; Gkika et al., 2015; De Feo and Ferrera, 2017; Diniz et al., 2017; Corbella et al., 2017; Lutterbeck et al., 2017; Kamble et al., 2017; Garfi et al., 2017). Although there are some contradictions, common views of the studies are that holistic environmental impacts of the natural-based systems are better than the energy-dependent ones; however, land occupation of these systems is a major concern in terms of not applicable in large scale.

In this study, wastewater treatment options, such as VLT, CWs, rotating biological contactor (RBC), conventional AST, membrane bioreactor (MBR), extended aeration (EA) and SP which can be used for the treatment of municipal wastewater from residential centers with less than 2000 p.e, are evaluated by using the LCA approach. The reason for the selection of natural-based treatment systems are the simplicity of the natural systems and the ability to be constructed with local possibilities in a very cost-effective way. In addition, energy-dependent systems are widely used in the world, and knowledge for their construction and operation exists.

10.2 MATERIALS AND METHODS

10.2.1 WASTEWATER TREATMENT ALTERNATIVES

A total of nine treatment alternatives were evaluated in this study. The short description and characteristics of these systems are given in the following sections. General information about the treatment alternatives is summarized in Table 10.1.

TABLE 10.1 General Information About Treatment Alternatives

	Treatment Alternatives								
	VLT	**CW-LS**	**CW-S**	**RBC**	**AST-LS**	**AST-S**	**SP**	**MBR**	**EA**
Land requirement	High (more than 20 m^2/p.e.)	Mid-high (about 5 m^2/p.e.)	High (about 9 m^2/p.e.)	Low (less than 1 m^2/p.e.)	Low (less than 1 m^2/p.e.)	Low (less than 1 m^2/p.e.)	Mid-high (about 2–4 m^2/p.e.)	Low (less than 1 m^2/p.e.)	Low (less than 1 m^2/p.e.)
Energy requirement	Very low (resulting from pumping, if needed)	Very low (resulting from pumping and main–taining)	Very low (resulting from pumping and maintaining)	Mid-Low (from shaft rotors and pre-treatment)	High	High	Low	Very high	Very high
Initial investment costs	Low (only occupied land cost)	Low (occupied land cost and cons–truction costs)	Low (occupied land cost and construction costs)	Mid (reinforced concrete structures, raw material requirements such as HDPE disks)	Mid-high (reinforced concrete structures, raw material require–ments and equipment costs)	High (reinforced concrete structures, raw material require–ments, equipment costs, phosphorus preci–pitation unit costs)	Mid-low (occupied land cost, raw material costs such as HDPE and con–struction costs)	Very high (reinforced concrete structures, membrane unit costs, raw material and equipment costs)	High (reinforced concrete structures especially bioreactor struc., equipment, and raw mat. costs.

	Treatment Alternatives								
	VLT	**CW-LS**	**CW-S**	**RBC**	**AST-LS**	**AST-S**	**SP**	**MBR**	**EA**
Recommended capacity	Less than 200 p.e. but it depends on the presence of suitable land	Less than 1000 p.e. but it depends on the presence of suitable land	Less than 500 p.e. but it depends on the presence of suitable land	Less than, 100 p.e	N/A	N/A	Less than 1000 p.e. but it depends on the presence of suitable land	N/A	N/A
Reuse of treated water	No effluent	Providing it is used for secondary treatment	Providing it is used for secondary treatment	Not suitable	Not suitable	It depends on disinfection level of the effluent	Not suitable	Suitable	It depends on disinfection level of the effluent

10.2.1.1 DETAILS OF VEGETATED LAND TREATMENT (VLT)

VLT is a basic treatment method in which wastewater is introduced to a vegetated area at a controlled rate. The methodology of land treatment is based on three methods: Slow rate, overland flow, and rapid filtration (US-EPA, 2006). The main aim, removal of the pollutants is achieved through the interaction of the plant root and soil matrix. Also, soil attenuation capacity has a significant role in the removal process (de Miguel et al., 2014). The main parts of the system are a primary settling tank and a vegetated area where water can infiltrate, evaporate, and can be uptaken by the plants (Madison et al., 1993; Kruzic, 1994).

In this study "slow-rate" land, treatment is evaluated. As summarized in Yildirim et al. (2012), agricultural crops (e.g., perennial grasses) and forests (e.g., *Eucalyptus camaldulensis*) are the most used vegetation types (Machado et al., 2007). This vegetation has individual advantages and disadvantages; for example, agricultural crops have greater nitrogen loadings, but a shorter application period. Forested systems have a longer application period and higher hydraulic loadings (McKim, 1982; Reed et al., 1991; Nuttler et al., 1996). The other advantages of this system are their low construction/operation/maintenance costs, low energy demand and operation dexterity (Martinez-Hernandez et al., 2018). The basic characteristics of VLT system used in this study are outlined in Table 10.2. Detailed information about hydrology, design criteria, and removal mechanisms of the VLT are discussed in detail in the study by Kruzic in 1994, US-EPA in 2006, and Crites et at. in 2014.

TABLE 10.2 Characteristics of Vegetated Land Treatment

Feature	Slow Rate
Minimum pretreatment	Primary sedimentation
Hydraulic loading rate (Lh)	0.5–6 m/a
Field area requirement	60–740 m^2/m^3/day
Disposition of applied wastewater	Evapotranspiration and percolation
Application techniques	Spray irrigation, surface or drip flow
Effluent discharge/receiving environment	No effluent is produced

Source: Taken from: Yildirim et al. (2012).

10.2.1.2 DETAILS OF CONSTRUCTED WETLAND (CW)

CW is also a natural treatment system evaluated in this study. They are designed to imitate natural wetlands in which plants and microorganisms are used to treat the incoming wastewater. It is an important eco-technological system for controlling water quality (Peterjohn et al., 1984; Kadlec et al., 1996; Kuusements et al., 1999; Teiter et al., 2005). Since 1950s, CWs have been used for reclamation of anthropogenic discharge such as wastewater, urban, and rural runoff, sewage treatment, and mining activities because of their nutrient capturing capacity, simplicity, low construction/operation/maintenance costs, low energy demand, process stability, little excess sludge production, effectiveness, and potential for creating biodiversity (Kaygusuz, 2004; Stefanakis et al., 2014;Tsihrintzis, 2017). CW can be divided into two major groups according to the flow pattern: Free water surface (FWS) and subsurface flow (SSF) (Mahmood et al., 2014). FWS type CWis evaluated in this study.

FWS type CW typically consists of a septic tank for the removal of primary sludge and basin or channels with a barrier to prevent seepage, soil to support the roots of the plants, and water at a relatively shallow depth flowing through the system (US-EPA, 1993). In FWS, the water surface is exposed to atmosphere. Although several macrophyte types can be found in FWS, emergent aquatic vegetations are dominant. The water depth ranges from a few centimeters to 0.8 m (Hammer, 1989; US-EPA, 1999; Kaygusuz, 2004; Crites et at., 2014). The basic characteristics of FWS-CW used in this study are outlined in Table 10.3. Detailed information about CW can be found in (Brix, 1993; US-EPA, 1993; Kadlec et al., 1996; US-EPA, 1999; Crites et at., 2014).

10.2.1.3 DETAILS OF ACTIVATED SLUDGE TREATMENT SYSTEMS (AST AND EA)

AST and EA are well-known and commonly used biological treatment method for municipal wastewater. AST and EA can be defined as the aeration of wastewater combined with organisms in order to develop microbiological flocs in which the organic constituents in wastewater are utilized. Treatment of wastewater is completed after settling of the flocs in the final clarifier and disinfection of effluent. In the case of discharging the effluent to a sensitive area, it may be necessary to add a chemical treatment unit for the removal of nutrients. The main difference of the EA from AST is that EA has the higher hydraulic retention time (HRT) and solids retention time (SRT) than the other. Additional information about activated sludge systems can be found in Medcalf et al. in 2003.

TABLE 10.3 Characteristics of the Constructed Wetland Systems

Flow Pattern	Free Water Surface (FWS)
Minimum pre-treatment	Primary sedimentation (septic tanks)
Annual loading rate	9–19 m/a
Field area requirement	20–140 m²/m³/d
Water depth	<10 cm (warm months)
	<45 cm (cold months)
Hydraulic retention time	5–14 d
Configuration	Multiple cells in series/parallel
Effluent discharge/receiving environment	Less sensitive water environments
	Sensitive water environments

Source: Kadlec et al. (1996); US-EPA (1999); Yildirim et al. (2012).

10.2.1.4 DETAILS OF ROTATED BIOLOGICAL CONTACTOR (RBC)

RBC is an attached growth treatment process consists of high-density polyethylene (HDPE) or polyvinyl chloride (PVC) disks, where microbiota can attach, mounted on a horizontal shaft. In this system, disks are partially submerged. During rotating the shaft, oxygen is transferred via adsorption by contacting air and bacteria, and then when the biomass passes through the liquid phase, nutrients, and organic pollutants can be absorbed. Once the biomass is overgrown, it is ruptured by the shearing forces exerted as the discs rotate, combined with the force of gravity, and excess biomass is removed in the clarifier. The basic characteristics of RBC used in this study are outlined in Table 10.4. Detailed information about RBC can be found in Clark in 1978 and Metcalf et al. in 2003.

TABLE 10.4 Typical Design Characteristics of Rotating Biological Contactors

Parameter	Unit	Treatment Level*		
		BOD Removal	BOD Removal and Nitrification	Separate Nitrification
Hydraulic loading	$m^3/m^2.d$	0.08–0.16	0.03–0.08	0.04–0.10
Organic loading	$gr.sBOD/m^2.d$	4–10	2.5–8	0.5–1.0
	$grBOD/m^2.d$	8–20	5–16	1–2
Max. 1st-stage org. loading	$gr.sBOD/m^2.d$	12–15	12–15	–
	$grBOD/m^2.d$	24–30	24–30	–

TABLE 10.4 *(Continued)*

Parameter	Unit	Treatment Level*		
		BOD Removal	BOD Removal and Nitrification	Separate Nitrification
NH_3 loading	grN/m².d	–	0.75–1.50	–
Hydraulic retention time	h	0.7–1.5	1.5–4	1.2–3
Effluent BOD	mg/L	15–30	7–15	7–15
Effluent NH_4-N	mg/L	–	<2	1–2

*Wastewater temperature above 13°C (55°F)

Source: Metcalf et al. (2003).

10.2.1.5 DETAILS OF STABILIZATION POND (SP)

SP is preferred in small communities for wastewater treatment. It consists of an open basin enclosed by soil (generally clay) embankments lined with HDPE, concrete or synthetic geofabrics (Verbyla et al., 2017). Since natural processes are imitated in this treatment, requirements of equipments and operation dexterity are low. However, reaction rate of the removal mechanism is slow because of no intervention. Therefore, very large areas are needed to reach desired treatment level (Arceivala, 1998; Qasim, 1999). Also, biological activities are easily affected by climate condition and effluent quality may be variable. The most common types of SPs are anaerobic ponds, facultative ponds, maturation or polishing ponds, aerated ponds, and high-rate algal ponds. Facultative ponds that are the most commonly applied are evaluated in this study. Detailed information about SPs can be found in von Sperling in 2007. The basic characteristics of SP used in this study are outlined in Table 10.5.

TABLE 10.5 The Basic Characteristics of Stabilization Ponds

Parameter	–
HRT, d	15–45
Water depth, m	1.5–2.0
L/B ratio	2 to 4
BOD_5 loading rate kg/ha.d	5–120
Soluble BOD_5 removal, %	85–95
Total BOD_5 removal, %*	70–90
COD, %	65–80

TABLE 10.5 *(Continued)*

Parameter	–
Ammonia, %	<50
Nitrogen, %	<60
Phosphorus, %	<35
Coliforms, %	90–99
Algae concentration, mg/L	20–80
Effluent TSS, mg/L	40–100

* Effluent BOD_5 removal efficiency is sometimes low due to algae concentration in the effluent.

Source: Von Sperling (2007).

10.2.1.6 DETAILS OF MEMBRANE BIOREACTOR (MBR)

MBR consist of a bioreactor that includes anaerobic, anoxic, or aerobic suspended growth and a microfiltration (MF) membrane pore size of which is 0.1–0.4 μm. MF membrane helps to se parate the sludge from the effluent by applying pressure (Brindle et al., 1996; Van Dijk et al., 1997; Metcalf et al., 2003; Lee et al., 2003). A significant advantage of the MBRs can be operated in high MLSS concentration (above10,000 mg/L). Also, the sludge can be retained in the bioreactor. Therefore, SRT, and HRT in the bioreactor can be controlled independently. HRT is very short and occupied area requirement of the MBR is low.

The major disadvantages of MBRs are initial investments, operation costs, and membrane fouling which significantly reduces membrane performance and lifespan, resulting in a significant increase in maintenance and operating costs. Detailed information about MBR can be found in Stephenson et al. (2000). The basic characteristics of MBR used in this study are outlined in Table 10.6.

TABLE 10.6 The Basic Characteristics of Membrane Bioreactors

Parameter	Unit	Range
Operation Data		
COD loading	kg/m^3d	1.2–3.2
MLSS	mg/L	5000–20,000
MLVSS	mg/L	4000–16,000
F/M	gCOD/gMLVSS.d	0.1–0.4
SRT	d	5–20

TABLE 10.6 *(Continued)*

Parameter	Unit	Range
HRT	h	4–6
Flux	L/m^2d	600–1100
Applied Vacuum	kPa	4–35
Dissolved oxygen	mg/L	0.5–1.0
Performance Data		
Effluent BOD	mg/L	<5
Effluent COD	mg/L	<30
Effluent NH_3	mg/L	<1
Effluent TN	mg/L	<10
Effluent turbidity	NTU	<1

Source: Adham et al. (2004); Hospido et al. (2012).

10.2.2 LCA PROCEDURE

LCA is a methodology that corresponds to the demands for selecting the best environmental friendly solution (Jensen et al., 1997). A LCA study includes four phases: goal and scope definition, inventory analysis, impact assessment, and interpretation. In the goal and scope definition phase, the purpose of the study, the functional unit (FU) to which all data/calculations is referred, and system boundaries are defined. The assumptions used in the study are also reported in this phase. The inventory analysis includes data collection and calculation procedures to quantify relevant inputs and outputs of a product system. These inputs and outputs may consist of resource consumption and releases to air, water, and soil-related to the system. In the impact assessment phase, the collected and processed data in the inventory analysis are evaluated by a systematic procedure and checking is realized to ascertain whether all requirements described in the goal and scope phase could be met or not.

The environmental impact assessment comprises of three phases: classification, characterization, and normalization. In the classification step, all released emissions are distributed into impact categories according to their environmental effects. Subsequently, emissions within each impact category are aggregated by using characterization factors that compared to the effect of a specific emission with a reference (Pre-Consultants, 2006).

Environmental impacts of the considered treatment alternatives are compared with each other with SimaPro 7.1 LCA software. The inventory data for treatment alternatives are obtained from *in-situ* measurements and laboratory analysis, by calculations and from the EcoInvent V.2 database, in which data from various treatment systems exists (Doka, 2009). For the impact assessment phase CML baseline 2000, which is developed by the Center of Environmental Science of Leiden University is used. The data derived from the EcoInventV.2 database are processed with CML baseline 2000 method and implicated by SimaPro 7.1 software (Yildirim et al., 2012). Selected impact categories in the evaluation of the scenarios are abiotic depletion (AD), acidification (A), eutrophication (E), global warming (GW), and toxicity (T).

In this section, the calculations of each system will be presented. The characteristics of raw wastewater as well as the discharge criteria are assumed to be the same for all alternatives (Table 10.7).

Principal assumptions and calculations based upon for each system are summarized below:

- The characteristics of raw wastewater as well as the discharge criteria are the same for all alternatives (Table 10.7).
- In VLT, wastewater is used for irrigation and no effluent is produced.
- Effluents from the CW, RBC, SP, and AST are discharged into water environments which are classified as "less sensitive" according to UWWT Directive (91/271/EEC). The main purpose is the removal of organic material.
- Effluents from the CW, AST, EA, and MBR are discharged alternatively into "sensitive" classified receiving water environments. In this case, according to the UWWT Directive, especially phosphorus should be removed in addition to the removal of organic material.
- Within the system boundaries of the LCA study all emissions (direct and indirect) during construction and the operation phase of the systems in a 20-year period are considered. Disassembly phase of the alternatives is excluded.
- In order to be able to compare the impacts of the systems, a FU is necessary. The FU, to which all emissions, material, and energy consumption data will be referred, is chosen as 1 population equivalent (p.e).
- The data required for assessing the impacts are obtained both from databases, literature, and existing treatment applications as well as from own calculations.

The details of the inventory data and assumptions regarding the treatment alternatives developed in this study are summarized below.

TABLE 10.7 Characteristics of Wastewater (91/271/EEC)

Parameter	Influent (mg/l)	Discharge Criteria According to UWWT Directive (mg/l)	
		Less Sensitive Areas	Sensitive Areas
COD	750	125	125
BOD_5	350	25	25
TSS	150	35	35
Tot P	10	–	1–2
Tot N	40	–	10–15

10.2.2.1 VEGETATIVE LAND TREATMENT (VLT)

In VLT systems wastewater is introduced to the area following the pre-treatment step in the septic tank (Figure 10.1). As originally evaluated in Yildirim et al. in 2012, VLT serves a population equivalent of 100 p.e.—based on a specific water consumption of 210 l/p.e.d.; the flow rate is calculated as 21 m^3/d. *Eucalyptus camaldulensis* which is well adapted to wastewater conditions and has high nutrient uptake capacity is chosen as biomass (Machado et al., 2007; Bhati et al., 2003).The nutrient uptake and hydraulic loading capacity of *Eucalyptus camaldulensis* have been increasing over the operation years together with improvement in their biomass (Tzanakakis et al., 2009). The surface area is calculated according to the annual hydraulic load that is determined by using the percolation rate of the soils and the evapotranspiration rate. These parameters are calculated according to the procedure described in US-EPA in 2006. The surface area required for the removal of the organic material is 2000 m^2 (calculated as 2026 m^2). The summary of the design parameters and calculations are given in Table 10.8. The flow chart of VLT is shown in Figure 10.1.

TABLE 10.8 Vegetated Land Treatment Design Parameters

Septic Tank Performance	
Volume	26.25 m^3 (hydraulic retention time: 30 hr)[1]
BOD_5 removal efficiency	50%[2]
Tot-P removal efficiency	15% [2]
Annual accumulated sludge amount*	64.9 L/p.e.yr. (80% Volatile Suspended Solids)
	34.4 kg/p.e.yr. stabilized sludge

TABLE 10.8 *(Continued)*

Septic Tank Performance	
VLT	
Flow rate (Q)	21 m^3/d = 7665 m^3/yr.
Evapotranspiration rate of *Eucalyptus camaldulensis*(ETc)	45 cm/yr.[1]
Precipitation (*P*)	500 cm/yr.[1]
Percolation rate (*Pr*)	15 cm/d[1]
Deep Percolation Rate (*Pw*)*	383.25 cm/yr.
Hydraulic Loading Rate (*Lh*)**	378.25 cm/yr.
Area requirements (A)***	2026 m^2≈2000 m^2

*$Pw = Pr \times Sf$ = 15 cm/d x 0.07 x 365 d/yr. = 383.25 cm/yr.; *Sf: 0,07 safety factor*) (US-EPA, 2006).

**$Lh = ETc - P + Pw$ (US-EPA, 2006).

***$A = Q/Lh$(US-EPA, 1999).

[1](Sunder, 1993).

Source: Taken from: Yildirim et al. (2012).

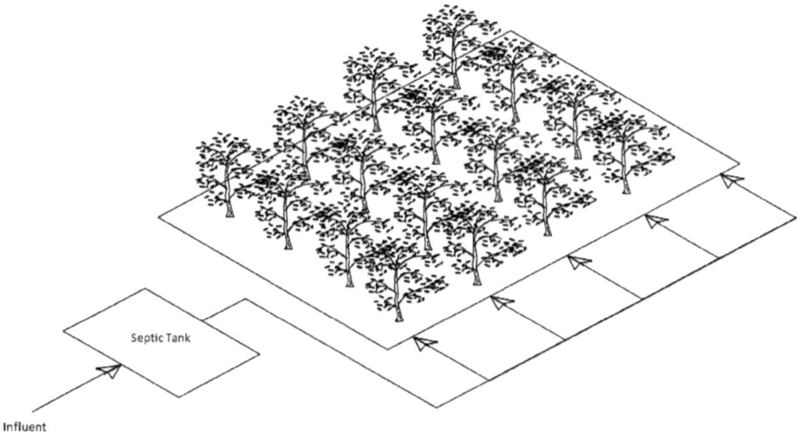

FIGURE 10.1 The flow diagram of VLT.

10.2.2.2 CONSTRUCTED WETLAND (CW)

This alternative is also evaluated originally in Yildirim et al. in 2012. The FWS type CW system also serves 100 p.e., and the flow rate is 21 m^3/d. Macrophytes in the wetland systems are selected as *Phragmites australis* (Machado et al.,

2007). It is assumed that these plants will be harvested every five years and disposed in the sanitary landfill. Annual growth of the macrophytes is assumed as 3 kg/m^2 and annual disposed green waste is calculated as 28 kg/p.e. (Vymazal et al., 2005). The capacity of the CW shows differences according to the required removal efficiency and specifications of the receiving water environment (less-sensitive/sensitive). For discharges into less sensitive areas, removal of organics will be sufficient and in this case, the required area of the CW will be less. In the case of sensitive areas, phosphorus compounds should be removed before discharging; therefore, the required area of the CW will increase substantially. For the design of the wetlands, the plug flow model suggested by Kadlec et al. in 1996 is used. The system components and the calculations are as follows: The two CW systems (for less-sensitive and sensitive areas) include four-cell septic tanks of total 25 m^3 volume (calculated as 26.25 m^3) as pre-treatment. Sludge accumulated in the septic tank is removed every year and is disposed in the sanitary landfill. Sludge accumulation is calculated as described in the study by Grayin (1995). According to this study, average annual sludge accumulation in the septic tank will be 64.9 L/p.e. In case 80% of this accumulated sludge will consist of volatile suspended solids (VSS), the amount of annual accumulated sludge is calculated as 34.4 kg/p.e. Regarding the wetland design in Table 10.3, it is foreseen that for discharges into less sensitive areas only organics will be removed, whereas for discharges into sensitive areas additional phosphorus removal will take place. It is assumed that in the pre-treatment unit 50% of the BOD_5 and 15% of Total Phosphorus in the influent will be removed. The calculation of the background concentration for BOD5 and Tot-P is given in Table 10.9.

Based on these data, wetland areas required for removing only organic compounds (discharges into less sensitive water bodies) is calculated as 573 m^2. On the other hand, the required area of wetlands planed for phosphorus removal is calculated as 933 m^2. This area is divided into six sub-wetlands in terms of operational concerns. The summary of the design parameters and calculations are given in Table 10.9. The flow diagram of the CW system is illustrated in Figure 10.2.

10.2.2.3 ACTIVATED SLUDGE TREATMENT (AST)

In the framework of this study, two alternative AST systems are investigated. The flow schemes of these systems are adapted from EcoInvent V.2 database. The first alternative is a conventional AST system consisting

TABLE 10.9 Constructed Wetland Design Parameters

Flow rate (Q)	21 m^3/d	
Annual growth rate of *Phragmites australis*	3 kg/m^2 a[1]	
Wetland Design		
	CW Design Criteria	
	For Less Sensitive	**For Sensitive**
Effluent Concentration from Septic Tank (C_i)		
(Influent concentration to wetland after pre-treatment)	175 mg/L BOD_5	8.5 mg/L Tot-P
Effluent Concentration from Wetland (C_e)	25 mg/L BOD_5	2 mg/L Tot-P
Background Concentration in Wetland (C_b)**	12.51 mg/L BOD_5	0.15 mg/L Tot-P
First Order Area Rate Constant (k_A)	0.093 m/d (for BOD_5; 20°C)	0.033 m/d (for Tot-P; 20°C)
Area Requirements (A)***	573 m^2	933 m^2

**Accumulated Sludge in septic tank: 64.9 L/p.e.yr * 0.20 * 2.65 kg/L = 34.397 kg/p.e.yr ≈34.4 kg/p.e.yr,* (Gray, 1995).

** *Wetland systems include various autotrophic and heterotrophic activities. The autotrophic ones can excess produced carbonaceous material that is buried as peat or is exported downstream to the next system. This production results in an internal release of particulate and dissolved biomass to the wetland water column, which is measured as non-zero of BOD, TSS, Tot-N, and Tot-P (background concentration)* (Vymazal et al., 2008).

*** *Plug-Flow k-C* model:* $A = -(Q/k_A).\ ln\ [(C_e - C_b)/(C_i - C_b)]$, (Kadlec et al., 1996).

[1](Gray, 1995).

[2](Bitton, 1994).

Source: Taken from: Yildirim et al. (2012).

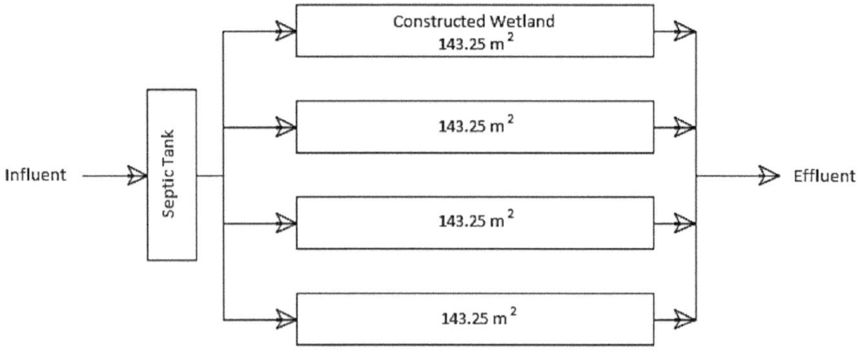

FIGURE 10.2 (A) The flow diagrams of CWs (less-sensitive).

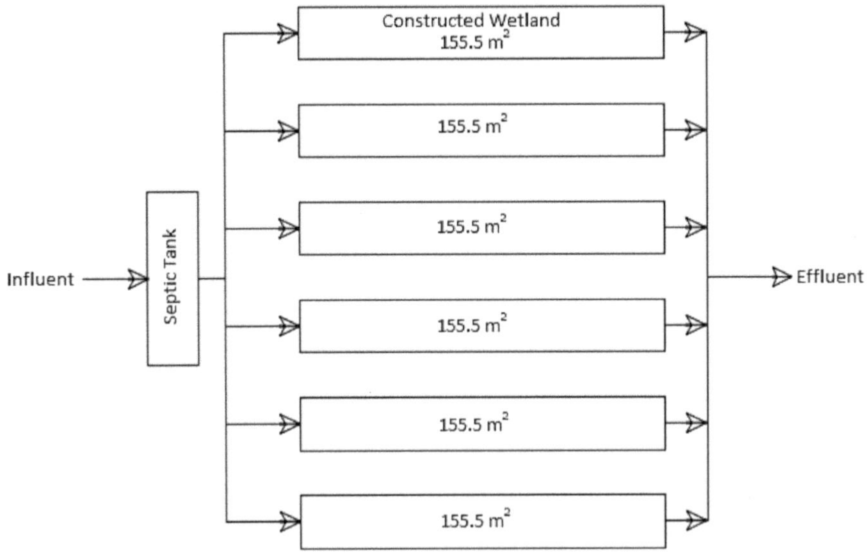

FIGURE 10.2 (B) The flow diagrams of CWs (sensitive).

of mechanical and biological treatment units including sludge digestion. In this alternative, biogas obtained from the digesting unit is used for producing electricity energy. Stabilized sludge in the digester unit is applied to agricultural soils. Wastewater effluent is discharged to a less sensitive area.

In the second alternative, nutrient removal is aimed prior to discharging of the treated water to sensitive areas. In order to comply with the requirements of the UWWT Directive, a chemical phosphorus precipitation unit is added.

All emissions, material, and energy consumption data for these two treatment alternatives, are obtained from EcoInvent V.2 database. It is reported that these inventory data are gathered from evaluations of 428 small scale treatment facilities (<2000 p.e.), by which the average capacity was 806 p.e. Flow diagrams of the AST systems are shown in Figure 10.3.

10.2.2.4 ROTATING BIOLOGICAL CONTACTORS (RBC)

RBC system considered in this study serves 2000 p.e. and the flow rate is 420 m^3/s. RBC is designed according to Hartman in the 1974 approach for removing BOD. The required disc area is calculated as 5547.5 m^2 and if 3 m. diameter

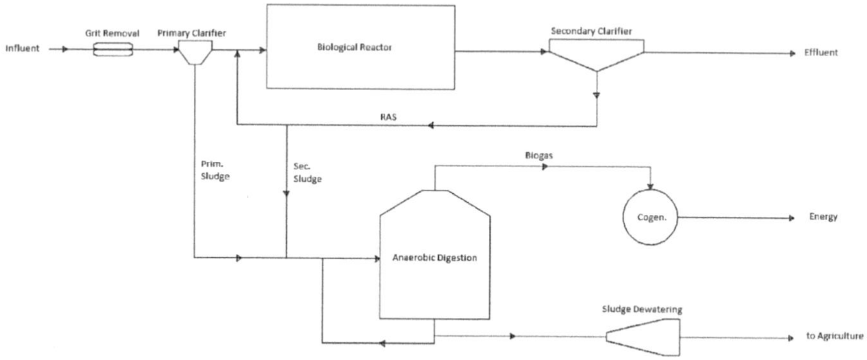

FIGURE 10.3 (A) The flow diagrams of ASTs (less-sensitive).

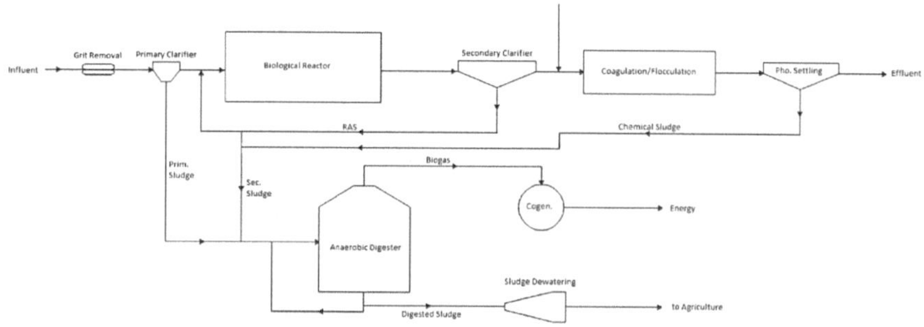

FIGURE 10.3 (B) The flow diagrams of ASTs (sensitive).

discs are used, 373 discs are needed. But, the number of discs is selected as 400 for ease for handling. It is accepted that RBC includes four parallel shafts and 100 discs are fixed on each of shafts. It is calculated that 25.44 t HDPE (disc thickness is 1 cm, density of HDPE is 0.96 t/m^3) and 1.81 t steel (shaft thickness is 2 cm, density of steel is 7.8 t/m^3) are needed for discs and shafts, respectively. Operational energy consumption is calculated as 24 kWh/d (according to Soyupakin (1985), 75 Wh energy is consumed for each meter of shaft). Excessive sludge is stabilized in anaerobic digester (AD) and produced methane is emitted to atmosphere. Methane and carbon dioxide productions in the AD are calculated as described in Metcalf et al. in 2003. According to this, produced methane and carbon dioxide are 1.871 kgCH_4/d and 2.182 kgCO_2/d, respectively. Also, carbon dioxide that is produced in the aerobic treatment via mineralization is calculated as 32.210 kgCO_2/d by using stoichiometric calculations. Flow diagram of the RBC system is shown in Figure 10.4.

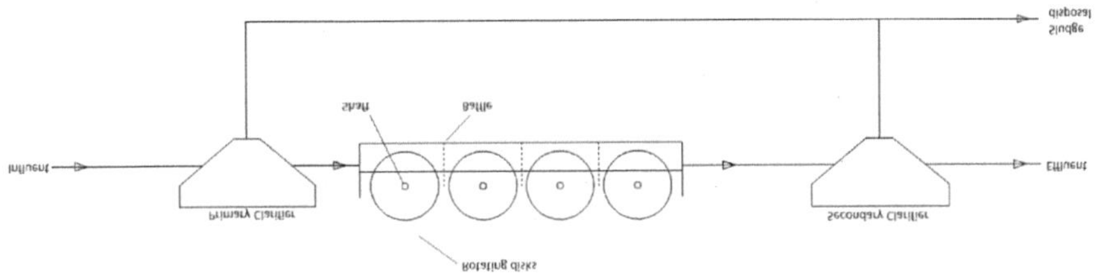

FIGURE 10.4 The flow diagram of RBC.

10.2.2.5 STABILIZATION PONDS (SP)

The *in-situ* data for SP is obtained from the WWTP systems of Harran University in Urfa, Turkey. This SP serves 190 p.e. and the flow rate is 40 m^3/d. length, width, and depth of the pond are 48 m, 18 m, and 1 m, respectively. The volume of the pond is 863 $m^3$and HRT is 19.8 d. Water quality data are collected *in-situ* measurements and laboratory analyzing. Greenhouse gas (GHG) emissions are calculated according to Ferrer in 2002 and 842.080 g CH_4/d is released. Stabilized sludge is calculated as described in Arceivala in 1998 and will be 0.016 m^3/d. The flow diagram of the SP system is shown in Figure 10.5.

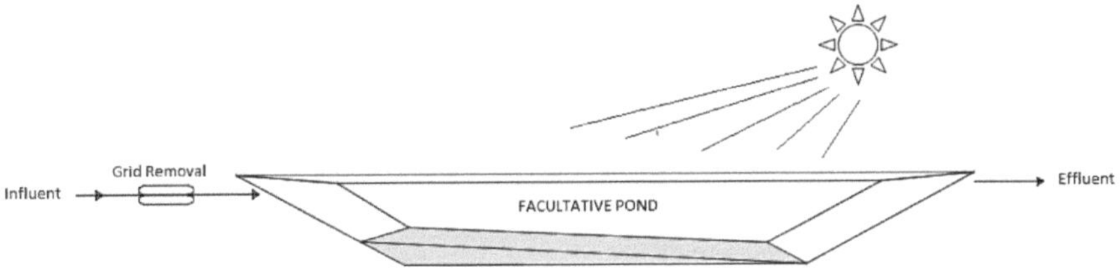

FIGURE 10.5 The flow diagram of SP.

10.2.2.6 EXTENDED AERATION (EA)

The main scenario for this alternative is the system evaluated in Lopsikin (2013). Data used in this study are modified: EA system developed in Lopsik in 2013 serves 1020 p.e. Flow rate is 214.2 m^3/d and the energy requirement of the facility is 160.65 kWh/d. In the data modification, primary settling and anaerobic reactor units are removed AD, and cogeneration units are added and stabilized sludge is used as fertilizer in agricultural applications. It is assumed that 75% of the volume of the sludge is stabilized and 55% of volatile solids in the sludge are removed in the AD. Thus, amount of the stabilized sludge is calculated as 0.226 kg/d (Y = 0.05, k_d = 0,025 d^{-1}). According to Ozturk et al. in 2005, TN, TP, and K ratios of the stabilized sludge weight can be approved as 3.3%, 2.3%, and 0.3%, respectively. If stabilized sludge is applied in agriculture, using fertilizers containing 7.458 g TN, 5.198 g TP, and 0.678 g K can be avoided. GHG emissions are calculated by using stoichiometric equations and IPCC in 2006. As the results of these calculations, 42.520 kg CO_2/d and 8.942 g N_2O/d are emitted in aerobic unit, 2.507 kg CO_2/d is emitted in AD. 2.151 kg CH_4/d are produced in AD and are used in cogeneration unit for producing energy. Hence, 10.943 kWh/d electricity and 3.044 kWh/d heat are produced, and 5.248 kg CO_2/d is released due to combustion of the methane (exergy of the cogeneration facility: 1 for electricity, 0.17 for heat; efficiency of the cogeneration facility: 0.33 for electricity, 0.54 for heat). Flow diagram of the EA system is shown in Figure 10.6.

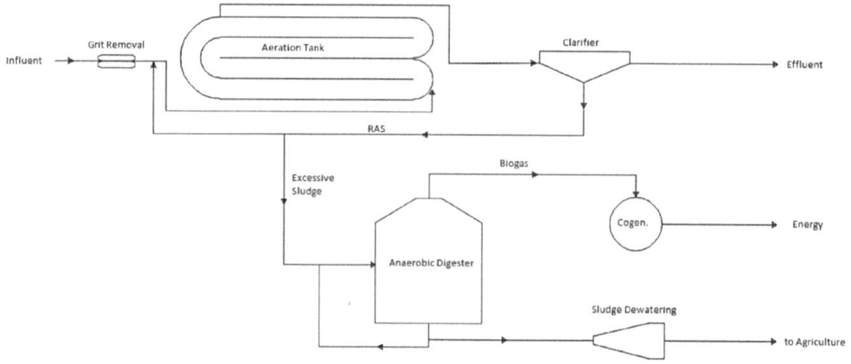

FIGURE 10.6 The flow diagram of EA.

10.2.2.7 MEMBRANE BIO-REACTORS (MBR)

For this work, a MBR process which serves 1000 p.e. is designed. To design MBR, flow rate, SRT, and MLSS are selected as 210 m^3/d, 20 d and, 10,000 mg/l, respectively. According to these assumptions, it is calculated that required bioreactor volume is 32 m^3, required membrane area is 100 m^2 and energy consumption of the MBR is 176.4 kWh/d. Produced excessive sludge in the MBR is sent to a central sludge processing facility where excessive sludge is stabilized and energy is obtained from biogas. Stabilized sludge is calculated as 0.292 kg/d and is applied in agriculture. Thus, using fertilizers containing 9.636 g TN, 6.716 g TP and 0.876 g K can be avoided according to Ozturk et al. in 2005. GHG emissions are calculated as 59.65 kg CO_2/d in aerobic units, 2.874 kg CO_2/d in AD and 5.968 kg CO_2/d from methane combustion by using stoichiometric equations. 2.466 kg CH_4/d are produced in AD and are used in cogeneration unit for producing energy. Therefore, 12.546 kWh/d electricity and 3.490 kWh/d heat are produced (exergy of the cogeneration facility: 1 for electricity, 0.17 for heat; efficiency of the cogeneration facility: 0.33 for electricity, 0.54 for heat). The flow diagram of the MBR system is shown in Figure 10.7.

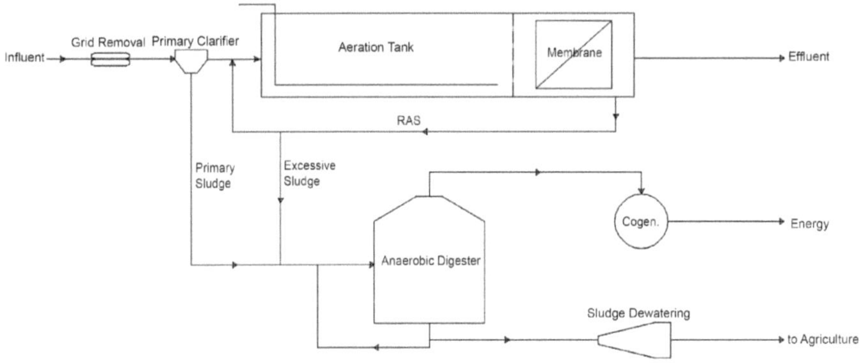

FIGURE 10.7 The flow diagram of MBR.

10.2.2.8 LCA STEP

In the framework of the impact assessment study, the emission data obtained from inventory steps are sorted into categories according to their effects. In order to evaluate the effects, these sorted emissions are converted to same dimension using characterization factors. In this study, the characterization factors and impact categories are selected according to CML2 baseline 2000, which is a LCA method included in SimaPro7.1 software. This method is one of the few which includes characterization factors for the emissions resulting from organic material, phosphorus, and nitrogen compounds (Doka, 2009). The default impact categories included in this method are shown in Table 10.10. The categories can be selected according to the characteristics of each study. In framework of this study, all of the default categories are taken into consideration.

TABLE 10.10 Characterization Factors of the Impact Categories

Impact Category	Unit
Abiotic depletion (AD)	kg (antimony eq.)
Acidification (A)	kg (SO_2 eq.)
Eutrophication (E)	kg (PO_4^{3-}eq.)
Global warming (GW)	kg (CO_2eq.)
Ozone layer depletion (OLD	kg (CFC-11 eq.)
Human toxicity (HT)	kg (1,4-DBC eq.)
Freshwater aquatic ecotoxicity (FWAE)	kg (1,4-DBC eq.)
Marine aquatic ecotoxicity (MAE)	kg (1,4-DBC eq.)
Terrestrial ecotoxicity (TE)	kg (1,4-DBC eq.)
Photochemical oxidation (PO)	kg (C_2H_4 eq.)

Source: Doka (2009).

10.3 RESULTS AND DISCUSSION

For the determination of the environmental impacts of the evaluated wastewater treatment alternatives, Life Cycle Impact Assessment, using data obtained from inventory databases as well as on-site measurements and own calculations is conducted. In the framework of this process impacts of each treatment alternative are calculated. The results of the characterizations step in which the impact potentials are calculated based on the LCI results are given in Table 10.11. These values are presented graphically in Figure 10.8. The maximum value of each impact category is marked as 100% and the other alternatives are calculated accordingly. Following this first step, the normalization step is also conducted. The normalization step provides an easier comparison of the impact categories by dividing the results into a selected reference value. Normalization provides a basis for comparing different types of environmental impact categories in which all impacts get the same unit. In this manner, it was possible to compare all the environmental impacts of the treatment options. In the normalization step, local specifications are taken into account for the evaluation of the local-dependent impact categories (eutrophication and toxicity categories). World normalization data listed in CML2 baseline 2000 are used as specific data for these impacts (Yildirim et al., 2012).

Normalization results presented in Figure 10.9 highlights the categories which are most affected by the treatment options. The most affected categories are *Eutrophication, Global Warming, and Toxicity* as for these categories the cumulative effects in a 20-year period are relatively high. Release of emissions and activities (such as raw material usage) affecting the other factors mainly occurred in the construction phase and environmental effects resulting from construction are relatively low taking into account the time period. The most effected categories are explained in detail in subsections.

TABLE 10.11 LCA Characterization Results of the Treatment Alternatives per 1 p.e. in 20 Years Period

Categories*	Unit**	VLT	CW-LS	CW-S	RBC	
AD	kg Sb eq	0.05	1.07	1.69	0.45	
A	kg SO_2eq	0.50	0.72	1.13	2.15	
E	kg PO_4^{3-}eq	9.74	18.21	12.45	34.76	
GW	kg CO_2eq	–1849.84	633.13	1008.10	346.46	
T	kg 1,4-DB eq	16326.40	30995.73	29430.68	53592.42	
Categories*	**Unit****	**SP**	**AST-LS**	**AST-S**	**EA**	**MBR**
AD	kg Sb eq	1.95	0.44	1.87	3.44	5.06
A	kg SO_2eq	1.44	3.79	4.67	1.99	3.04
E	kg PO_4^{3-}eq	16.16	31.70	19.20	14.91	15.53
GW	kg CO_2eq	1471.86	404.66	671.59	795.01	1170.93
T	kg 1,4-DB eq	94468.61	87510.43	145305.57	129657.70	153146.25

***AD:** Abiotic Depletion, **A:** Acidification, **E:** Eutrophication, **GW:** Global Warming, **T:** Toxicity

****Sb eq:** Antimony equivalent, **SO_2eq**: Sulphur dioxide equivalent, **PO_4^{3-}eq:** Ortophosphate equivalent, **CO_2eq:** Carbon dioxide equivalent, **1,4-DB eq:** 1,4-Dichlorobenzene equivalent

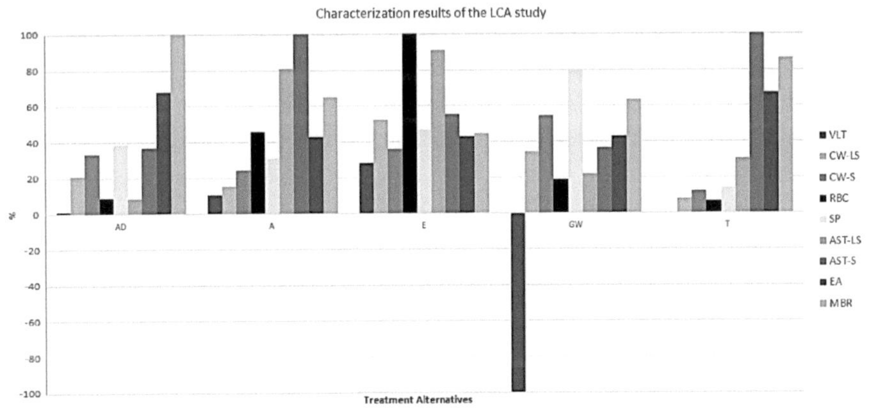

FIGURE 10.8 Characterization results of the LCA study.

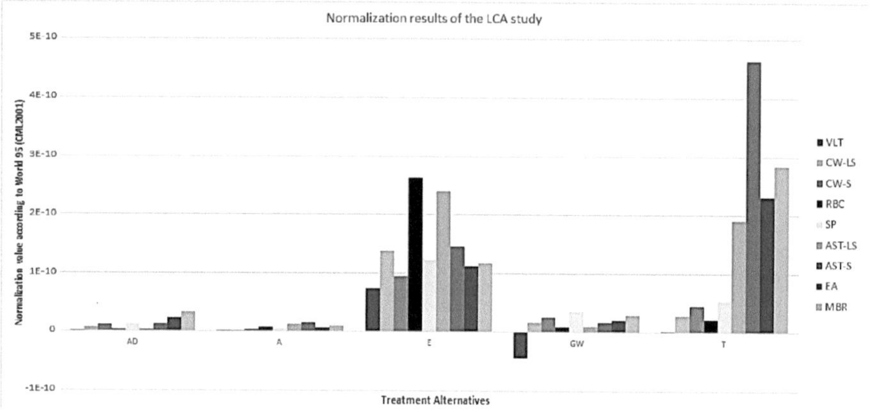

FIGURE 10.9 Normalization results of the LCA study.

10.3.1 ABIOTIC DEPLETION (AD)

MBR is the most resource depleting alternative as the energy consumption during the operation stage is relatively high (0.84 kWh/m^3 for 1 p.e.). EA is also an energy-consuming alternative followed by AST-S. In the case of discharges into sensitive areas, chemicals for phosphorus precipitation are needed resulting in indirect emissions during production. VLT option is the least resource consuming as a raw material requirement of the VLT is very low as these materials are used only in the construction phase of the primary treatment facilities (Figure 10.8). In natural systems, the required area per p.e is very much higher than the others. This is necessary in order to achieve the planned treatment targets. In the framework of this study, it is calculated that about 20 m^2/p.e. for VLT is needed. Area demand of CW is lower, 4–6 m^2/p.e. for discharges into less sensitive areas and 8–10 m^2/p.e. for discharges into sensitive environments. Also, area demand of SP is 4–5 m^2/p.e. Area requirements for other alternatives are significantly lower. 1–0.4 m^2/p.e. for facilities which serve 1000 to, 100,000 p.e., respectively as there is no correlation between area demand per capita and the capacity of the facility.

10.3.2 ACIDIFICATION

Compounds of SO_2, NO, N_2O, NO_x, NH_3, HCl, and HF cause acidification. These compounds are generally occurred from combustion of fossil fuels. The other significant source of the compounds is biological activities that are

NH_3 producing during degradation of organic-N and N_2O emissions causing from incomplete denitrification. NH_3 of these compounds has the most acidification potential (1.88 kg SO_2eq/kg NH_3). According to characterization result, AST-S, AST-LS, and MBR have the most effect on the acidification due to indirect emissions releasing from energy consumption during construction and operation phases of the treatment alternatives, respectively. Additionally, released emissions on production of the chemicals that are used in phosphorus precipitation in the AST-S have contributed to acidification category. NO_2 emissions are high in the AST-LS, SP, and RBC because nitrification cannot occur. In CW-S, N_2O producing can cause from incomplete denitrification because CW-S has plug-flow regime and organic materials requiring for denitrification are consumed along the flow direction of CW.

10.3.3 EUTROPHICATION

This study assumes that treated wastewater from VLT will not be discharged into surface waters; consequently, there is no direct contribution to the eutrophication category. RBC and AST-LS have the greatest impact on the eutrophication category as phosphorus removal is limited to biological uptake and post-settling efficiency. In AST-S, EA, and MBR alternatives, the amount of phosphorus in the wastewater may be reduced below the permitted limits by means of, e.g., chemical precipitation. However, part of the phosphorus compounds in the sludge may enter surface waters with agricultural runoff after being introduced to the soil. Theoretically, the removal efficiency of the CW increases parallel with the size of the plant. The area requirement of the CW varies based on the limits of the target parameters (P, N, BOD, TSS). If phosphorus compounds are to be removed, the required CW area will increase, and it is expected that other contaminants in the wastewater will remain below the permitted limits as the area required for their removal is less than that of phosphorus. Conversely, if it is targeted to remove predominantly organic matter (discharging into less sensitive areas), the area requirement of the CW will be relatively low and the nutrient concentrations in the effluents will remain above allowable limits because there is not enough area to remove these nutrients is available (Yildirim et al., 2012).

10.3.4 GLOBAL WARMING

The production of GHGs in the SP is very high, as the deeper parts of the SP are under anaerobic conditions. Although some of the produced CH_4 is consumed in the water column, the emitted CH_4 is higher than the other alternatives. As organic matter, under anaerobic conditions, mainly mineralizes, CO_2 emissions are also high.

The VLT has the lowest GWP due to the negative contribution of GHGs resulting from the CO_2 fixation capacity of *Eucalyptus camaldulensis.*

RBC and AST-LS show no significant effects on GW. Their contribution to the GW comes from indirect emissions that are used in the production of energy and direct emissions from the production of CH_4 from excessive sludge.

The contribution of MBR, EA, AST-S, and AST-LS on GW are derived from indirect emissions released by the generation of consumed energy. But the methane produced by the digestion of the sludge is used for power generation and thus the GW of these alternatives is reduced slightly. In addition, AST-S has an additional contribution to the GW category due to released emissions in the production of chemicals used for phosphorus precipitation.

GHGs produced in the CW are higher than some of the alternatives. There are conflicting studies in the literature about the global warming potential (GWP) of CWs. Dixon et al. (2003) reported that most of the emissions in the reed systems are emitted during the construction phase and that the CO_2 emissions during the operational phase are negative. Machado et al. (2007) found that the CWs' contributions to global warming and acidification during the operational and maintenance phases are negative (Machado et al., 2007). On the other hand, Søvik et al. (2006) and Ström et al. (2007) investigated the GHG contribution of the CW to the atmosphere and found in their *in-situ* measurements that the CWs made a positive contribution to GHG (Søvik et al., 2006; Ström et al., 2007).

The calculations of these studies prove to be acceptable. Because CW has a plug-flow regime, the first part of CW consumes organic material. Therefore, it may not be possible to meet the demand for organic material in the subsequent anoxic zone. In fact, the following denitrification process cannot be completed and the increase in N_2O will positively contribute to the GHGs of CW.

10.3.5 TOXICITY

In this study, it is assumed that the consumed energy in the WWTP is produced from fossil fuels in terms of reflecting county conditions. Since toxic chemicals (dioxins, PAHs, etc.), are releasing due to combustion of the fossil fuels, energy-dependent alternatives are the most affecting options in toxicity categories. Also, indirect toxic emissions are emitted from producing chemicals that are used for phosphorus precipitation in AST-S.

10.4 CONCLUSIONS

For small communities with <2000 p.e., there are several treatment options and these systems will become more important in the near future. In this study, nine different WWTPs are evaluated, which are suitable for small residential areas. For evaluating individual systems for their environmental impact in this work, LCA methodology was used.

To be able to choose the best alternative, detailed knowledge about the pollution potential of wastewater, as well as the introductory standards will be available. Economic efficiency is another important parameter that was not taken into account in this work.

In order to be able to compare the nine alternative treatment systems, it was assumed that the wastewater to be treated has the same properties and the quality of the treated wastewater meets the same standards.

Based on the calculated environmental impact, the VLT system is the most appropriate option as shown in Figures 10.8 and 10.9. This alternative has the lowest environmental impact for all categories. As a result of the CO_2 uptake, capacity of biomass depleted in the system the GWP of these systems is negative. The disadvantage of these systems, however, is the fact that it cannot be used in large capacities because the space requirement p.e. is very high (about 20 m^2/p.e.). In addition, in closed water catchment areas where groundwater is at risk of contamination, they cannot be used.

The other natural system, the CW, is more suitable in areas of higher groundwater vulnerability. They have an impermeable floor made of clay and HDPE layer on top. Unlike VLT, the CW requires a receiving (water) environment. CW has more GWP than VLT. The main drawback is also the relatively high land requirement: For discharges into less sensitive areas, the specific land requirement is calculated as 4–6 m^2/p.e. The required area increases to 8–10 m^2/p.e. if the treated wastewater is discharged to a sensitive area.

SP showed relatively good results in terms of environmental performance, with the exception of GW. However, the main disadvantages of choosing the SP as an alternative are that biological activities in SP are easily affected by climatic conditions and the required area is high.

Generally, the high space requirements of these treatment alternatives are a major factor that is very important especially for coastal areas. This issue can play a key role in the final decision.

Indirect emissions from RBC are lower than energy-dependent technologies due to their low energy requirements. But, the treatment process of RBC cannot achieve to remove nitrogen and phosphorus compounds and its effects cause direct emissions, the nitrogen and phosphorus compounds on the eutrophication category is very high. Therefore, RBC should not be preferred, especially in sensitive areas.

The energy-dependent processes have the highest environmental impact in construction and operating phases compared to the other natural systems. On the other hand, AST systems have the lowest area requirements and specific area requirements decrease with increasing capacity. While the greenhouse potentials of alternatives may be high due to their energy use, these potentials are lower than expected, as the biogas produced from sludge is converted into energy.

In summary, for small communities with less than 2000 p.e. and wastewater discharges into less sensitive environments VLT and CW are the most environmentally friendly options. When discharging in vulnerable areas and/or where there is not enough space, energy-consuming treatment alternatives may be considered.

KEYWORDS

- **abiotic depletion**
- **activated sludge treatment**
- **eutrophication**
- **free water surface**
- **global warming**
- **high-density polyethylene**

REFERENCES

Adham, S., & Decaloris, J. F., (2004). *Optimization of Various MBR Systems for Water Reclamation-Phase III Final Technical Report*. Bureau of Reclamation Denver Federal Center, 01-FC-81-0736. USA.

Arceivala, S. J., (1998). *Wastewater Treatment for Pollution Control.* McGraw-Hill Publishing Co.; 2nd edition.

Bhati, M., & Singh, G., (2003). Growth and mineral accumulation in *Eucalyptus camaldulensis* seeding irrigated with mixed industrial effluents. *Biosource Technol., 88*, 221–228.

Bitton, G., (1994). *Wastewater Microbiology*. John Wiley and Son's, Inc. Publication.

Bounds, T. R., (1997). In: Bedinger, M. S., Johnson, A. I., & Fleming, J. S., (eds.), *Design and Performance of Septic Tanks, Site Characterization and Design of Onsite Septic Systems*. American Society for Testing Materials, Philadelphia.

Brindle, K., & Stephenson, T., (1996). The application of membrane biological reactors for the treatment of wastewaters. *Biotechnology and Bioengineering, 49*, 601.

Brix, H., (1993). Wastewater treatment in constructed wetlands: System design, removal processes, and treatment performance. In: Moshiri, A. G., (ed.), *Constructed Wetlands for Water Quality Improvement* (pp. 9–22). CRC Press, Boca Raton, Florida.

Clark, J. H., Moseng, E. M., & Asano, T., (1978). Performance of a rotating biological conductor under varying wastewater flow. *Jour. of Water Pol. Cont. Fed., 50*(5), 896–911.

Corbella, C., Puigagut, J., & Garfí, M., (2017). Life cycle assessment of constructed wetland systems for wastewater treatment coupled with microbial fuel cells. *Science of the Total Environment, 584, 585*, 355–362.

Crites, R. W., Reed, S. C., Middlebrooks, E. J., Bastian, R. K., & Reed, S. C., (2014). *Natural Systems for Waste Management and Treatment* (2nd edn.). CRC Press, Taylor and Francis Group, Boca Raton, Florida.

Dalemo, M., (1995). *The Modeling of an Anaerobic Digestion Plant and a Sewage Plant in the ORWARE Simulation Model, Report 213*. The Swedish University of Agricultural Sciences, Department of Agricultural Engineering, Uppsala, Sweden.

De Feo, G., & Ferrera, C., (2017). A procedure for evaluating the most environmentally sound alternative between two on-site small-scale wastewater treatment systems. *Journal of Cleaner Production, 164*, 124–136.

De Miguel, A., Meff, R., Leal, M., González-Naranjo, V., Martínez-Hernández, V., Lillo, J., Martín, I., et al., (2014). Treating municipal wastewater through a vegetation filter with a short-rotation poplar species. *Ecological Engineering, 73*, 560–568.

Diniz, G. S., Tourinho, T. C. O., Silva, A. F., & Chaloub, R. M., (2017). Environmental impact of micro algal biomass production using wastewater resources. *Clean Technç. Environ. Policy, 19*, 2521–2529.

Dixon, A., Simon, M., & Burkitt, T., (2003). Assessing the environmental impact of two options for small scale wastewater treatment: Comparing a reed bed and an aerated biological filter using a life cycle approach. *Ecol. Eng., 20*, 297–308.

Doka, G., (2009). *Wastewater Treatment, Eco Invent Report No: 13, Swiss Centre for Life Cycle Inventories*. Dübendorf, Switzerland.

Downing, A. L., Painter, H. A., & Knowles, G., (1964). Nitrification in the activated sludge process. *J. Inst. Sew. Purif.*, no. 130.

Emmerson, R. H. C., Morse, G. K., Lester, J. N., & Edge, D. R., (1995). Life-cycle analysis of small-scale sewage treatment processes. *Int. Water Environ. Man., 9*(3), 317–325.

EU Council Directive 91/271/EEC. *Urban Wastewater Treatment Directive*.

Ferrer, A. V. M., (2002). *Algae and Duckweed-Based Pond Reactors: An Evaluation and Comparison of CH_4 Gas Emissions* (p. 42). MSc Thesis Institute UNESCO-IHE, Delft.

Foley, J., Haas, D., Hartley, K., & Lant, P., (2010). Comprehensive life cycle inventories of alternative wastewater treatment systems. *Water Res., 44*(5), 1654–1666.

Gallego, A., Hospido, A., Moreira, M. T., & Feijoo, G., (2008). Environmental performance of wastewater treatment plants for small populations. *J. Resour. Conserv. Recycl., 52*(I.6), 931–940.

Garfí, M., Flores, L., & Ferrer, F., (2017). Life cycle assessment of wastewater treatment systems for small communities: Activated sludge, constructed wetlands, and high rate algal ponds. *Journal of Cleaner Production, 161*, 211–219.

Gkika, D., Gikas, G. D., & Tsihrintzis, V. A., (2015). Environmental footprint of constructed wetlands treating wastewater. *Journal of Environmental Science and HealthPart A, 50*, 631–638.

Gray, N. F., (1995). The influence of sludge accumulation rate on septic tank design. *Environmental Technology, 16*(I. 8/1), 795–800.

Guinee, J. B., (2001). *Life Cycle Assessment, an Operational Guide to the ISO Standards, Final Report*. Centre of Environmental Science, Leiden University (CML).

Hammer, D. A., (1989). *Constructed Wetlands for Wastewater Treatment* (p. 831). Lewis Publishers, Inc., Chelsea, MI.

Hospido, A., Moreira, M. T., & Feijoo, G., (2008). A comparison of municipal wastewater treatment plants for big centers of population in Galicia (Spain). *Int. J. LCA, 13*(1), 57–64.

Hospido, A., Moreira, M. T., Fernandez-Couto, M., & Feijoo, G., (2004). Environmental performance of a municipal wastewater treatment plant. *Int. J. LCA, 9*(4), 261–271.

Hospido, A., Moreira, M. T., Martin, M., Rigola, M., & Feijoo, G., (2005). Environmental evaluation of different treatment processes for sludge from urban wastewater treatments: Anaerobic digestion versus thermal processes. *Int. J. LCA, 10*(I. 5), 336–345.

Hospido, A., Sanchez, I., Rodriguez-Garcia, G., Iglesias, A., Buntner, D., Reif, R., Moreira, M. T., & Feijoo, G., (2012). Are all membrane reactors equal from an environmental point of view? *Desalination*, 285, 263–270.

Houillion, G., & Jolliet, O., (2005). Life cycle assessment of processes for the treatment of wastewater urban sludge: Energy and global warming analysis. *J. Clean. Prod., 13*, 287–299.

IPCC (Intergovernmental Panel on Climate Change), (2006). In: Eggleston, H. S., Buendia, L., Miwa, K., Ngara, T., & Tanabe, K., (eds.), *Guidelines for National Greenhouse Gas Inventories, the National Greenhouse Gas Inventories Programme*. IGES, Japan.

Jensen, A. A., Hoffman, L., Møller, B. T., Schmidt, A., Christiansen, K., Elkington, J., & Van, D. F., (1997). *Life Cycle Assessment (LCA): A Guide to Approaches, Experiences, and Information Sources, Report to the Environmental Agency, Environmental Issues Series, No. 6*. Copenhagen.

Kadlec, R. H., & Knight, R. L., (1996). *Treatment Wetlands*. Boca Raton, FL: CRC Press.

Kamble, S. J., Chakravarthy, Y., Singh, A., Chubilleau, C., Stark, M., & Bawa, I., (2017). A soil biotechnology system for wastewater treatment: Technical, hygiene, environmental LCA and economic aspects. *Environ. Sci. Pollut. Res., 24*, 13315–13334.

Kaygusuz, E. A., (2004). *Domestic Wastewater Treatment in Pilot-Scale Constructed Wetlands Implemented in the Middle East Technical University, PhD Thesis.* Natural and Applied Science Ins., METU, Ankara, Turkey.

Kruzic, A. P., (1994). Natural treatment systems. *Water Environ. Res., 66*(4), 357–361.

Kuusemets, V., & Mander, U., (1999). Ecotechnological measures to control nutrient losses from catchments. *Water Sci. Technol., 40*(10), 195–202.

Lee, W., Kang, S., & Shin, H., (2003). Sludge characteristics and their contribution to microfiltration in submerged membrane bioreactors. *J. Membr. Sci., 216*, 217–227.

Li, Y., Lou, X., Huang, X., Wang, D., & Zhang, W., (2013). Life cycle assessment of a municipal wastewater treatment plant: A case study in Suzhou, China. *Journal of Cleaner Production, 57*, 221–227.

Lopsik, K., (2013). Life cycle assessment of small-scale constructed wetland and extended aeration activated sludge wastewater treatment system. *Int. J. Environ. Sci. Technol., 10*, 1295–1308.

Lorfano, G., & Brown, J., (2010). Wastewater management through the ages: A history of mankind. *Science of the Total Environment, 408*, 5254–5264.

Lutterbeck, C. A., Kist, L. T., Lopez, D. R., Zerwes, F. V., & Machado, E. L., (2017). Life cycle assessment on integrated wastewater treatment systems with constructed wetlands in rural areas. *Journal of Cleaner Production, 148*, 527–536.

Machado, A. P., Urbano, L., Brito, A. G., Janknecht, J., Salas, J. J., & Nogueria, R., (2007). Life cycle assessment of wastewater treatment options for small and decentralized communities. *Water Sci. Tech., 56*(3), 15–22.

Madison, M., & Henderson, M., (1993). Zero discharge all-weather land application with soil storage. *Proceeding of the 66th Annual Conference of the Water Environment Federation.* Anaheim, CA.

Mahmood, Q., Pervez, A., Zeb, B. S., Zaffar, H., Yaqoob, H., Waseem, M., Afsheen, Z., & Afsheen, S., (2013). Natural treatment systems as sustainable ecotechnologies for the developing countries. *Bio. Med. Research International*. ID 796373, 19.

Martínez-Hernández, V., Leal, M., Meffe, R., De Miguel, A., Alonso-Alonso, C., De Bustamante, I., Lillo, J., et al., (2018). Removal of emerging organic contaminants in a poplar vegetation filter. *Journal of Hazardous Materials, 342*, 482–491.

McKim, H. L., (1982). *Wastewater Application in Forest Ecosystems, Report 82–19*. USA CRREL, Hanover, NH.

Medcalf, & Eddy, Inc, (2003). *Wastewater Engineering, Treatment, and Reuse* (4th edn.). McGraw Hill Publishing.

Mels, A. R., Nieuwenhuijzen, A. F., Van, D. G. J. H. J. M., Klapwijk, B., De Koning, J., & Rulkens, W. H., (1998). *Sustainability Criteria as a Tool in the Development of New Sewage Treatment Methods, Options for Closed Water Systems: Sustainable Water Management Congress*. Wageningen.

Molinos-Senante, M., Garrido-Baserba, M., Reif, R., Hernandez-Sancho, F., & Poch, M., (2012). Assessment of wastewater treatment plant design for small communities: Environmental and economic aspect. *Science of the Total Environment, 427, 428*, 11–18.

Molinos-Senante, M., Gómez, T., Garrido-Baserba, M., Caballero, R., & Sala-Garrido, R., (2014). Assessing the sustainability of small wastewater treatment systems: A composite indicator approach. *Science of the Total Environment, 497, 498*, 607–617.

Munoz, I., Rieradevall, J., Torrades, F., Peral, J., & Domenech, X., (2005). Environmental assessment of different solar driven advanced oxidation processes. *J. Solar Ener., 79*, 369–375.

Neumayr, R., Dietrich, R., & Steinmüller, H., (1997). Life cycle assessment of sewage sludge treatment. *Proceedings of the 5th SETAC Annual Conference*. Brussels.

Nutter, W., Philpott, L., & Morris, L. A., (1996). Long-term environmental impacts of municipal wastewater irrigation to forests at Clayton Country, Georgia. *Proceedings of Land Application of Wastes in Australia and New Zealand: Research and Practice.* Australian Conference.

Ødegaard, H., (1995). An evaluation of cost efficiency and sustainability of different wastewater treatment processes. *VATTEN, 51*, 291–299.

Ontiveros, G. A., & Campanella, E. A., (2013). Environmental performance of biological nutrient removal process from a life cycle perspective. *Bioresource Technology, 150*, 506–512.

Öztürk, İ., Timur, H., & Koşkan, U., (2005). Atıksu Arıtımının Esasları: Evsel, Endüstriyel Atıksu Arıtımı Ve Arıtma Çamurlarının Kontrolü, Çevreve Orman Bakanlığı Yayınları.

Pasqualino, J. C., Meneses, M., Abella, M., & Castells, F., (2009). LCA as a decision support tool for the environmental improvement of the operation of a municipal wastewater treatment plant. *Environ. Sci. Technol., 43*(9), 3300–3307.

Peterjohn, W. T., & Correll, D. L., (1984). Nutrient dynamics in an agricultural watershed: Observations on the role of a riparian forest. *Ecology, 65*(5), 1466–1475.

Pre Consultants, (2006). *Sima Pro: Introduction to LCA with SimaPro.* PRe-Consultants, Amersfoort, Netherlands.

Qasim, S. R., (1999). *Wastewater Treatment Plants: Planning, Design, and Operation.* CRC Press.

Reed, S. C., & Bastian, R. K., (1991). Potable water via land treatment and AWT. *Water Environ. Technol., 3*(8), 40–47, WEF, Alexandria, VA.

Roeleveld, P. J., Klapwijk, A., Eggels, P. G., Rulkens, W. H., & Starkenburg, W., (1997). Sustainability of municipal wastewater treatment. *Water Sci. Technol., 35*(10), 221.

Sonesson, U., Dalemo, M., Mingarini, K., & Jönsson, H., (1997). ORWARE: A simulation model for organic waste handling systems. Part 2: Case study and simulation results. *J. Resour. Conserv. Recycl., 21*, 39–54.

Søvik, A. K., Augustin, J., Heikkinen, K., Huttunen, J. T., Necki, J. M., Karjalainen, S. M., & Kløve, B., et al., (2006). Emission of the greenhouse gases nitrous oxide and methane from constructed wetlands in Europe. *J. Environ. Qual., 35*, 2360–2373.

Soyupak, S., (1985). Tatil Köyleri, Düşük Nüfuslu Meskûn Bölgelerve Sitelerden Gelen Atıksuların Arıtılmasıiçin Bir Yöntem: Dönen Biyolojik Reaktör Sistemi. Türkiye Mühendislik Haberleri Dergisi, Sayı: 315, İnşaat Mühendisleri Odası, Ankara.

Stefanakis, A., Akratos, C., & Tsihrintzis, V. A., (2015). *Vertical Flow Constructed Wetlands: Eco-Engineering Systems for Wastewater and Sludge Treatment* (2015th edn.). Elsevier Publishing, Amsterdam, Netherlands.

Stephenson, T., Judd, S., Jefferson, B., & Brindle, K., (2000). *Membrane Bioreactors for Wastewater Treatment.* IWA Publishing, London, UK.

Ström, L., Lamppa, A., & Cristensen, T. R., (2007). Greenhouse gas emissions from a constructed wetland in southern Sweden, *Wetlands Ecol. Man., 15*, 43–50.

Suh, Y. J., & Rousseaux, P., (2002). An LCA of alternative wastewater sludge treatment scenarios. *J. Resour. Conserv. Recycl., 35*(I. 3), 191–200.

Sunder, S. S., (1993). The ecological, economic, and social effects of eucalyptus. *FAO Regional Office for Asia and the Pacific, Proceedings-Regional Expert Consultation on Eucalyptus, I.* Bangkok, Thailand.

Teiter, S., & Mander, U., (2005). Emission of N_2O, N_2, CH_4, and CO_2 from constructed wetlands for wastewater treatment and from riparian buffer zones. *Eco. Eng., 25*(I.5/1), 528–541.

Tillman, A. M., Lundström, H., & Svingby, M., (1996). *Life Cycle Assessment of Alternative Wastewater Systems in Bergsjönand Hamburgsund.* Appendix; Report 1996:1b; Technical environmental planning, Chalmers University of Technology: Goteborg, Sweden.

Tsihrintzis, V. A., (2017). The use of vertical flow constructed wetlands in wastewater treatment. *Water Resources Management, 31*(10), 3245–3270.

Tzanakakis, V. A., Paranychianakis, N. V., & Angelakis, A. N., (2009). Nutrient removal and biomass production in land treatment systems receiving domestic effluent. *Eco. Eng., 35*, 1485–1492.

US-EPA, (1993). *Subsurface Flow Constructed Wetlands for Wastewater Treatment: A Technology Assessment, US-EPA Report.* EPA 832-R-93-008.

US-EPA, (1999). *Free Water Surface Wetlands for Wastewater Treatment: A Technology Assessment*. Arizona, USA.

US-EPA, (2006). *Land Treatment of Municipal Wastewater Effluents, Process Design Manual*. EPA/625/R-06/016. Cincinnati, Ohio, USA.

Van, D. L., & Roncken, G. C. G., (1997). Membrane biorector for wastewater-treatment: The state of the art and new developments. *Water Sci. and Tech., 35*, 53.

Verbyla, M., Von, S. M., & Maiga, Y., (2017). Waste Stabilization Ponds. In: Rose, J. B., & Jiménez, C. B., (eds.), *Global Water Pathogens Project*. http://www.waterpathogens.org (accessed on 11 June 2020). Haas, C., Mihelcic, J. R., & Verbyla, M. E., (eds.), *Part 4 Management of Risk from Excreta and Wastewater*. http://www.waterpathogens.org/book/waste-stabilization-ponds (accessed on 11 June 2020). Michigan State University, E. Lansing, MI, UNESCO.

Von, S. M., (2007). *Waste Stabilization Ponds*. IWA Publishing, Alliance House, 12 Caxton Street, London, UK.

Vymazal, J., & Kröpfelova, L., (2005). Growth of *Phragmites australis* and *Phalaris arundinacea* in constructed wetlands for wastewater treatment in the Czech Republic. *Eco. Eng., 25*, 606–621.

Vymazal, J., & Kröpfelova, L., (2008). *Wastewater Treatment in Constructed Wetlands with Horizontal Sub-Surface Flow*. Springer Science + Business Media B.V.

Yildirim, M., & Topkaya, B., (2012). Assessing environmental impacts of wastewater treatment alternatives for small-scale communities. *Clean-Soil, Air, and Water, 40*(2), 171–178.

Zhang, Q. H., Wang, X. C., Xiong, J. Q., Chen, R., & Cao, B., (2010). Application of life cycle assessment for an evaluation of wastewater treatment and reuse project-case study of Xi'an, China. *Biosource Technology, 101*, 1421–1425.

Zhang, Z., & Wilson, F., (2000). Life-cycle assessment of a sewage-treatment plant in South-East Asia. *Water Environ. J., 14*(1), 51–56.

Index

D

E

P

T

Y

Z